Springer Atmospheric Sciences

More information about this series at http://www.springer.com/series/10176

Ewa Łupikasza

The Climatology of Air-Mass and Frontal Extreme Precipitation

Study of meteorological data in Europe

Springer

Ewa Łupikasza
Faculty of Earth Sciences
University of Silesia
Sosnowiec, Poland

ISSN 2194-5217 ISSN 2194-5225 (electronic)
Springer Atmospheric Sciences
ISBN 978-3-319-81047-8 ISBN 978-3-319-31478-5 (eBook)
DOI 10.1007/978-3-319-31478-5

Printed on acid-free paper

This Springer imprint is published by Springer Nature
The registered company is Springer International Publishing AG Switzerland

Preface

This book consists of eight chapters. The executive summary followed by description of the sources from which meteorological and synoptic data were taken, the meteorological data selection and homogenisation procedure, as well as the characteristics of the spatial and temporal coverage of the data selected, are described in Chap. 1, *Introduction*. The diversity of definitions of an extreme event translates into a diversity of existing extreme event indices, including those applied to extreme precipitation. The definitions of extreme events and key indices of precipitation extremes are discussed in Chap. 2, *Definitions and Indices of Precipitation Extremes*. This chapter, which provides a background for further deliberations, discusses the amount and frequency of extreme precipitation in Europe, taking into account "absolute" extremes, and goes on to describe and explain the choice of the definition of extreme precipitation in this study. Chapter 3, *Origin-Based Types of Extreme Precipitation*, presents a classification of the origin-based extreme precipitation types based on fundamental dynamic processes leading to precipitation formation and the spatial and seasonal variability in the occurrence of the different types identified. Chapter 4, entitled *Regionalization of Extreme Precipitation Types Occurrence in Europe*, synthesizes the spatial and seasonal variability in the occurrence of individual extreme precipitation types by regionalising them, and thus to identify groups of stations characterised by similar structure of extreme air-mass and frontal precipitation occurrence. The chapter also includes the characteristics of the regional groups identified. In Europe, weather conditions, including the occurrence and amount of precipitation, depend significantly on the intensity of westerly air flow, which is described by the North Atlantic Oscillation Index. Chapter 5, *Relation Between Extreme Precipitation Occurrence and North Atlantic Oscillation*, describes the impact of westerly air flow on extreme precipitation occurrence in Europe. Chapter 6, *Air-Mass and Frontal Extreme Precipitation Occurrence in Circulation Types*, analyses the relationship between the occurrence of extreme air-mass and frontal precipitation and atmospheric circulation. To this end, a methodologically uniform continent-wide catalogue, or strictly speaking, catalogues, of synoptic situations were developed.

Sosnowiec, Poland Ewa Łupikasza

Acknowledgements

The research was partly financed by the National Research Centre, Ministry of Science and Higher Education within the project no. N 306 2439 39 "Synoptic background of extreme precipitation in Europe and its long-term variability."

This publication has been financed from the funds of the Leading National Research Centre (KNOW) received by the Centre for Polar Studies for the period 2014–2018.

I am also immensely grateful to Prof. Joanna Wibig, Prof. Elwira Żmudzka and Prof. Zbigniew Ustrnul for their comments on the manuscript, although any errors are my own and should not tarnish the reputations of these esteemed persons.

I would like to express my appreciation to Prof. Tadeusz Niedźwiedź and Prof. Jacek Jania for disscussions and support which enabled me to publish this book. I also would like to thank Paweł Pilch for linguistic support.

Thank you

Contents

Chapter 1
Introduction

Abstract Extreme climatic and weather events are imperishable subjects of climatological studies. Although the understanding of these rare and unexpected phenomena has improved over the past decades, the factors governing their occurrence and magnitude still need to be studied. This chapter starts with the executive summary including the most important conclusions on origin-based extreme precipitation types (air-mass and frontal precipitation) and their relationship to atmospheric circulation which are discussed in this book. It also reviews a vast body of literature on synoptic contributions to the development of extreme precipitation. It presents sources of meteorological and synoptic data used throughout this study and discusses the quality of weather data and the criteria used for sectioning precipitation series. The climatology of frontal precipitation and air-mass precipitation presented in this volume has been developed using daily totals recorded at 513 European weather stations during the period spanning December 1950 to February 2008. A large set of synoptic charts covering the same period has also been used. The majority (64 %) of the individual precipitation series has been found to be complete, that is, either free from gaps or where any gaps account for no more than 1 % of days in the study period. This chapter also includes the executive summary, which brings together the most important conclusions from the comprehensive synoptic/climatic analysis of extreme precipitation taking into account both circulation types and atmospheric fronts in Europe. The analysis has demonstrated the existence of regional groups with seasonal patterns of occurrence of extreme precipitation types. A regional and seasonal variability in the relationship between the occurrence of origin-based types of extreme precipitation and atmospheric circulation was also discovered.

Keywords Air-mass precipitation • Frontal precipitation • Synoptic charts • Europe

People have realised the enormous importance of precipitation in their lives since the earliest civilisations. The Egyptians and Sumerians undertook the first hydroengineering projects around 3200 BC. Even though humanity recognised the importance of precipitation as a source of life-giving water, questions about its origins only began much later. The earliest attempts to explain the underlying mechanisms were made by Greek philosophers. Despite the scarce knowledge available at the

E. Łupikasza, *The Climatology of Air-Mass and Frontal Extreme Precipitation*,
Springer Atmospheric Sciences, DOI 10.1007/978-3-319-31478-5_1

time, the attempts to solve the mystery were surprisingly successful. Anaximander, a contemporary of Thales, explained rain as a product of the humidity pumped up from Earth by the sun. He believed that hail is frozen rain and that snow originates after smaller crystals of ice cluster in the air (Strangeways 2007).

The earliest references to precipitation measurements come from India, from the fourth century BC, and from Palestine, from the second century BC. The earliest accounts about the use of rain gauges originate from China (1247 AD). Regular precipitation records began in Korea in 1441 during the rule of King Sejong and continued until 1907 (Strangeways 2007). In Europe, no precipitation measurements had been carried out until the seventeenth century. The first attempt to construct a rain gauge was made by Benedetto Castelli in Italy in 1639. In 1660, Sir Christopher Wren designed the first British rain gauge (Strangeways 2007). The earliest regular precipitation measurements in Europe were made in 1715 in Hoofdorp-Zwannenburg (the Netherlands), in 1725 in Padua (Italy), in 1739 in Uppsala (Sweden), and in 1739 in Gdańsk and Żagań (Poland). In the early nineteenth century, precipitation was already being measured in many European cities, including Warsaw (from 1803), Prague (1804), Copenhagen (1805), Jena (1827), Dresden (1828), Helsinki (1844), and Berlin (1847) (von Rudloff 1967).

Thanks to regular records, T. Bergeron explained the precipitation formation mechanisms quite well as early as 1935. Still, the relationship between the occurrence of precipitation and its determinants requires further research. Invariably, for millennia, rainfall has determined people's living conditions and many processes in the geographic environment, which follows from the fact that precipitation is an element of the hydrological cycle with crucial importance for the environment. As Okołowicz (1969) observed, precipitation is the last link of the water cycle that is of interest to atmospheric sciences, and, at the same time, the first link of the section of the cycle which is of interest to hydrologists. Thus, the high importance of precipitation is related to its importance for the dynamics of Earth's ecosystems, both natural and human managed (Weltzin et al. 2003; Weltzin and McPherson 2003). The element linking precipitation and ecosystems is soil moisture, which depends on the amount of rainwater reaching the surface, on the physical properties of the surface itself, and on the season of the year.

According to Knapp et al. (2002), changes in such a spatially and temporally diversified environmental factor (precipitation) may have a greater impact on ecosystems than changes in the other, relatively stable, factors, such as CO_2. Hence, investigating the spatial and temporal variability of precipitation is, from the perspective of its impact on ecosystems, of crucial importance (Knapp et al. 2002). Identifying both the temporal and spatial differentiation in precipitation requires, above all, determining the nature of the relationship between its occurrence and determinants, notably atmospheric circulation.

Given the current climate change, particular importance is attached to extreme climatic and meteorological events, including extreme precipitation. Existing knowledge about its occurrence is incomplete because, being a "nonstandard" phenomenon, such precipitation is accidental and hard to predict. Knowledge about the conditions underlying extreme phenomena is important because such events often have disastrous consequences and, in extreme cases, lead to human death. As Jentsch et al. (2006) observed—and as has not changed much since—scientists are unable to predict accurately the development of extreme events, and, above all, pinpoint their location. The low predictability of extreme events and their abruptness limit people's preparedness for such phenomena and their ability to protect themselves against their negative consequences. Thus, extreme events are among the most difficult research areas in climatology. The results of climate research into the link between the occurrence of extreme events and atmospheric circulation are helpful in studies aimed at modelling of the occurrence of precipitation extremes (Hellström 2005). These results are also crucial for impact studies in such areas as flood risk estimation (Beven 1993), structural engineering (Cowpertwait 1994), soil erosion rate (Favis-Mortlock and Boardman 1995), and changes in the quality of water resources (Wilby 1993).

1.1 Executive Summary

This study discusses spatial and seasonal variability in the occurrence of origin-based extreme precipitation in Europe and its associations with atmospheric circulation. A vast set of daily precipitation records was used from 513 weather stations in Europe covering the period from December 1950 to February 2008. At more than 56 years, this relatively long study period offered a sufficiently large sample of the precipitation events in question for the results to be reliable.

The considerable degree of variation found in European precipitation regimens, as defined by their totals and concentration seasons, demands the definitions of extreme precipitation events to be regionalised. This study adopted the statistical definition of extreme events. An event was identified as extreme if its frequency of occurrence was above the 95th percentile (95P) of the statistical frequency distribution. Threshold values corresponding to the 95P were derived separately for each station and each month using records of precipitation of 1 mm or more in the standard period 1961–1990. The spatial and seasonal variability of the daily totals of 95P precipitation (i.e., the threshold values for the identification of days with extreme precipitation) is associated with the variability of monthly precipitation totals. The overall range of variability in the 95P precipitation spans the daily totals of 4.2 mm (January) and 114.4 mm (October). The highest values of 95P were recorded in the Mediterranean (above 40 mm in its western and central parts in autumn) and along the western Scandinavian coast, especially its southwestern section (above 25 mm in winter). Elsewhere the highest seasonal threshold values of extreme precipitation were identified in summer (15–30 mm) and the lowest in winter (5–10 mm in Central Europe and Eastern Europe; 10–15 mm in Western Europe). The spatial and seasonal variability of 95P daily totals is a result of the combined circulation and altitude effects.

Two major origin-based types of precipitation were identified: air-mass (type A) and frontal (type F). These types were distinguished considering basic processes leading to the development of precipitation (free convection, forced convection, and less vertical overrunning movements, such as observed on the warm front) and their identification potential on synoptic maps. The types were further subdivided taking into account dynamic processes depending on the type of the weather front, and this produced precipitation associated with the passage of different fronts (type Ff), precipitation associated with the passage of a cold front (type Fc), precipitation associated with the passage of a warm front (type Fw), precipitation associated with an occluded front (type Fo), precipitation associated with a stationary front (type Fs), and precipitation associated with a discontinuity line (type Fd).

One of the fundamental differences between air-mass precipitation and frontal precipitation is the size of the precipitation zone. An analysis of the spatial distribution of the major origin-based types of extreme precipitation (type A and type F) demonstrated that during the whole year air-mass precipitation was simultaneously (i.e., on the same precipitation day) recorded by a much smaller number of stations than was frontal precipitation. Two or fewer stations simultaneously recording air-mass precipitation accounted for about 38–40 % of days with extreme precipitation whereas the same figure for frontal precipitation was just 15 %.

The frequency of origin-based precipitation types in Europe follows clear spatial and seasonal regularities. These regularities depend on the varying pace of the cyclone life cycle during the year and on ground relief. Ground relief also impacts the pace of cyclogenesis, thus influencing the spatial variability of origin-based types of extreme precipitation.

The average proportion of air-mass precipitation (type A) in the overall number of days with extreme precipitation ranged from 14 % in winter to 25 % in summer. In all seasons, type A extreme precipitation is the most frequent in the south of the continent. In summer, when conditions for the development of convection in this area are favourable, it accounts for more than 50 % of extreme precipitation. This frequency declines northwards. Elsewhere in the continent it only reaches 10–20 % of extreme precipitation (ExP) in summer, while in other seasons its share typically drops to no more than 10 % ExP. Type A precipitation is somewhat more frequent in upland and mountainous areas and along the southeastern North Sea coast (~30–50 % ExP). This difference is explained by orographic barrier effect or by a sea–land transition effect that boosts atmosphere dynamics and leads to convective movement which produces heavy precipitation.

In Europe, most days with extreme precipitation are associated with the occurrence of weather fronts. In every season, the average proportion of frontal precipitation (type F) in the overall number of days with extreme precipitation was several times higher than that of air-mass precipitation. Type F precipitation accounts for the largest portion of extreme precipitation in winter (on average, 86 % ExP), as cyclonic activity increases on the continent, and for the lowest in summer (on average, 75 % ExP), when conditions favourable for free convection are far more frequent. At the scale of the whole of Europe the lowest frequency of frontal precipitation is found in southern Europe, especially in summer, when this type accounts for approximately 10 % ExP. Both the highest frequencies of air-mass precipitation and the lowest frequencies of frontal precipitation that characterise this area in summer are an effect of the influence of the northern edge of the subtropical anticyclone zone. In winter, the frequency of frontal precipitation in southern Europe increases in response to the area's intensive seasonal cyclonic activity. Frontal precipitation accounts for between 50 and 80 % of extreme precipitation along the southeastern North Sea coast as well as along the northern and the western Scandinavian coast in summer and on the Polish Baltic coast in autumn. The lower frequency of type F extreme precipitation and higher frequency of type A along the southeastern North Sea coast in summer are partly linked to a strong breeze effect in that area.

Extreme precipitation associated with the passage of a cold front (type Fc) is the most frequent in summer, when Europe receives cooler air from over the Atlantic Ocean. In all seasons, the highest frequencies of this origin type of extreme precipitation occurs in Western Europe (~30–50 % ExP in summer and 30–40 % in other seasons), along a belt extending from the northwestern tip of the Iberian Peninsula to France and the Alps, as well as in southeastern Europe. The higher number of days with type Fc extreme precipitation in Western Europe's mountainous areas is linked to the orographic barriers of the Massif Central and the Alps that slow down and intensify the oncoming cold fronts. The summer maximums of type Fc extreme precipitation in the southeastern part of the continent are linked with a vast low-pressure system centred to the southeast of the Caspian Sea, which causes advection of continental air over the heated ground surface and with a strong anticyclone activity, as either the Azores' high extending deep into the continent with its wedge or smaller highs forming over the continental Europe cause an inflow of air from the northwest. In Central Europe, the frequency of type Fc precipitation reaches 20–30 % ExP only in summer; in other seasons, lower figures are recorded.

Extreme precipitation associated with a warm front (type Fw) occurs in Europe less frequently than that from cold-front precipitation. Its contribution to the overall number of days with extreme precipitation peaks in winter (on average, 13 % ExP), when the ground is the coldest, and in spring, when the temperature of the landmass, cooled down during winter, remains lower than that of the surrounding seas. In general, the frequency of type Fw precipitation increases towards the continental part of Europe, and the rate of this change is the highest in winter and the lowest in summer. In winter, the highest frequencies of type Fw precipitation exceed 20 % ExP in the northern part of southern Europe, in Eastern Europe, at isolated stations of Central Europe and the Scandinavian Peninsula and, in spring, also in Eastern Europe.

Extreme precipitation associated with the passage of various weather fronts (type Ff) is the most frequent of all frontal precipitation types. On average it accounts for between about 24 % ExP in summer and about 37 % ExP in winter, when the speed of front travel across Europe is the highest. The main features of the spatial distribution of type Ff remain largely unchanged during the year. Type Ff extreme precipitation most frequently occurs: in Western Europe, including the Atlantic islands; the western continental coast between the Bay of Biscay and Bay of Finland; and along the western Scandinavian coast, especially its southern section. In this area the proportion of type Ff precipitation in winter ranges from 50 to 60 % ExP, whereas in summer it rarely exceeds 40 % ExP. In winter, the zone of high frequency of type Ff precipitation reaches deeper into the continent up to and including Central Europe. The autumn/winter maximums of extreme precipitation associated with the passage of various weather fronts deeper into the continent are explained by the speed of atmospheric fronts that is higher than in the warm half of the year. This effect is a result of higher pressure gradients and of intensive cyclogenesis which, in turn, is caused by stronger thermal gradients in this half of the year in the northern hemisphere. In summer, Mediterranean stations recorded no extreme precipitation associated with the passage of various fronts during the study period.

The proportion of ExP associated with an occluded front (type Fo) in the overall number of days with extreme precipitation follows a regular zonal pattern across Europe. During the year, main features of this pattern vary little, as the proportion of type F precipitation remains relatively stable at about 20 % ExP. It is the most frequent in northern Europe (more than 40 % ExP), except the western Scandinavian coast, where its proportion in the overall number of days with extreme precipitation during a year does not exceed 20 % ExP. The central part of the southern tip of the Scandinavian Peninsula is especially prone to the occurrence of type Fo precipitation, which accounts for 50 % ExP. This local peak of type Fo occurrence has an orographic background. Upon reaching the Scandinavian Mountains, weather fronts slow down, which accelerates the occlusion development process. The frequency of type Fo precipitation gradually decreases from the north to the south. In summer, many southern European stations did not record any type F precipitation during the study period.

Extreme precipitation associated with a stationary front (type Fs) and a discontinuity line (type Fd) is among the least frequent in Europe. Type Fs accounts on average for between about 3 % ExP (in spring and in autumn) and 4 % ExP (in summer) annually, whereas type Fd does not exceed 1 % ExP even in a single season. Type Fs precipitation is the most frequent in Western Europe and in Central Europe (~10–20 % ExP at individual stations), but numerous stations in southern and northern Europe record no such events. Summer is the peak season of type Fd precipitation, which is rare in winter. In summer, type Fd precipitation is concentrated mainly in Western Europe and in Central Europe.

The identification of strong spatial and seasonal variabilities in the occurrence of the origin-based precipitation types inspired a further study intended to obtain a more synthetic picture. Using cluster analysis of *k*-mean method and a careful selection of the grouping variables, six groups of stations with different structure of the occurrence of origin-based extreme precipitation types were identified (Table 1.1).

Although the selected groups of stations fail to meet the spatial continuity criterion required to become true regions, they represent regional features of such a regional structure and hence they have been called regional groups (GR). Considering the enormous spatial and temporal variability of precipitation, which is even more true of extreme precipitation, it must be noted that the spatial regularity in the distribution of the regional groups proves an existence of factors that govern the occurrence of origin-based extreme precipitation types in Europe. Some of these groups include weather stations scattered across the study area, but rather than random their

Table 1.1 Seasonal distribution of regional groups of origin-based types of extreme precipitation

GR	Spring	Summer	Autumn	Winter
GR1	Southern Europe between the Atlantic Ocean and the Caspian Sea and southern part of Central Europe	Northern part of southern Europe	Western Scandinavian coast, coastal stations in Western Europe and Central Europe (Baltic coast) and in mountains	Northern Europe and western coast of the Scandinavian Mountains
GR2	Western, Central and Eastern Europe and western Scandinavian coast	Southern Europe	Northern part of southern Europe and southern part of Western and Central Europe	Northern part of southern Europe, Eastern Europe
GR3	Western Scandinavian coast and northern part of south-western Europe	Western Scandinavian coast, mountain stations and Central European coastal stations, isolated stations in Eastern Europe	Western Scandinavian coast, northern part of south-western Europe	Scandinavian coast, northern part of south-western Europe
GR4	Northern Europe	Northern part of Western, Central and Eastern Europe and south-western part of the Scandinavian Peninsula	Western Europe, Northern part Central Europe and Eastern	Western and Central Europe
GR5	Eastern and Western part of southern Europe	Southern part of Western and Central Europe and the northern part of southern Europe	Southern Europe	Scattered stations in southern, Central and northern Europe
GR6	Southern part of the Scandinavian Peninsula	Northern Europe	Northern Europe (lee side of the Scandinavian Mountains, Atlantic islands)	Northern Europe on the lee side of the Scandinavian Mountains, Iceland and Great Britain

GR regional groups

distribution is an effect of specific local influences on the processes leading to the development and controlling the amount of extreme precipitation.

These regional groups synthesise the most important features of the spatial and seasonal variability of origin-based types of extreme precipitation (structure of ExPT occurrence). In the regional groups of stations located in northern Europe (GR6 in all seasons), the largest proportion of extreme precipitation is associated with occluded fronts (~50 % ExP in winter, in autumn and in spring and ~40 % ExP in summer). In summer, air-mass precipitation is the dominant origin-based type in three regional groups covering southern Europe (GR1, GR2, GR3). In the southernmost regional group, GR2, type A precipitation accounts for almost 70 % of extreme precipitation. In autumn, air-mass precipitation is the most frequent origin-based type in two regional groups (GR1, GR5), while in spring and in winter it is only in one of them (GR5). In autumn, one of these groups includes stations in the southern part of the continent, but in this group type A precipitation accounts for a considerably lower proportion of extreme precipitation (just short of 40 % ExP) than in summer. In this regional group much ExP is also associated with cold fronts (more than 25 % ExP). The other group identified in autumn featuring the highest frequency of type A precipitation, just like similar regional groups identified in spring and in winter, covers coastal and mountain stations. An important role in the shaping of precipitation in these groups is filled by local factors, mainly ground relief and dynamic process at the contact of land and ocean. In every season, the highest European frequencies of precipitation associated with a cold front are found within regional groups covering stations in the southern parts of Western, Central, and Eastern Europe, but this type accounts for the highest proportion of extreme precipitation (just short of 30 % ExP) only in one regional group identified in summer (GR5). In the other groups, the maximum frequencies are accounted for by precipitation associated with the passage of various fronts (type Ff). Type Ff precipitation in general accounts for the largest proportion of extreme precipitation in regional groups covering Western and Central Europe, as well as the northern and central parts of Eastern Europe. This type reaches its highest frequency (~50 % ExP) in the winter group GR4 that covers Western and Central Europe. In general, precipitation associated with the passage of various fronts accounts for the largest part of extreme precipitation in winter across Europe, when the rate of cyclone movement in the area is the highest. In this season, type Ff accounts for the largest proportion of extreme precipitation in four regional groups of stations.

The frequency of extreme precipitation in Europe depends significantly on atmospheric circulation. In many parts of Europe, the occurrence of extreme precipitation is linked to the intensity of the zonal circulation measured with the NAO index, which was analysed by comparing empirical distribution functions of NAO on days with ExP and on days without ExP. Both the strength and the spatial extent of these relationships are seasonal in nature, with the peak in winter, and vary depending on the origin-based type of precipitation in question. In autumn, the influence of NAO on ExP occurrence is weaker than in winter, but clearly stronger than in summer and in spring. In winter, across most of Europe (including northern, Western, and Central Europe), a majority of extreme precipitation coincided with the positive NAO phase and with an inflow of moist air masses associated with it over the parts of Europe involved. The majority of precipitation events recorded at that time in Western Europe are associated with a cold front (Fc), while in the south of the Scandinavian Peninsula these are associated with an occluded front (type Fo). In the southern part of the continent the air advection from over the Atlantic Ocean and the occurrence of extreme precipitation are associated primarily with a negative NAO phase. This extreme precipitation is associated with the passage of various fronts (type Ff) or with an occluded front (type Fo). In autumn, the nature of the relationship between ExP and NAO changes: most of the extreme precipitation recorded in Central and

Western Europe, as well as in the western part of the Iberian Peninsula, is observed during NAO–, while that along the western Scandinavian coast and on the Atlantic islands is related to NAO+. The negative NAO phase in Western and Central Europe, as well as in the southern part of Eastern Europe, mostly coincides with type Ff and type Fo precipitation. NAO+ favours the occurrence of these two precipitation types in the northwestern part of Europe, including the Atlantic islands and the western Scandinavian coast.

The influence of synoptic-scale circulation on the occurrence of extreme precipitation varies among precipitation types. The frequency of various origin-based ExP types in Europe is linked to the direction of air advection and to the type of the pressure system, but varies from region to region. All origin-based types of extreme precipitation (both air-mass and frontal) occur in both cyclonic and anticyclone situations, with the former accounting for a higher frequency.

Relationships between air-mass precipitation and atmospheric circulation are characterised by less regional variability than these between circulation and the occurrence of frontal precipitation. Extreme air-mass precipitation displays the strongest link with atmospheric circulation in areas including southern Europe, mountainous areas, the southeastern North Sea coast, and the western slopes of the Scandinavian Mountains. Type A precipitation in southern Europe occurs during air advection from the southeastern sector (situation E+SEc) during the whole year. Elsewhere in Europe this type is normally observed during air advection from the northern sector. Along the western Scandinavian coast its occurrence is particularly favoured by situations with advection from the northwestern sector (situation W+NWc), while in mountainous areas deeper in the continent it occurs the most frequently during air advection from the northwestern sector (situation N+NEc), especially in summer.

The occurrence of extreme frontal precipitation (type F) is associated both with the direction of air advection and with the type of the pressure system. This relationship remains stable between seasons in anticyclone systems, whereas in cyclones there is visible seasonality. The strongest dependencies between the frequency of type F precipitation and anticyclone situations was found in the Alps, in the central part of the southern tip of the Scandinavian Peninsula, and on the lee side of the Scandinavian Mountains. In the Alps the occurrence of type F precipitation, just as type A, is favoured by the northern advection, especially type W+NWa. In the central part of the southern tip of the Scandinavian Peninsula and on the lee side of the Scandinavian Mountains, type F precipitation displays the strongest connection with southern advection, especially from the southeast (situation S+SEa). A conditional probability analysis of the occurrence of type F precipitation also suggests that in autumn and in winter along the western Scandinavian coast this type is the most likely in situations N+NEa and W+NWa, although in the southern part of Norway during the whole year in situation S+SWa.

The relationship between frontal precipitation and cyclonic situations is much more complicated. In northern Europe (except the western coast of the Scandinavian Mountains) and in Eastern Europe, extreme frontal precipitation has the strongest link with air advection from the southeast (situation E+SEc). In Central Europe the highest frequency and the highest probability of the occurrence of type F precipitation coincide with air advection from the northeastern sector (situation N+NEc). In Western Europe, its occurrence is favoured by air advection from the northwestern sector (situation W+NWc). All these relationships gain in strength in autumn and in winter. Throughout the year, type F precipitation in the British Isles and in the southern part of the Scandinavian Peninsula coincides mostly with air advection from the southwest (situation S+SWc). In the British Isles, the probability of type F precipitation in situation E+SEc is comparable to that in situation S+SWc.

The breaking down of extreme frontal precipitation into the types of fronts that produced it helped to better define the relationship between its occurrence and atmospheric circulation, especially in cyclonic synoptic situations. This relationship between precipitation associated with a cold front (type Fc) and anticyclonic situations is weak, but discernible in specific areas, such as the Alps and, in certain seasons, also in the Iberian Peninsula and along the western Scandinavian coast. The Alpine precipitation associated with a cold front is favoured by northern air advection, especially situation W+NWa, regardless of the season. Along the western Scandinavian coast the frequency of type Fc precipitation remains low in all anticyclone situations, but the analysis of conditional probability demonstrated that its occurrence in situation N+NEa is more likely than that in other anticyclone situations, especially in winter and in autumn. Along the western Iberian coast the situation W+NWa (in summer and in winter) coincides with increased frequencies of type Fc precipitation, although this relationship is not confirmed by the results of conditional probability analysis.

Extreme precipitation of type Fc displays a stronger relationship with cyclonic situations. In southeastern Europe the occurrence of this type is normally favoured by advection from the western sector. Around France this type coincides with advection from the northwestern sector (situation W+NWc, especially in winter and in autumn) whereas along the western Iberian coast its probability is similar or, at some stations, slightly higher in situation S+SWc. Advection from the northern sector favours the occurrence of type Fc extreme precipitation in the southern part of Eastern Europe. In the Alps, type Fc is linked to air advection from the northern sector in both cyclonic and anticyclonic situations, although its probability in lows is higher.

In Europe, precipitation generated by warm fronts (type Fw) is relatively rare. It displays a clear relationship with atmospheric circulation in cyclonic situations; this is the strongest in Eastern Europe and slightly weaker in a broadly defined Central Europe that reaches up to the Alps and the western part of Eastern Europe. In Eastern Europe it coincides with southeastern advection (situation E+SEc) throughout the year and especially in winter. In the central part of Europe this type is normally linked with air flowing from the northeast (situation N+NEc), but with seasonal variations: in spring, this is true of the territory of Poland, in summer in western part of Eastern Europe, while in autumn in southern Poland and in areas to the south of it between the Alps and the Black Sea.

Precipitation associated with the passage of various fronts (type Ff) is the most frequently occurring origin-based precipitation type, and for this reason the relationship between its frequency and atmospheric circulation follows a similar pattern to that of frontal precipitation (type F), especially in cyclonic situations. However, because type Ff precipitation accounts for the largest proportion of extreme precipitation in Western Europe, its occurrence in winter is favoured by advection from the northwestern sector (situation W+NWc). The frequency of type Ff precipitation in anticyclone situations is lower than that of type F precipitation, but its spatial variability is similar. The only difference concerning Ff type is a lack of the relationship between northern air advection and this type in the Alps in autumn and in spring.

Precipitation generated on occluded fronts (type Fo) accounts for the largest proportion of extreme precipitation in northern Europe. The area the most prone to occurrence of this precipitation type is the central part of the southern tip of the Scandinavian Peninsula. There, type Fo precipitation is favoured by air advection from the southwestern sector in both cyclonic and anticyclone situations (situations S+SWc and S+SWa). Elsewhere in Europe type Fo is only related to cyclonic situations. In northern Europe, especially on the lee side of the Scandinavian Mountains, in the British Isles, and in Eastern Europe, the occurrence of ExO type Fo, especially in autumn and in winter, is linked to southeastern air advection. In summer, along the western Scandinavian coast and the southeastern North Sea coast, most of the extreme precipitation is related to occluded fronts during air advection from the northern sector (situations N+NEc and W+NWc). Extreme precipitation of type Fo displays the strongest relationship with the type of the pressure system, which most frequently is the cyclonic system.

This study provides the first pan-European comprehensive synoptic and climatic analysis of extreme precipitation to take into account both circulation types and atmospheric front types. Six regional groups featuring seasonally different patterns of the occurrence of origin-based extreme precipitation types were identified. Regional and seasonal variability of the relationship between the occurrence of origin-based types of extreme precipitation and atmospheric circulation was also found.

1.2 Motivation and Scope

Although the key determinants of the occurrence and amount of precipitation are generally known, the nature of this relationship, which is subjected to both spatial and seasonal variability, still requires detailed research, in particular in the area of extreme precipitation. As is shown in the next section (Sect. 1.2), existing knowledge about the relationships between extreme precipitation and atmospheric circulation is, for the most part, regional in nature. As a rule, analysing the dependence of extreme precipitation events on atmospheric circulation involves determining the relationships between the occurrence of such events and model synoptic situations, which tend to be defined in different ways depending on the region. Another approach to the issue involves identifying the synoptic situations that generate extreme precipitation. Very little attention has been devoted to the role of atmospheric fronts in shaping precipitation. The problem has been addressed by research conducted for weather forecasting purposes, mainly through case studies. Climatology-oriented studies are few and are based on local research (e.g., Twardosz 2005; Raddatz and Hanesiak 2008; Mätlik and Post 2008). Recently, three global-scale publications dedicated to climatology of atmospheric fronts have been released (Berry et al. 2011a, b; Catto et al. 2012), yet the automatic front location method used in those studies provides very general information about the occurrence of fronts.

To date, there have been no wide-ranging temporal or spatial climatology studies to explain the nature of the relationships between the occurrence and amounts of precipitation and the associated types of weather fronts. There is no doubt as to the impact of atmospheric processes associated with weather fronts on extreme precipitation (Kašpar 2003; Berry et al. 2011b), especially at moderate latitudes, where this element of atmospheric circulation is a key weather-forming factor (Berry et al. 2011a, b; Catto et al. 2012). To date, the problem, which is of crucial importance in climatology, has not been addressed on a continental or long-term scale, probably because of the ambiguity of the criteria that would allow weather fronts to be automatically mapped, all the more so that the conditions in which fronts are actually formed do not always correspond to those criteria (Yarnal 2000; Barry and Carleton 2001). As a consequence, such studies require considerable efforts and resources.

This study discusses the climatology of extreme air-mass and frontal precipitation and its relationship to atmospheric circulation on a continental scale. The issues discussed in this study supplement existing knowledge about the relationships between high precipitation and weather fronts, which are a major component of atmospheric circulation. To date, climatology literature has not discussed the problem in such broad terms, that is, at a continental scale. There have been very few studies on the relationship between extreme precipitation and atmospheric circulation taking into account weather fronts (see Sect. 1.2: Synoptic background of extreme precipitation in Europe in the light of literature). The sources from which meteorological and synoptic data were taken, the meteorological data selection and homogenisation procedure, as well as the characteristics of the data selected, are described in Sect. 1.3: Meteorological and synoptic data: sources, selection procedure, and homogeneity.

The diversity of definitions of an extreme event translates into a diversity of existing extreme event indices, including those applied to extreme precipitation. This book consists of eight chapters. The definitions of extreme events and key indices of precipitation extremes are discussed in Chap. 2: Definitions and Indices of Precipitation Extremes. This chapter, which provides a background for further deliberations, discusses, by using selected extreme event indices, the amount and frequency of extreme precipitation in Europe, taking into account "absolute" extremes, and goes on to describe and explain the choice of the definition of extreme precipitation in this study.

Chapter 3, Origin-Based Types of Extreme Precipitation, describes the resultant classification of the types of extreme precipitation origin based on fundamental dynamic processes leading to precipitation formation and on the presence or absence of weather fronts on days with extreme precipitation. This part of the study also presents the spatial and seasonal variability in the occurrence of extreme precipitation according to the different types identified.

The apparent spatial and seasonal variability in the occurrence of individual extreme precipitation types on the continent motivated me to synthesise the results by regionalising them, and thus to identify groups of stations characterised by similar structure of extreme air-mass and frontal precipitation occurrence, discussed in Chap. 4, entitled Regionalization of Extreme Precipitation Types Occurrence in Europe. The chapter also includes the characteristics of the regional groups identified.

In Europe, weather conditions, including the occurrence and amount of precipitation, depend significantly on the intensity of westerly air flow, which is described by the North Atlantic Oscillation (NAO) Index. Chapter 5, Relation Between Extreme Precipitation Occurrence and North Atlantic Oscillation, describes the impact of westerly air flow on extreme precipitation occurrence in Europe. For this purpose, at each of the meteorological stations under study the probability density function of daily values of the NAO Index on days without precipitation was compared with that on days with extreme precipitation. Moreover, conditional probability of extreme precipitation occurence in positive and neagtive NAO phases was calculated including for extreme precipitation origin-based types.

The distribution of weather fronts is determined by spatial changes in atmospheric pressure at sea level, which, in turn, determine the directions of the movement of air masses differing in thermodynamic properties. Chapter 6, Air-Mass and Frontal Extreme Precipitation Occurrence in Circulation Types, analyses the relationship between the occurrence of extreme air-mass and frontal precipitation and atmospheric circulation. To this end, a methodologically uniform continent-wide catalogue or, strictly speaking, catalogues of synoptic situations were developed, for each grid point spaced at 2.5° in longitude and latitude between 30°N and 80°N and between 30°W and 70°E. The uniform classification of synoptic situations on a continent-wide scale, prepared for geographic grid points, takes into account local characteristics of atmospheric circulation, which are of crucial importance for extreme precipitation. The relationships between the occurrence of each type of extreme precipitation and atmospheric circulation were defined by analysing the frequency of these types in high-pressure and low-pressure systems, and in the individual synoptic situations defined. To complement the frequency analysis, the conditional probability of the occurrence of the precipitation types under study in the synoptic situations was also estimated. In Sects. 6.2 and 6.3 a simplified typology of synoptic situations (see Sect. 6.1) was used because of the relatively small number of extreme precipitation types at the individual stations.

The changing nature of atmospheric circulation during the year required adopting a seasonal approach in the research, in which the year was subdivided into standard climatic seasons (spring: March, April, May; summer: June, July, August; autumn: September, October, November; winter: December, January, February). For the most part, the results of calculations are presented in maps presenting the spatial variability of the precipitation characteristics analysed. The maps are supplemented with histograms showing the percentage of the stations where the values of the characteristics under study fell within the respective ranges adopted for them. The class ranges in the histograms that correspond to the precipitation index value ranges in the spatial distribution maps, vary in width. As the maps were being prepared, a number of tests were conducted, showing that making use of different-width ranges reflects in a better way the spatial variability of the occurrence of the extreme precipitation types under investigation. For most of the precipitation characteristics, the key descriptive statistics were calculated, including continent-wide arithmetic means. It must be stressed that, even though continental means constitute very general information about the precipitation characteristics under study, when accompanied by other statistics describing the precipitation characteristics structure more accurately, they present well the diversity of the occurrence of the extreme precipitation types in Europe across the seasons.

The statistical methods used in this study are discussed in detail in the initial section of each chapter, which should facilitate understanding the research results presented in the successive parts of the book.

Atmospheric droughts, which are treated by many climatologists as opposite phenomena to extremely high precipitation, do not fall within the scope of this study. The occurrence of droughts depends not only on absence of precipitation, but also on other meteorological elements (Klementová and Litschman 2001; Brunovský et al. 2009). Extremely low precipitation, which occurs relatively frequently, does not satisfy the extremity criteria specified by the statistical definition of extreme events adopted in this study. Undoubtedly, droughts are an important characteristic of the precipitation regime, yet given the underlying assumptions of this study, they do not fall within the scope of the study.

1.3 Synoptic Background of Extreme Precipitation in Europe in the Light of Literature

As early as 1937, A.C.W. Baskin stated, in his book entitled *Factors Governing Different Types of Precipitation*, that the causes of precipitation have always interested man because they determine man's prosperity, contentment, and existence. Baskin's (1937) view is confirmed by very extensive scientific literature about precipitation, which could hardly be listed in a single study. This chapter discusses selected and, as the author believes, crucial studies addressing the dynamic mechanisms leading to extreme precipitation, as well as studies on the relationship between extreme precipitation occurrence and atmospheric circulation. Such studies are relatively few among all the studies about extreme precipitation, which usually address precipitation trends on different temporal and spatial scales (e.g., in Europe and its parts: Groisman et al. 1999; Heino et al. 1999; Zolina et al. 2004, 2005; Moberg and Jones 2005; Klein Tank et al. 2009; Kürbis et al. 2009; Łupikasza et al. 2011, in Spain: Beguería et al. 2010, in Croatia: Gajić-Čapka and Cindrić 2011, in the Iberian Peninsula: Garciá et al. 2007; Rodrigo 2009; López-Moreno et al. 2010, in Germany:

Hundecha and Bárdossy 2005; Trömel and Schönwiese 2007, in the Czech Republic: Kyselý et al. 2007; Kyselý 2009, in Poland: Lupikasza 2010b, in the Mediterranean Region: Norrant and Douguédroit 2006; Ramos and Martinez-Casanovas 2006, in the United Kingdom: Osborn and Hulme 2002; Maraun et al. 2008; Burt and Ferranti 2012, in Italy: Pavan et al. 2008).

The earliest studies about the origin of precipitation were dedicated to determining its organisation in mesoscale cyclonic systems. Even though the general principles governing the distribution of clouds, precipitation, and other weather components within low-pressure systems have been known from the earliest days of synoptic meteorology (Abercromby 1887), learning the detailed structure of cyclonic systems became possible only thanks to the use of radars in meteorology. Thus, intensive studies in the area were carried out in the 1960s and the early 1970s. The distribution of precipitation zones within frontal systems in the West Coast of the United States was first described in 1962 (Nagle and Serebreny 1962). Two years later, Elliott and Hovind (1964) published the results of their research dedicated to identifying convective precipitation zones within the fronts passing in the area of the Californian coast. The key pioneer studies of the structure of frontal systems and the precipitation zones occurring within them include Kreitzberg (1964), Elliott and Hovind (1965), Nozumi and Arakawa (1968), Kreitzberg and Brown (1970), Austin and Houze (1972), and Marwitz (1972). An important result of the last of these studies is the discovery that the precipitation occurrence structure in each of the cyclonic (thunderstorm) systems and the nature of the precipitation are similar. The conclusion is important because it allowed the low-pressure system pattern identified to be used for more in-depth studies of precipitation mechanisms at scales smaller than the synoptic scale.

Later studies into precipitation-producing mechanisms focused on refining and verifying the prognostic models applied to heavy precipitation representing a threat to humans. They were dedicated to analysing the development of convective systems and the nature of associated precipitation (Kane et al. 1987; Market et al. 2003) or to researching such processes as the formation of convective or *stratiformis* precipitation (Anagnostou and Kummerow 1997; Houze 1997; Bolliger et al. 2004; Mohr 2004; Anip and Market 2007; Langer and Reimer 2007). The foregoing studies demonstrate that the most intensive precipitation is associated with the mature stage of the development of a convective system, and appears in areas with inflowing warm and moist air. Its character differs depending on the mechanism initiating the convection, which develops in a sector of warm air and is associated with a warm or stationary front. Convection associated with cold fronts results in higher precipitation over vaster areas than convection in a sector of warm air in a typical low. Smaller convection systems, which occur in summer, represent free convection, whereas larger ones, which occur in spring, are associated with large-scale atmospheric processes (Kane et al. 1987). Precipitation efficiency in convective systems is positively correlated with humidity in the layer below the condensation level, change in wind direction with altitude, and convective inhibition (Market et al. 2003). It is higher than in *stratiformis* systems and is characterised by annual variability with a maximum in summer and minimum in winter (Anip and Market 2007).

In 1997, R. Houze published the results of his research on the origins of rainfall within the intertropical convergence zone. He made, as he himself states, a paradoxical discovery that the convective cells that occur there also produce *stratiformis* precipitation, which is associated with the so-called old convection. In the same year, Anagnostou and Kummerow (1997), using satellite photographs, developed a scheme for classifying clouds producing convective and *stratiformis* precipitation. The model allowed distinguishing areas characterised by a high (70 %), moderate (40–70 %), and low (40 %) probability of the occurrence of *stratiformis* precipitation. Bolliger et al. (2004) carried out similar research to distinguish convective precipitation areas from *stratiformis* precipitation areas based on the cloud regimes in the Alps. However, the researchers did not obtain satisfactory results, finding that most of the cases were hybrid situations. Similar studies, including the use of satellite imagery, were conducted by Langer and Reimer (2007) in Germany. However, the satellite imagery did not prove to be accurate enough to allow convective and *stratiformis* precipitation areas to be recognized. The diurnal course of convective precipitation in desert areas, as studied by Mohr (2004), is highly diversified, which is attributable to the high variability in the occurrence and development of organised convective systems.

In Europe, the Alps and the Mediterranean Basin are the areas most exposed to extreme precipitation, and, as a result, they have been a main focus of research on precipitation origins. The mesoscale convective systems that cause intensive precipitation in the northern Alps are characterised by linear organisation. A specific feature of the systems is an area of *stratiformis* precipitation stretching behind the system line (Hagen et al. 2000). The convective systems with cold fronts, which are slowed down by the orographic barrier created by the mountain range in question, move over the Alps from the west eastwards (Steinacker 1981; Frei et al. 2000; Hagen et al. 2000; Kljun et al. 2001; Pradier et al. 2004). Such situations, enhanced by inflowing moist air from the southern sector, contribute to the occurrence of high orographically intensified flood-producing precipitation, especially on northern slopes of the Alps (Sénési et al. 1996; Buzzi et al. 1998; Massacand et al. 1998; Buzzi and Foschini 2000; Ferretti et al. 2000; Schneidereit and Schär 2000; Houze 2001; Rotunno and Ferretti 2001; Pradier et al. 2004). However, the occurrence of extreme precipitation in the Alps is associated with a range of synoptic situations and with land features. The situations include cyclonic troughs in the upper layers of the troposphere, Genoa lows, reactivated under the influence of the mountain range in question, incoming fronts from over the Atlantic and convection of unstable air masses, in which orography is particularly significant (Tibaldi et al. 1990; Cacciamani et al. 1995; Grazzini 2005; Hoinka et al. 2006). The synoptic situations most conducive to the occurrence of extreme precipitation in the Alps occur most frequently in autumn (57 % of the cases), with their frequency in the other seasons reaching 30 % in summer, 18 % in spring, and 1 % in winter (Hoinka et al. 2006).

The research of extreme precipitation in the Mediterranean Basin included both case studies focusing on isolated precipitation events or several selected events (Dutton and Dougherty 1979; Watson et al. 1982; Millán et al. 2005; Milelli et al. 2006; Federico et al. 2008a; Nuissier et al. 2008) and research spanning longer periods, of at least a dozen or so years, thus acquiring the nature of climatology studies (Llasat and Puigcerver 1997; Mizrahi 2000; Tolika et al. 2007; Lorenzo et al. 2008; Federico et al. 2008b; Fragoso and Gomes 2008; Toreti et al. 2010). The foregoing case studies (dating back to the 1970s, 1980s, and 1990s) focussed on analysing convective precipitation. Present-day research aims towards identifying synoptic situations producing extreme precipitation.

Llasat and Puigcerver (1997) determined the share of convective precipitation in overall precipitation totals in Catalonia. Llasat and Puigcerver identified convective precipitation using an intensity-based criterion, yet they did not use the threshold applied in Southern Europe of 50 mm*h^{-1} (Dutton and Dougherty 1979; Watson et al. 1982), above which precipitation is considered to be convective. Instead, they differentiated between nonconvective precipitation, low-intensity convective precipitation (approximately 0.8 mm*min^{-1}), moderate- and high-intensity convective precipitation (greater than 0.8 mm*min^{-1}), and storm precipitation. In Catalonia, convective precipitation accounts for approximately 25 % of precipitation between January and April. In the months that follow, its share increases, ranging between 70 and 83 %, with a maximum in August. Convective precipitation represents more than half of annual precipitation (59 %). The distribution of precipitation in the area is highly influenced by orography and topography, as well as the distance from the sea (Llasat and Puigcerver 1997). The results of similar studies in Catalonia completed by Llasat et al. (2005), based not only on instrumental records but also on radar data, indicate that there is less convective precipitation than identified in earlier studies, because it represents merely 8 % of precipitation events in Catalonia, its share in annual precipitation total reaches approximately 36.5 %, and it predominantly occurs in autumn. Of this 8 %, precipitation resulting from low-intensity convection represents 3.8 %, that associated with moderate convection is 2.9 %, whereas that arising from strong convection is 1.3 %. Precipitation resulting from strong convection lasts less than 1 h, whereas that associated with less intensive convection is longer in duration. Flood-producing precipitation in the area is usually associated with moderate convection, lasting from several hours to several days (Llasat et al. 2005).

In the Valencia Region in the Iberian Peninsula, three types of precipitation were differentiated depending on origin (Millán et al. 2005). These types include precipitation associated with the Atlantic systems of fronts, which appears on the windward (western) side of the mountains in the region, convective precipitation associated with the Iberian Peninsula thermal low, which appears on the eastern side of the mountain ranges and is highly enhanced by the orographic effect, and precipitation associated with advection of the air from the east, from over the Mediterranean Sea, which covers the eastern coast of the Iberian Peninsula at times when a high occurs over Central Europe (Pastor et al. 2001; Millán et al. 2005; Milelli et al. 2006; Nuissier et al. 2008). In turn, flood-producing precipitation in the northwestern part of the Mediterranean Basin is caused by the deep trough extending from the British Isles towards the Iberian Peninsula, leading to an inflow of moist air from the south and southeast, enhanced by orographic effects (Milelli et al. 2006). As is noticed by Federico et al. (2008a, b), high rainfall in the Mediterranean area may appear in different synoptic situations, also those not typically causing extreme precipitation events. Recent studies show that in many such cases, the source of humidity needed for them to come into being was located outside the Mediterranean Basin, as a result of which its transport was crucially important for such precipitation to occur (Krichak et al. 2004; Turato et al. 2004; Federico et al. 2008a). The advection of moist and warm air from the south associated with the presence of a dynamic low from the Bay of Cádiz also drives extreme precipitation in southern Portugal (Fragoso and Gomes 2008). As is shown by studies in the western part of the Iberian Peninsula, which is no longer treated as part of the Mediterranean Region, the largest proportion of intensive rainfall in Galicia, which lies in the northwest of Spain, occurs in cyclonic situations involving an inflow of air from the west and southwest with a centre of a low lying over the area (Lorenzo et al. 2008). The tracks of low-pressure systems crossing the area are highly correlated with the NAO Index, especially in winter (Zorita et al. 1992; Esteban-Parra et al. 1998; Rodriguez-Puebla et al. 2001; García et al. 2005; Trigo 2006).

In the central part of the Mediterranean Region, that is, in Calabria, which lies in southern Italy, extreme precipitation is determined by 11 synoptic types, the most important of which are associated with cyclones moving over the Mediterranean Sea from the west eastwards when extreme precipitation is recorded in the west of Calabria, and with cyclones moving from the north to the south when precipitation occurs in eastern Calabria. In such a case, convective precipitation is associated with the low-pressure systems occurring in the area of the Gulf of Genoa and the Balkans. In the area, orography is also important in the forming of extreme precipitation, especially at times of advection of air from the south, when low-pressure systems move from Gibraltar towards the centre of the Mediterranean Region (Federico et al. 2008b). In northern Italy, the formation of extreme rainfall is highly influenced by meridional flow (Busuioc et al. 2008). In the eastern part of the Mediterranean Sea, where extreme precipitation is mainly recorded in cyclonic situations, the highest frequency of such precipitation characterises days with air flowing from the WSW direction (Tolika et al. 2007). Toreti et al. (2010) differentiated three types of synoptic situations fostering extreme precipitation in the western part of the Mediterranean Basin and two situations leading to intensive precipitation in the east of the Basin. The airflow in the types of situations characteristic of the western and central parts of the Mediterranean Region is determined by the atmospheric centres of action located between Spain and northwest Africa (negative) and the northern part of the Scandinavian Peninsula, as well as over the Gulf of Lion and the Balearic Sea (negative) and over the northern part of the Atlantic (positive). The third type of situations consists of the following three atmospheric centres of action: a trough stretching from the UK towards Africa and two centres marked by positive anomalies in the subtropical part of the north Atlantic and northeast Europe (Toreti et al. 2010). Intensive precipitation in the eastern part of the area depends on the presence of a trough stretching between from the Baltic region towards northeast Africa, which is associated with pressure centres over the North Atlantic and the east part of the Mediterranean Sea. Another situation producing extreme precipitation events results from a negative pressure anomaly with the centre over the Aegean

Sea and a positive anomaly located over the North Sea (Eshel and Farrell 2000; Toreti et al. 2010).

In the central part of east France, which lies in the border area of the Mediterranean region, the relationships between extreme precipitation occurrence and atmospheric circulation are similar to the western part of the Mediterranean area, especially in the Iberian Peninsula. The occurrence of extreme precipitation phenomena in that area is associated with the inflow of air from the south caused by the low-pressure area located near the Bay of Biscay, and by the inflow of air from the west (oceanic inflows). The foregoing situations have crucial importance for producing heavy precipitation in the cold half of the year. By contrast, in summer convection appearing in the presence of small pressure gradients is important (Mizrahi 2000).

The origins of flood-producing rainfall were also investigated on the British Isles (Browning and Hill 1984; Collier and Hardaker 1996; Wheeler 1997; Wilby 1998; Hand et al. 2004; Little et al. 2008; Maraun et al. 2010). The results of the former research (Collier and Hardaker 1996) indicate that mesoscale convective complexes, which are characterised by the co-occurrence of convective and *stratiformis* precipitation, are rare in the United Kingdom and usually associated with deep inland convection (Browning and Hill 1984). Wilby (1998) examined the relationships between extreme rainfall and circulation types in central and southern England, taking into account weather fronts. In those areas, extreme precipitation occurs mainly in cyclonic situations (61 %). Approximately half as many extreme precipitation events are associated with the types characterised by a pronounced airflow (32 %) and only 7 % with anticyclonic types. Weather fronts were present during 85 % of the cases examined (Wilby 1998). The studies by Hand et al. (2004) and Little et al. (2008) are particularly valuable, given their long-term—and thus climatology—nature, which allows drawing universal conclusions. As is demonstrated in the study by Hand et al. (2004), which comprised 50 extreme precipitation events recorded on the British Isles in the twentieth century, most of them were convective in nature, 30 cases; 15 cases were classified as frontal, and 5 as orographic. Extreme precipitation associated with free convection is recorded, above all, in the central and southern areas of Great Britain, and in summer, also in lowland areas (Little et al. 2008). All the cases recognised as frontal occurred in coastal areas (Hand et al. 2004). However, later research showed that frontal precipitation may extend over the entire country (Little et al. 2008). Among the frontal cases, the highest precipitation totals were associated with a stationary front, with the highest proportion of frontal precipitation associated with a warm front or occlusion having the nature of a warm front (80 % of cases). A large proportion of the extreme events studied occurred in the presence of a low-pressure system with its centre to the south or east of the precipitation area. As a rule, orographic precipitation was associated with a high-pressure system in the area of the Bay of Biscay or Spain and the resultant advection of air from the west or southwest (Hand et al. 2004; Hanna et al. 2008; Little et al. 2008). In the central part of the east coast of England, most of the precipitation appears during cyclonic weather types and during advection of air from the north towards the southeast (Ross et al. 2006).

There are relatively few publications investigating the origins of extreme precipitation or examining the relationships between its occurrence and atmospheric circulation elsewhere in Europe. Using the Lamb classification (1972), Hellström (2005) determined the relationships between the occurrence of extreme precipitation in Sweden and atmospheric circulation. Most of the extreme precipitation in that part of Europe occurs during cyclonic weather types and is fostered by advection of air from the south. A vast majority of extreme precipitation associated with advective circulation types was accompanied by atmospheric fronts (90 % of events), and this was also true for events recorded during anticyclonic weather types, where fronts were present during 75 % of the events (Hellström and Malmgren 2004).

Estonian research of synoptic conditions leading to very heavy precipitation, in excess of 50 mm*24 h^{-1}, proves that approximately 88 % of such events are associated with low-pressure systems and fronts, and merely 12 % of them are convective in nature (Mätlik and Post 2008). Usually, such low-pressure systems arrive in Estonia from the south, with their source areas being the Black Sea or the Mediterranean Region; a much smaller proportion of them originate over the Baltic. Precipitation-bearing fronts flow onto Estonia from the west and northwest. Fronts inflowing from the north rarely result in extreme precipitation (Mätlik and Post 2008).

In Belgrade, high daily precipitation totals are mainly associated with cold weather fronts. Such precipitation originates during unstable atmospheric conditions within cold air masses and in situations when a low builds up over the Mediterranean Sea with the centre located over the southwest coast of the Black Sea (Unkašević and Radinović 2000).

In Austria, the relationship between the occurrence of extreme precipitation and atmospheric circulation differs from region to region. Each of the seven synoptic situations distinguished by Seibert et al. (2007) is conducive to the occurrence of extreme precipitation in different parts of the country. In a situation of a low-pressure gradient, extreme precipitation is more or less uniformly distributed across Austria, with a slight maximum in the southeast regions. The northwesterly airflow across the troposphere levels leads to extreme precipitation, mainly in the northern regions. The fast northwesterly airflow with northerly upper winds drives extreme events associated with orographic uplift in the southern part of central Austria and in the northern and western areas of the country. In Central Austria and in the northern and eastern regions, extreme precipitation is associated with the low that is located to the east of Austria and moves along the Vb track, driving northerly airflow; this leads to the accumulation of moist air masses (German *Stau*) on Alpine slopes, producing intensive rainfall several days long in the mountains and their foreland. The situations, known in Austria as the southerly 'Stau,' westerly 'Stau,' and Vb, generate the highest precipitation: the southerly 'Stau' in the south of Austria, to the north of the main Alpine range and in the eastern region, the westerly 'Stau,' in the west, and Vb in north and east Austria (Seibert et al. 2007).

In Poland, the synoptic conditions of extreme precipitation were studied on a countrywide scale by Ustrnul and Czekierda (2001, 2009) and Lupikasza (2010a).

Many such studies are regional in scope (Morawska-Horawska 1971; Kwiatkowski 1984; Kożuchowski 1986; Cebulak 1992; Bogucka 1998; Wibig and Fortuniak 1998; Kossowska-Cezak and Mrugała 1999; Niedźwiedź 1999; Siwek 2010). In Poland, the highest daily precipitation is recorded during the inflow of air from the northeast sector. The probability that such phenomena will occur is the highest during such air advection and during a trough and a centre of a low-pressure system (Ustrnul and Czekierda 2001, 2009). The flood-producing precipitation recorded in the Carpathian Mountains and the Sudety Mountains results from the occurrence of relatively shallow, stationary, low-pressure systems over southeast Poland, Hungary, or western Ukraine, which cause an inflow of cool and moist air from the northeast sector. Usually, such lows reach Central Europe from over the Adriatic along the Vb track, which was delineated as early as the nineteenth century by van Bebber (1891). The orographic barrier created by the Sudety Mountains and the Carpathians, similarly to the Alps, intensifies cold air-mass accumulation processes, leading to intensive precipitation lasting from 3 to 5 days (Niedźwiedź 1999, 2003a, b; Ustrnul and Czekierda 2001, 2009). Such low-pressure systems are responsible for flood-producing precipitation not only in Poland but also in other countries of Central Europe. A similar situation, that is, a low-pressure system lying, this time, over the border area between Saxony and Poland, which had moved from over England across France, Italy (where it intensified) and further north, caused flood-producing rainfall in 2002 in the Ore Mountains. In addition, on those days, a N–S-oriented stationary front was present over Germany, along which very warm and moist air masses flowing from the south met very cold air incoming from Scandinavia (James et al. 2004).

In their study of cyclonic precipitation trends in Europe, Karagiannidis et al. (2012) prove that high cyclonic precipitation changes correspond to changes in all extreme precipitation events. Interestingly, they also found that the frequency of days with extreme rainfall grows with altitude, which relationship—an increase in extreme precipitation—begins to be clearly observable only above 700 m a.s.l. (Karagiannidis et al. 2012). Other studies of the relationships between precipitation and local conditions in areas with complex land relief, carried out in the Alps and on the British Isles, demonstrate that in addition to the strong relationship with altitude (Sevruk 1997; Frei and Schär 1998; Brunsdon et al. 2001; Bongioannini et al. 2005), precipitation is also strongly related with topography (Sevruk 1997; Frei and Schär 1998; Brunsdon et al. 2001), the prevailing wind direction (Sevruk 1997), circulation of the atmosphere (Frei and Schär 1998), and distance from the sea (Brunsdon et al. 2001).

Importantly, atmospheric circulation in Europe is characterised by intensive zonal flow, which is best described by the NAO (North Atlantic Oscillation) Index (Hurrell and Van Loon 1997). Most research of the relationship between precipitation and atmospheric circulation, as expressed by the NAO Index, has addressed monthly or seasonal precipitation (not daily precipitation as in this study); however, the huge significance of westerly flow for the weather conditions in Europe requires taking into account the key studies concerning that problem: Bojariu and Gimeno (2003), Yu and Zhou (2004), Wang and You (2004), Antunes et al. (2006), Jaagus

(2006), Feidas et al. (2007), Fukutomi et al. (2007), Santos et al. (2007), Mares et al. (2009), Andrade et al. (2011), Black (2012), and Jones et al. (2013), including Polish studies into the relationship between the occurrence of precipitation and the NAO Index and other circulation indices: Kożuchowski and Marciniak (1988), Twardosz (1998, 1999), Wibig (1999, 2001), Kirschenstein (2004), Niedźwiedź and Twardosz (2004), and Niedźwiedź et al. (2009). Studies of the relationship between daily precipitation extremes and the NAO Index are extremely rare and have been conducted in the Iberian Peninsula (Valencia et al. 2012; Ramis et al. 2013). Ramis et al. (2013) did not find any strong relationships among the variables they studied. Extreme precipitation total in the Iberian Peninsula is determined to a greater extent by local circulation (local circulation indices: WeMOi, the western Mediterranean Oscillation; IBEI, Iberian Index) than by NAO (Ramis et al. 2013).

The early 2010s saw the publication of three large-scale global studies dedicated to the climatology of weather fronts (Berry et al. 2011a), the trends in their frequency (Berry et al. 2011b), and the amount of precipitation generated at weather fronts (Catto et al. 2012). These studies, which covered the areas of the globe between 60°N and 60°S, relied on an objective weather front location method, based on data obtained from ERA-40 reanalyses. Basically, the results obtained by the authors prove the regularities, which are known from the literature, concerning the frequency of the occurrence of fronts on Earth. The automatic front location method supplies very general information about the occurrence of fronts. The authors criticise analysing weather fronts based on—as they claim—highly subjective synoptic charts (Berry et al. 2011a). However, as it turns out, they verify the results obtained by the objective method by comparing them with the very same subjective synoptic charts. At the same time, they find the results of front location by the subjective and objective methods to be highly consistent, which proves, in fact, that the information about front location and movement shown by synoptic charts is very valuable. In addition, the authors identify a drop (by 10–20 % in the years 1989–2009) in the frequency of the occurrence of fronts along the tracks of North Atlantic lows (Berry et al. 2011b). Their analysis of the relationship between precipitation and weather fronts, spanning 1997–2008, demonstrates that at moderate latitudes most precipitation is associated with weather fronts (68 % on average), and that in extreme cases frontal precipitation represents 90 % of the rainfall. The authors also find that, over land at moderate latitudes of the Northern Hemisphere, warm fronts generate more precipitation than cold fronts (Catto et al. 2012).

This literature review demonstrates that no research has been undertaken to date to investigate the origins of extreme precipitation, whether air-mass or frontal, and its relationship with atmospheric circulation in climatological terms and at a scale of the continent. In the current era of development and automation of tools allowing fast analysis of huge amounts of data, research involving effort-intensive data collection methods is undertaken less and less frequently. Such studies include identifying extreme precipitation types (air-mass, frontal), which requires information about the occurrence and types of weather fronts. To date, researchers have not developed a reliable automatic method for identifying weather fronts and their direction. The automatic atmospheric front location method (Berry et al. 2011a) relies on a number of assumptions that even though necessary are yet subjective, but do not allow precise and unambiguous front type identification.

1.4 Meteorological and Synoptic Data: Sources, Selection Procedure, and Homogeneity

Homogeneous meteorological data are required to conduct reliable analysis of climatic conditions and their changes. The availability of meteorological datasets, especially daily records, is still limited. Existing online databases contain records from thousands of stations, yet only few of them offer long, gap-free series of homogeneous observation data. Preparing a suitable database that would forms a basis for further climatology analyses is one of the most painstaking stages of climatology research.

1.4.1 Sources of Meteorological and Synoptic Data

Meteorological Data For the most part, daily precipitation totals, which form the basis of this study, come from online meteorological databases (daily data) and synoptic databases (subdaily data) and from archives made available by the national meteorological services of selected European countries:

- European Climate Assessment & Dataset (ECA&D), http://ecad.knmi.nl/.
- Global Historical Climatology Network – Daily (GHCN – Daily), (http://www.ncdc.noaa.gov/oa/climate/ghcn-daily/).
- Hydrometeorology data service system CliWare, http://cliware.meteo.ru/meteo.
- Archives of DWD (Deutscher Wetterdienst), http://www.dwd.de/.
- Archival data made available on the website of the Dutch weather service KNMI (Koninklijk Nederlands Meteorologisch Instituut), http://www.knmi.nl/klimatologie/.
- e-klima meteorological database of the Norwegian weather service NMI (Norwegian Meteorological Institute), http://eklima.met.no.
- Internet weather service containing synoptic (sub-daily) data OGIMET, www.ogimet.com.
- Archives of the Armagh Observatory in the United Kingdom (Butler et al. 1998), http://www.arm.ac.uk/.

Most of the data, except that for Poland, the Netherlands, Norway, Russia, and the Armagh station in Ireland, comes from the ECA&D database, which contains observation data series from 3640 measuring points (as of June 2012). Data from Norwegian stations were obtained from the extensive e-klima meteorological database, which comprises records from 650 stations, both those functioning as part of the NMI and other institutions conducting regular meteorological observations and measurements in Norway. The data for the Dutch and German stations were obtained from the websites of the national weather services of those countries, and data for Russia were taken from the CliWare database.

Notwithstanding the unquestionable value and usefulness of the records from the foregoing databases, when selecting the data, a number of shortcomings and inaccuracies were detected as regard the metadata published in the respective databases and its quality. The most serious shortcoming of the extensive, global GHCN-Daily database, which contains records from 43,000 weather stations, is the lack of reliable information about the method for calculating daily precipitation totals. For this reason, the daily precipitation from this database was only used to fill in the missing data following prior identification of daily precipitation totals. As regard ECA&D, the database is to be criticised for the procedure used to recalculate the daily precipitation totals for the purpose unifying them within the database as a whole; usually the recalculations spanned the last 10 years only. As a result, at some stations the precipitation days changed within the time series. However, the detailed description of each observation series published in ECA&D allows the days to be identified with accuracy. For most of the stations, the daily precipitation totals in the ECA&D database are provided to an accuracy of 1 mm.

The synoptic data in the OGIMET database span a relatively short recent period, since 28 September 1999 until the present day. The database, which contains data from synoptic weather stations worldwide, is an extremely valuable source of information about all meteorological elements. Data in the OGIMET database are available via synoptic reports, which need to be decoded to obtain specific values of the meteorological elements. The daily totals obtained from the OGIMET database were used to obtain missing data and to verify data and were helpful in identifying precipitation day. Verification of precipitation data for the recent years of the research period has been facilitated by the use of the information about meteorological phenomena recorded in the reports as 'present weather' (real-time observations) and past weather (W_1W_2).

Data for Polish stations come, in part, from the database of extreme and hydrological events in Poland, compiled as a result of the research project PBZ-KBN-086/P04/2003 entitled "Extreme meteorological and hydrological events in Poland. (Assessment of events and predicting their impact on the human environment)." Some of the data were taken from *Roczniki Meteorologiczne* yearbooks (1954–1965), and *Opady Atmosferyczne* yearbooks (1954–1981), which are published by the Institute of Meteorology and Water Management (IMGW), and the meteorological database compiled by the staff of the Meteorological Observatory of the University of Silesia in Katowice (Poland), based on daily bulletins of the State Hydrological and Meteorological Institute (PIHM) and daily bulletins of IMGW.

The first stage of the selection involved choosing the stations located within the study area, and then selecting, from among those stations, the data series that met the time criterion, that is, spanned the period between December 1950 and February 2008. The thus selected data were subject to further selection.

Synoptic Data The types of extreme precipitation (frontal, air-mass precipitation) were determined on the basis of surface synoptic charts for the period between 1950 and February 2008. For each day in the study period, synoptic situation charts were collected for at least two measuring times, depending on the period and source: 06 UTC and 18 UTC or 12 UTC and 06 UTC or 00 UTC and 12 UTC. Very many of the synoptic charts were the very well prepared maps of the German weather service (Deutscher Wetterdienst), published in *Europäischer Wetterbericht* in 1976–2000 (times: 00 UTC, 12 UTC), and maps made available by DWD via its internet archives, which span the period since 2003 until the present day (http://www.wetter3.de/Archiv/archiv_dwd.html; times: 00 UTC, 06 UTC, 12 UTC, 18 UTC). Use was also made of charts published by the French weather service (Météo-France) in *Météorologie Nationale Bulletin Quotidien d'Études* in the years 1951–1975 (times: 06 UTC, 18 UTC), and charts of the Polish Institute of Meteorology and Water Management issued in 1975–1979 (times: 00 UTC and 12 UTC), as well as charts published in *Codzienny Biuletyn Meteorologiczny* of IMGW spanning the years 1980–2007 (time: 00 UTC). Many of the synoptic charts used were taken from the Weather Chart Archives of the Climatology Department of the University of Silesia in Katowice. The missing charts were obtained from the library of Zentralanstalt fur Meteorologie und Geodynamik in Vienna. Overall, 45,592 charts were pooled in the study.

1.4.2 Selection and Quality Control of Daily Precipitation Series

Criteria for Data Selection In choosing the weather stations to be used in this study, the precipitation series for the stations located in Europe were selected. Following this, the stations with insufficiently long observation series, that is, shorter than the study period of December 1950–February 2008, were eliminated. At many European stations, such as most of the Norwegian sites, regular instrumental measurements only started in 1961. For this reason, account was also taken of the stations for which precipitation series began between December 1950 and January 1961, as well as the stations where precipitation records were taken at least until the end of 2000.

In the next stage, the data were reviewed for missing values. The percentage of the days with missing data was calculated throughout the study period, that is, between December 1950 and February 2008, and in a standardised 30-year period spanning 1961–1990, which was adopted as the reference period for calculating the threshold values needed for distinguishing days with extreme precipitation. Of the 956 precipitation series meeting the these time criteria, the following series satisfying each of the following conditions were selected:

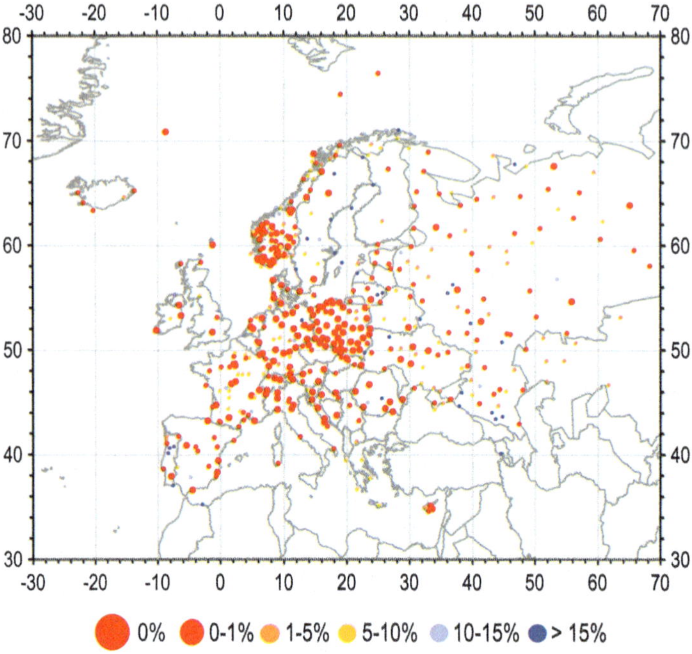

Fig. 1.1 Location of meteorological stations and gaps (% of days) in daily precipitation series over December 1950–February 2008

- In the reference period of 1961–1990 used for determining the threshold values, the level of missing data did not exceed 3 %.
- Throughout the study period (December 1950–February 2008), the level of missing data did not exceed 20 %. Less strict criteria for missing data were applied by Busuioc et al. (2008), and Mätlik and Post (2008), who adopted a threshold of 30 % of days with missing data.

The resultant selection of data included 513 best-quality time series of daily precipitation totals in Europe, the distribution of which is presented in Fig. 1.1.

Homogeneity of Daily Precipitation Data The data published in the ECA&D database, which is crucial for this study, are verified on an ongoing basis by four statistical homogeneity tests (Wijngaard et al. 2003; Klok and Klein Tank 2009): Standard Normal Homogeneity Test SNH (Alexandersson 1986), Buishand Range test (BHR) (Buishand 1982), Petitt test (PET) (Pettit 1979), and Von Neumann Ratio test (VON) (von Neumann 1941). As regard precipitation data, the annual number of days with precipitation of >1 mm was tested. When selecting precipitation data the series marked with the code corresponding to the best-quality data, that is, homogeneous data, were considered. After each update, the data published in the GHCN database are verified by a multi-step QA homogenization procedure (Automated Quality Assurance Procedures) (Durre et al. 2010; Menne et al. 2012).

Overall, the procedure consists of 20 tests, 12 of which are applied with respect to precipitation. The tests are used to check data for outliers identified by several statistical methods, detect series of repeating values and unnaturally long precipitation-free consecutive days, compare the values of corresponding climate characteristics from neighbouring stations, and detect duplicating dates. The QA was described in detail by Durre et al. (2010). However, it must be remembered that assessing the homogeneity of precipitation data, especially at a daily resolution, is virtually impossible because of the enormous spatial and temporal variability of precipitation. Currently, there is still no statistical method available for reliable assessment of homogeneity, not only of precipitation datasets, but also other daily meteorological data (Klok and Klein Tank 2009). As stated by Klok and Klein Tank (2009), even when a meticulous statistical homogeneity testing procedure is applied, data errors cannot be excluded (http://eca.knmi.nl). Therefore, the quality of the data selected was subjected to an additional check, which involved, above all, comparing precipitation series coming from various databases (mainly from ECA&D, GHCN–Daily, and OGIMET) for the same stations. Based on fixed-time synoptic data from the OGIMET database, daily precipitation totals were calculated using two 'precipitation days': between 06 UTC on the day when a precipitation total was recorded and 06 UTC on the next day, and between 18 UTC on the previous day and 18 UTC on the day when a precipitation total was recorded. The differences between the corresponding totals obtained from the foregoing databases were calculated, following which results greater than 1 mm were verified. By comparing data coming from the different databases, it was possible to eliminate error values and control and identify the daily precipitation total for each record in the database created. Erroneous data correction based on statistical testing is exemplified by the 42-mm storm precipitation with snowfall and granular snowfall recorded on 27 February 2004 at the Lerwick station (UK), which was corrected to 5 mm. The homogeneity of data from Polish stations was verified using the Home_R package developed as a result of Action COST – ES0601 (*Advances in homogenisation methods of climate series*, http://www.homogenisation.org/ version October 2011) and the AnClim application developed in the Czech Republic (Stěpánek 2010; Stěpánek et al. 2012). It was necessary to determine the method for calculating daily precipitation totals because of the need to match correctly the relevant weather charts to the days with extreme precipitation. Based on this, the precipitation types were determined (frontal, air-mass).

1.4.3 Spatial Distribution and Temporal Availability of the Daily Precipitation Series

This study is based on time-series of daily precipitation totals between December 1950 and February 2008, coming from 513 weather stations in Europe (Fig. 1.1). The choice of data was dictated, above all, by the availability and quality of precipitation series.

The distribution of the weather stations within the study area is uneven, with the highest density in southern Norway and in Central Europe, and the lowest in Eastern Europe and the Mediterranean Region. The lowland nature of the eastern part of the continent translates into lower spatial variability in precipitation in that part of Europe. Consequently, the number of stations representing the area used in the study is sufficient to reflect the spatial variability of precipitation, as opposed to Southern Europe, which is highly diversified in landform terms and for which data are, unfortunately, unavailable. It must by noted that weather stations in eastern Europe are evenly spaced. Finland and the Kola Peninsula are particularly poorly represented by the data.

The series of daily precipitation totals in the period between December 1950 and February 2008 are complete for 38 % of the stations, and 26 % of the stations show a level of missing data below 1 %. Thus, it may be concluded that 64 % of the stations located mainly in Western Europe and Central Europe provide full observation series. For 11 % of the stations, the data gaps range between 1 and 5 % of the days. Thirteen percent of the stations show gaps between 5 and 10 % of the days. A relatively small group (5 % of the stations) are stations with missing data between 10 and 15 % of the days, and for 7 % of the stations the gaps are higher than 15 %.

The temporal changes in the number of weather stations considered in the study are shown in Fig. 1.2. In the period under investigation, the lowest number of days with missing precipitation data was between the early 1960s and 2000. Many weather stations in Europe started to take meteorological records only in the 1950s, as a result of which the amount of data for that decade is markedly lower. This was also the case after 2000, which is attributable to growing restrictions in access to meteorological data. The precipitation series in the reference period of 1961–1990, which is used in this study to calculate the threshold values for identifying days with extreme precipitation, are much more complete (Fig. 1.3). Most of the stations (76 %) offer full precipitation data series, with 16 % of them having gaps of not more than 1 %, and 8 % with gaps between 1 and 5 %.

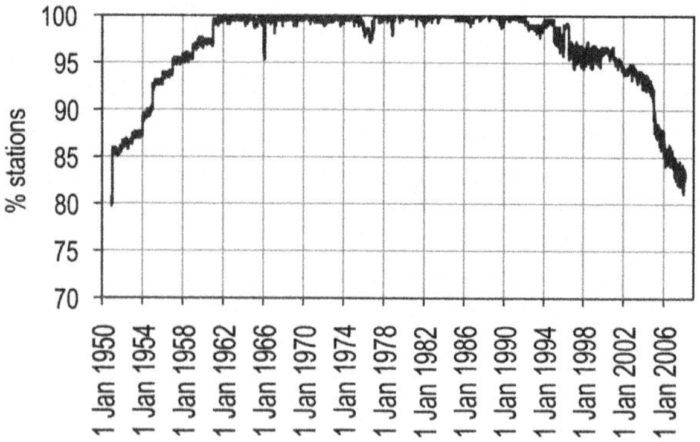

Fig. 1.2 Temporal availability of daily precipitation data in the period of December 1950–February 2008

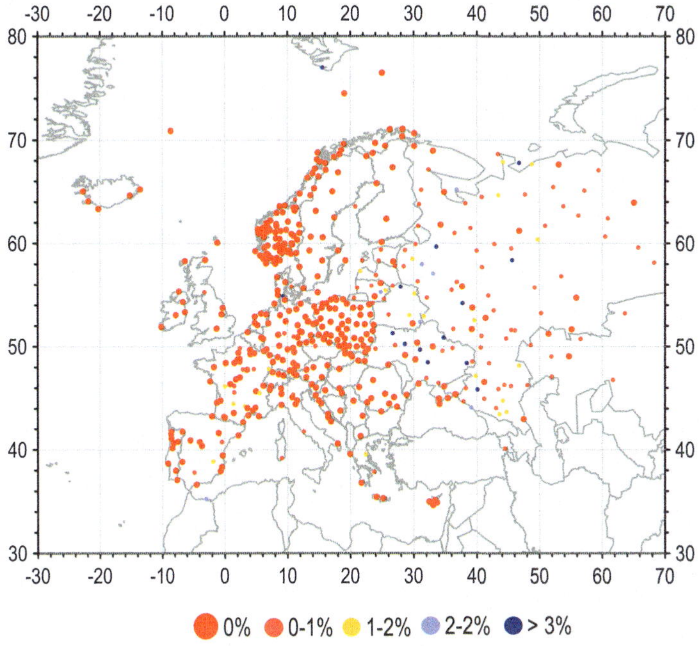

Fig. 1.3 Gaps (% of days) in daily precipitation series in the 1961–1990 normal period

1.5 Conclusion

This chapter provides the motivation and scope of the study and presents a vast body of literature on synoptic contributions to the development of extreme precipitation. It presents sources of meteorological and synoptic data used throughout this study, discusses the quality of weather data and the criteria used for sectioning precipitation series, and includes an executive summary of the results.

The climatology of extreme air-mass and frontal precipitation and its relationship to atmospheric circulation on a continental scale is presented. The issues supplement existing knowledge about the relationships between high precipitation and weather fronts, which are a major component of atmospheric circulation. This book consists of eight chapters. The sources from which meteorological and synoptic data were taken, the meteorological data selection and homogenisation procedure, as well as the characteristics of the data selected, are described in Sect. 1.3, Meteorological and synoptic data: sources, selection procedure and homogeneity. The definitions of extreme events and key indices of precipitation extremes are discussed in Chap. 2, which also presents the amount and frequency of extreme precipitation in Europe, taking into account "absolute" extremes, and goes on to describe and explain the choice of the definition of extreme precipitation in this study. Chapter 3, Origin-Based Types of Extreme Precipitation, describes the resultant classification of the types of extreme precipitation origin based on fundamental dynamic processes leading to precipitation formation and on the presence or absence of weather fronts on days with extreme precipitation and also presents the spatial and seasonal differences in the occurrence of origin-based extreme precipitation types. Chapter 4, entitled Regionalization of Extreme Precipitation Types Occurrence in Europe, includes the characteristics of the regional groups identified differing in the structure of the occurrence of origin-based extreme precipitation types. Chapter 5, Relation Between Extreme Precipitation Occurrence and North Atlantic Oscillation, describes the impact of westerly air flow on extreme precipitation occurrence in Europe. Chapter 6, Air-Mass and Frontal Extreme Precipitation Occurrence in Circulation Types, analyses the relationship between the occurrence of extreme air-mass and frontal precipitation and atmospheric circulation types.

A vast set of daily precipitation records was used from 513 weather stations in Europe covering the period from December 1950 to February 2008. At more than 56 years this relatively long study period offered a sufficiently large sample of the precipitation events in question for the results to be reliable. The majority (64 %) of the individual precipitation total data series has been found to be complete, that is, either free from gaps or where any gaps account for no more than 1 % of days in the study period. The lowest number of days with missing precipitation data was between the early 1960s and 2000. Many weather stations in Europe started to take meteorological records only in the 1950s, as a result of which the amount of data for that decade is markedly lower; this was also true after 2000, which is attributable to growing restrictions in access to meteorological data. The precipitation series in the reference period of 1961–1990, which is used in this study to calculate the threshold values for identifying days with extreme precipitation, are much more complete. Most of the precipitation series (76 %) is complete or almost complete with gaps of not more than 1 % (16 % of stations). The distribution of the weather stations within the study area is uneven, with the highest density in southern Norway and in Central Europe, and the lowest in Eastern Europe and the Mediterranean Region.

The results of the research point to the following significant spatial variabilities in the occurrence of origin-based types of extreme precipitation in Europe:

- All origin-based extreme precipitation types occurring in anticyclonic systems and in areas where precipitation development is enhanced by the orographic effect are associated with advection from the northern sector.
- The highest frequency of precipitation associated with a cold front is recorded in Western Europe (France and the Benelux countries) and in southwestern Europe, where it is associated with advection from the western sector, as well as in the southern part of Eastern Europe, where it coincides with advection from the northern sector.
- Precipitation associated with a warm front is the most frequent in the northern part of Southern Europe in winter and in Eastern Europe in winter and in spring during advection from the southeastern sector.
- Precipitation associated with the passage of various fronts is the most frequent in Western and Central Europe (from northern France to Poland) during air advection from the northwestern sector.

- Precipitation associated with an occluded front is observed mainly in northern Europe, especially in the southern part of the Scandinavian Peninsula on the eastern side of the Scandinavian Mountains, in association mainly with advection from the southeast.

This study does not exhaust the research potential offered by the meteorological and synoptic database collected for the purpose. Many of the areas addressed here require a more in-depth approach supplemented by additional weather data. These areas include the amount of precipitation generated on various weather fronts that could be analysed in the light of the convection indices, which goes beyond the scope of this volume. Another obvious step in a research into origin-based types of extreme precipitation would be to cover its temporal change.

Climatic research that covers relationships between the occurrence of extreme events and atmospheric circulation can be useful in studies intended to develop reliable scenarios of future change in extreme climatic events. It can also provide crucial information for impact studies, such as estimations of flood risks, rates of soil erosion, and changes in the quality of water resources.

References

Abercromby R (1887) On the relation between tropical and extra-tropical cyclones. Proc R Soc Lond 43:1–30. doi:10.1098/rspl.1887.0108
Alexandersson H (1986) A homogeneity test applied to precipitation data. J Climatol 6:661–675
Anagnostou EN, Kummerow CH (1997) Stratiform and convective classification of rainfall using SSM/I 85–GHz brightness temperature observations. J Atmos Ocean Technol 14:570–575
Andrade C, Santos JA, Pinto JG, Corte-Real J (2011) Large-scale atmospheric dynamics of the wet winter 2009–2010 and its impact on hydrology in Portugal. Climate Res 46:29–41
Antunes S, Pires O, Rocha A (2006) Detecting spatio-temporal precipitation variability in Portugal using multichannel singular spectra analysis. Int J of Climatol 26:2199–2212
Anip MHM, Market PS (2007) Dominant factors influencing precipitation efficiency in a continental mid-latitude location. Tellus 59A:122–126
Austin PM, Houze RA (1972) Analysis of the structure of precipitation patterns in New England. J Appl Meteorol 11:926–935
Barry RG, Carleton AM (2001) Synoptic and dynamic climatology. Routledge, London
Baskin AWC (1937) Factors governing different types of precipitation. Thesis (M.S. in Meteorology), California Institute of Technology
Beguería S, Angulo-MartÐnez M, Vicente-Serrano SM, López-Moreno JI, El-Kenawyb A (2010) Assessing trends in extreme precipitation events intensity and magnitude using non-stationary peaks-over-threshold analysis: a case study in northeast Spain from 1930 to 2006. Int J Climatol 31:2102–2114
Berry G, Jacob C, Reeder M (2011a) Recent global trends in atmospheric fronts. Geophys Res Lett 38:L21812
Berry G, Reeder JM, Jacob CH (2011b) A global climatology of atmospheric fronts. Geophys Res Lett 38:L04809
Beven K (1993) Riverine flooding in a warmer Britain. Geogr J 159:157–161
Black E (2012) The influence of the North Atlantic oscillation and European circulation regimes on the daily to interannual variability of winter precipitation in Israel. Int J Climatol 32:1654–1664

Bogucka M (1998) Extreme rainfalls in Poland in the last 30 years. In: Proceedings of the 2nd European conference on applied climatology, Österreichische Beiträgezu Meteorologie und Geophysik, Zentralanstalt für Meteorologie und Geodynamik (ZAMG), Vienna. Publication 384, ECAC98 CD–RO

Bojariu R, Gimeno L (2003) The role of snow cover fluctuations in multiannual NAO persistence. Geophys Res Lett 30:1156. doi:10.1029/2002GL015651

Bolliger M, Binder P, Roesli HP, Wanner H (2004) Satellite- and radar-based investigations of heavy precipitation systems on the southern side of the European Alps. Meteorol Atmos Phys 87:219–234

Bongioannini Cerlini P, Emanuel KA, Todini E (2005) Orographic effect on convective precipitation and space-time rainfall variability: preliminary results. Hydrol Earth Syst Sci 9(4):285–299

Browning KA, Hill FF (1984) Structure and evolution of a mesoscale convective system near the British Isles. Q J R Meteorol Soc 110:897–913

Brunovský P, Lapin M, MelicherčĐk I, SomorčĐk J, Ševčovič D (2009) Variability of K-day extreme precipitation totals and other K-day extreme events. J Hydrol Hydromech 57(4):250–263

Brunsdon C, McClatchey J, Unwin DJ (2001) Spatial variations in the average rainfall–altitude relationship in Great Britain: an approach using geographically weighted regression. Int J Climatol 21:455–466

Buishand TA (1982) Some methods for testing the homogeneity of rainfall records. J Hydrol 58:11–27

Burt TP, Ferranti EJS (2012) Changing patterns of heavy rainfall in upland areas: a case study from northern England. Int J Climatol 32:518–532

Busuioc A, Tomozeiub R, Cacciamanib C (2008) Statistical downscaling model based on canonical correlation analysis for winter extreme precipitation events in the Emilia-Romagna region. Int J Climatol 28:449–464

Butler CJ, Coughlin ADS, Fee DT (1998) Precipitation at Armagh Observatory 1838–1997. Biol Environ Proc R Ir Acad 98B(2):123–140

Buzzi A, Foschini L (2000) Mesoscale meteorological features associated with heavy precipitation in the southern Alpine region. Meteorol Atmos Phys 72:131–146

Buzzi A, Tartaglione N, Malguzzi P (1998) Numerical simulations of the 1994 Piedmont flood: role of orography and moist processes. Mon Weather Rev 126:2369–2383

Cacciamani C, Battaglia F, Patruno P, Pomi L, Selvini A, Tibaldi S (1995) A climatological study of thunderstorm activity in the Po Valley. Theor Appl Climatol 50:185–203

Catto JL, Jakob C, Berry G, Nicholls N (2012) Relating global precipitation to atmospheric fronts. Geophys Res Lett 39:L10805

Cebulak E (1992) Wpływ sytuacji synoptycznych na maksymalne opady dobowe w dorzeczu Górnej Wisły (The influence of synoptic situations on the maximum daily precipitation in the upper Vistula basin). Folia Geogr Ser Geogr Phys 23:81–95

Collier GC, Hardaker PJ (1996) Estimating probable maximum precipitation using a storm model approach. J Hydrol 183:277–306

Cowpertwait PS (1994) A generalized point process model for rainfall. Proc R Soc Lond A447:23–37

Durre I, Menne MJ, Gleason BE, Houston TG, Vose RS (2010) Comprehensive automated quality assurance of daily surface observations. J Appl Meteorol Climatol 49:1615–1633

Dutton EJ, Dougherty HT (1979) Year-to-year variability of rainfall for microwave applications in the USA. IEEE Trans Commun COM-27 5:829–832

Elliott RD, Hovind EL (1964) On convection bands within Pacific Coast storms and their relation to storm structure. J Appl Meteorol 3:143–154

Elliott RD, Hovind EL (1965) Heat, water and vorticity balance in frontal zones. J Appl Meteorol 4:196–211

Eshel G, Farrell B (2000) Mechanisms of Eastern Mediterranean rainfall variability. J Climate 57:3219–3232
Esteban-Parra MJ, Rodrigo FS, Castro-Diez Y (1998) Spatial and temporal patterns of precipitation in Spain for the period 1880–1992. Int J Climatol 18:1557–1574
Europäischer Wetterbericht (European Meteorological Bulletin) (1976–2000) Offenbach, Germany
Favis-Mortlock DT, Boardman J (1995) Nonlinear responses of soil erosion to climate change: a modelling study on the UK South Downs. Catena 25:365–388
Federico S, Avolio E, Bellecci C, Lavagnini A, Colacino M, Walko RL (2008a) Numerical analysis of an intense rainstorm occurred in southern Italy. Nat Hazards Earth Syst Sci 8:19–35
Federico S, Avolio E, Pasqualoni L, Bellecci C (2008b) Atmospheric patterns for heavy rain events in Calabria. Nat Hazards Earth Syst Sci 8:1173–1186
Feidas H, Noulopoulou CH, Makrogiannis T, Bora-Senta E (2007) Trend analysis of precipitation time series in Greece and their relationship with circulation using surface and satellite data: 1955–2001. Theor Appl Climatol 87:155–177
Ferretti R, Low-Nam S, Rotunno R (2000) Numerical simulations of the Piedmont flood of 4–6 November 1994. Tellus 52A:162–180
Fragoso M, Gomes PT (2008) Classification of daily abundant rainfall patterns and associated large-scale atmospheric circulation types in Southern Portugal. Int J Climatol 28:537–544
Frei C, Schär C (1998) A precipitation climatology of the Alps from high-resolution rain-gauge observations. Int J Climatol 18:873–900
Frei CH, Davies HC, Gurtz J, Schär CH (2000) Climate dynamics and extreme precipitation and flood events in Central Europe. Integr Assess 1:281–299
Fukutomi Y, Masuda K, Yasunari T (2007) Cyclone activity associated with the interannual seesaw oscillation of summer precipitation over northern Eurasia. Global Planet Change 56:387–398
Gajić-Čapka M, Cindrić K (2011) Secular trends in indices of precipitation extremes in Croatia 1901–2008. Geofizika 28:293–312
GarcĐa NO, Gimeno L, De La Torre L, Nieto R, AĐel JA (2005) North Atlantic Oscillation (NAO) and precipitation in Galicia (Spain). Atmosfera 18:25–32
Garciá JA, Cruz Gallego M, Serrano A, Vaquero JM (2007) Trends in block-seasonal extreme rainfall over the Iberian peninsula in the second half of the twentieth century. J Climate 20:113–130
Grazzini F (2005) Predictability of a large-scale flow conducive to extreme precipitation over the western Alps. Meteorol Atmos Phys 95:123–138
Groisman PY, Karl TR, Easterling DR, Knight RW, Jamason PF, Hennessy KJ, Suppiah R, Page CM, Wibig J, Fortuniak K, Razuvaev V, Douglas A, Rorland E, Zhai P-M (1999) Changes in probability of heavy precipitation: important indicators of climatic change. Clim Change 42:243–283
Hagen M, Schiesser H-H, Dorninger M (2000) Monitoring of mesoscale precipitation systems in the Alps and the Northern Alpine Foreland by radar and rain gauges. Meteorol Atmos Phys 72:87–100
Hand WH, Fox NI, Collier CG (2004) A study of twentieth-century extreme rainfall events in the United Kingdom with implications for forecasting. Meteorol Appl 11:15–31
Hanna E, Mayes J, Beswick M, Prior J, Wood L (2008) An analysis of the extreme rainfall in Yorkshire, June 2007, and its rarity. Weather 63(9):253–260
Heino R, Brazdil R, Forland E, Tuomenvirta H, Alexandersson H, Beninston M, Pfister C, Rebetez M, Rosenhagen G, Rosner S, Wibig J (1999) Progress in the study of climatic extremes in northern and central Europe. Clim Change 42:151–181
Hellström C (2005) Atmospheric conditions during extreme and non-extreme precipitation events in Sweden. Int J Climatol 25:631–648
Hellström C, Malmgren B (2004) Spatial analysis of extreme precipitation in Sweden 1961–2000. Ambio 33:187–192

Hoinka KP, Schwierz C, Martius O (2006) Synoptic-scale weather patterns during Alpine heavy rain events. Q J R Meteorol Soc 132:2853–2860
Houze R (1997) Stratiform precipitation in regions of convection: a meteorological paradox? Bull Am Meteorol Soc 78(10):2179–2196
Houze RA Jr (2001) Orographic control of precipitation: what are we learning from MAP? MAP Newsl 14:3–5
Hundecha Y, Bárdossy A (2005) Trends in daily precipitation and temperature extremes across Western Germany in the second half of the 20th century. Int J Climatol 25:1189–1202
Hurrell JW, Van Loon H (1997) Decadal variations in climate associated with the North Atlantic Oscillation. Clim Change 36:301–326
Jaagus J (2006) Climatic changes in Estonia during the second half of the 20th century in relationship with changes in large-scale atmospheric circulation. Theor Appl Climatol 83:77–88
James P, Stohl A, Spichtinger N, Eckhardt S, Forster C (2004) Climatological aspects of the extreme European rainfall of August 2002 and a trajectory method for estimating the associated evaporative source regions. Nat Hazards Earth Syst Sci 4:733–746
Jentsch V, Kantz H, Albeverio S (2006) Extreme events: magic, mysteries, and challenges. In: Albeverio S, Jentsch V, Kantz H (eds) Extreme events in nature and society. Springer, Berlin, pp 1–18
Jones MR, Fowler HJ, Kilsby CG, Blenkinsop S (2013) An assessment of changes in seasonal and annual extreme rainfall in the UK between 1961 and 2009. Int J Climatol 33:1178–1194
Kane RJ, Chelius CR, Fritsch JM (1987) Precipitation characteristics of mesoscale convective weather systems. J Clim Appl Meteorol 26:1345–1357
Karagiannidis AF, Karacostas T, Maheras P, Makrogiannis T (2012) Climatological aspects of extreme precipitation in Europe, related to mid-latitude cyclonic systems. Theor Appl Climatol 107:165–174
Kašpar M (2003) Objective frontal analysis techniques applied to extreme/nonextreme precipitation events. Stud Geophys Geod 47:605–631
Kirschenstein M (2004) Rola cyrkulacji atmosferycznej w kształtowaniu opadów w północno-zachodniej Polsce (The role of atmospheric circulation in shaping precipitation in north-western Poland). Wydawnictwo Pomorskiej Akademii Pedagogicznej w Słupsku, Słupsk
Klein Tank AMG, Zwiers FW, Zhang X (2009) Guidelines on analysis of extremes in a changing climate in support of informed decisions for adaptation, Climate Data and Monitoring WCDMP-No. 72. World Meteorological Organization, Geneva, p 52
Klementová E, Litschman T (2001) Vyśledky hodnotenia sucha v oblasti Hurbanova. Sucho, hodnocei a predika. Pracovni seminar, CHMU, Brno 19 November 2001
Kljun N, Sprenger M, Schär C (2001) Frontal modification and lee cyclogenesis in the Alps: a case study using the ALPEX reanalysis data set. Meteorol Atmos Phys 78:89–106
Klok EJ, Klein Tank AMG (2009) Updated and extended European dataset of daily climate observations. Int J Climatol 29:1182–1191
Knapp AK, Fay PA, Blair JM, Collins SL, Smith MD, Carlise JD, Harper CW, Danner BT, Lett MS, McCarron JK (2002) Rainfall variability, carbon cycling and plant species diversity in a mesic grassland. Science 298:2202–2205
Kossowska-Cezak U, Mrugała S (1999) Opady atmosferyczne o anomalnej wysokości (na przykładzie Warszawy i Lublina) (Precipitation of an anomalous amount (in Warsaw and Lublin as an example). Prz Geofizyczny 44(1-2):39–51
Kożuchowski K (1986) Zmienność opadów w Polsce w okresie 1881–1980 (Variation in precipitation in the years 1881–1980 in Poland). Acta Univ Lodziensis Folia Geogr 48:158
Kożuchowski K, Marciniak K (1988) Variability of mean monthly temperatures and semi-annual precipitation totals in Europe in relation to hemispheric circulation patterns. Int J Climatol 8:191–199
Kreitzberg CW (1964) The structure of occlusion, as determined from serial ascents and vertically directed radars. Air Force Cambridge Res Lab Rep 64-20:1–121

Kreitzberg CW, Brown HA (1970) Mesoscale weather systems within an occlusion. J Appl Meteorol 9:419–432

Krichak SO, Alpert P, Melina D (2004) The role of atmospheric processes associated with Hurricane Olga in the December 2001 floods in Israel. J Hydrometeorol 5:1259–1270

Kürbis K, Mudelsee M, Tetzlaff G, Brázdil R (2009) Trends in extremes of temperature, dew point, and precipitation from long instrumental series from central Europe. Theor Appl Climatol 98:187–195

Kwiatkowski J (1984) Relationship between atmospheric precipitation in the Polish Sudety Mts. and its foreland and circulation factors (in Polish). PAN, Komisja Nauk o Ziemi, Wrocaw

Kyselý J (2009) Trends in heavy precipitation in the Czech Republic over 1961–2005. Int J Climatol 29:1745–1758

Kyselý J, Picek J, Huth R (2007) Formation of homogeneous regions for regional frequency analysis of extreme precipitation events in the Czech Republic. Stud Geophys Geod 51:327–344

Lamb HH (1972) British Isles weather types and register of the daily sequence of circulation patterns 1861–1971. Geophys Memories 16(116):85

Langer I, Reimer E (2007) Separation of convective and stratiform precipitation for a precipitation analysis of the local model of the German Weather Service. Adv Geosci 10:159–165

Little MA, Rodda HJE, McSharry PE (2008) Bayesian objective classification of extreme UK daily rainfall for flood risk applications. Hydrol Earth Syst Sci Discuss 5:3033–3060

Llasat MC, Puigcerver M (1997) Total rainfall and convective rainfall in Catalonia, Spain. Int J Climatol 17:1683–1695

Llasat MC, Rigo T, Ceperuelo M, Barrera A (2005) Estimation of convective precipitation: the meteorological radar versus an automatic rain gauge network. Adv Geosci 2:103–109

López-Moreno JI, Vicente-Serrano SM, Angulo-MartÐnez M, BeguerÐa S, Kenawya A (2010) Trends in daily precipitation on the northeastern Iberian Peninsula, 1955–2006. Int J Climatol 30:1026–1041

Lorenzo MN, Taboadab JJ, Gimenoa L (2008) Links between circulation weather types and teleconnection patterns and their influence on precipitation patterns in Galicia (NW Spain). Int J Climatol 28:1493–1505

Lupikasza E (2010a) Relationships between occurrence of high precipitation and atmospheric circulation in Poland using different classifications of circulation types. Phys Chem Earth 35:448–455

Lupikasza E (2010b) Spatial and temporal variability of extreme precipitation in Poland in the period 1951–2006. Int J Climatol 30:991–1007. doi:10.1002/joc.1950

Łupikasza E, Hänsel S, Matschullat J (2011) Regional and seasonal variability of extreme precipitation trends in southern Poland and central-eastern Germany 1951–2006. Int J Climatol 15:2249–2271

Maraun D, Osborn TJ, Gillett NP (2008) United Kingdom daily precipitation intensity: improved early data, error estimates and an update from 2000 to 2006. Int J Climatol 28:833–842. doi:10.1002/joc.1672

Maraun D, Rust HW, Osborn TJ (2010) Synoptic airflow and UK daily precipitation extremes development and validation of a vector generalized linear model. Extremes 13:133–153

Mares C, Mares I, Stanciu A (2009) Extreme value analysis in the Danube lower basin discharge time series in the twentieth century. Theor Appl Climatol 95:223–233

Market PS, Allen S, Scofield R, Kuligowski R, Gruber A (2003) Precipitation efficiency of warm-season Midwestern mesoscale convective systems. Weather Forecast 18:1273–1285

Marwitz JD (1972) Precipitation efficiency of thunderstorms on the high plains. J Res Atmos 6:367–370

Massacand AC, Wernli H, Davies HC (1998) Heavy precipitation on the Alpine southside: an upper-level precursor. Geophys Res Lett 25:1435–1438

Mätlika O, Post P (2008) Synoptic weather types that have caused heavy precipitation in Estonia in the period 1961–2005. Estonian J Eng 14(3):195–208

Météorologie Nationale Bulletin Quotidien Détudes, 1951–1975, Secrétariat Détat aux transports, Secréttariat Général a Laviation Civile, Paris
Menne MJ, Durre I, Vose RS, Gleason BE, Houston TG (2012) An overview of the global historical climatology network-daily database. J Atmos Ocean Technol 29:897–910
Milelli M, Llasat MC, Ducrocq V (2006) The cases of June 2000, November 2002 and September 2002 as examples of Mediterranean floods. Nat Hazards Earth Syst Sci 6:271–284
Millán MM, Estrela MJ, Miró J (2005) Rainfall components: variability and spatial distribution in a Mediterranean area (Valencia region). J Climate 18:2682–2705
Mizahari F (2000) Heavy daily precipitation distribution in east-central France and west European meteorological patterns. Theor Appl Climatol 66:199–210
Moberg A, Jones PD (2005) Trends in indices for extremes in daily temperature and precipitation in Central and Western Europe, 1901–1999. Int J Climatol 25:1149–1171
Mohr KI (2004) Interannual, monthly, and regional variability in the wet season diurnal cycle of precipitation in Sub-Saharan Africa. J Climate 17:2441–2453
Morawska-Horawska M (1971) Meteorologiczne przyczyny powodzi w Polsce Południowej w lipcu 1970 (Meteorological causes of the flood in the Southern Poland in July 1970). Prz Geol 16(24):299–315
Nagle RE, Serebreny SM (1962) Radar precipitation echo and satellite cloud observations of a maritime cyclone. J Appl Meteorol 1:279–295
Neumann J (1941) Distribution of the ratio of the mean successive difference to the variance. Ann Math Stat 12:367–395
Niedźwiedź T (1999) Rainfall characteristics in Southern Poland during the severe flooding event of July 1997. Stud Geomorphol Carpatho-Balc 33:5–25
Niedźwiedź T (2003a) Extreme precipitation in Central Europe and its synoptic background. Global Change 10:15–29
Niedźwiedź T (2003b) Extreme precipitation events on the northern side of the Tatra mountains. Geogr Pol 76(2):15–23
Niedźwiedź T, Twardosz R (2004) Long-term variability of precipitation at selected stations in central Europe. Global Change IGBP 11:73–100
Niedźwiedź T, Twardosz R, Walanus A (2009) Long-term variability of precipitation series in east central Europe in relation to circulation patterns. Theor Appl Climatol 98:337–350
Norrant C, Douguédroit A (2006) Monthly and daily precipitation trends in the Mediterranean (1950–2000). Theor Appl Climatol 83:89–106
Nozumi Y, Arakawa H (1968) Prefrontal rainbands located in the warm sector of subtropical cyclones over the oceans. J Geophys Res 73:787–492
Nuissier O, Ducrocq V, Ricard D, Lebeaupina C, Anquetinb S (2008) A numerical study of three catastrophic precipitating events over southern France. I: Numerical framework and synoptic ingredients. Q J R Meteorol Soc 134:111–130
Okołowicz W (1969) Klimatologia Ogólna (General climatology). PWN, Warszawa
Osborn TJ, Hulme M (2002) Evidence for trends in heavy rainfall events over the UK. Philos Trans R Soc Lond A360:1313–1325
Pastor F, Estrela MJ, Peñarrocha D, Millán MM (2001) Torrential rains on the Spanish Mediterranean coast: modeling the effects of the sea surface temperature. J Appl Meteorol 40:1180–1195
Pavan V, Tomozeiu R, Cacciamani C, di Lorenzo M (2008) Daily precipitation observations over Emilia-Romagna: mean values and extremes. Int J Climatol 28:2065–2079. doi:10.1002/joc.1694
Pettit AN (1979) A non-parametric approach to the change-point problem. Appl Stat 28:126–135
Pradier S, Chong M, Roux F (2004) Characteristics of some frontal stratiform precipitation events south of the alpine chain during MAP. Meteorol Atmos Phys 87:197–218
Raddatz RL, Hanesiak JM (2008) Significant summer rainfall in the Canadian Prairie Provinces: modes and mechanisms 2000–2004. Int J Climatol 28:1607–1613
Ramis C, Homar V, Amengual A, Romero R, Alonso S (2013) Daily precipitation records over mainland Spain and the Balearic Islands. Nat Hazards Earth Syst Sci 13:2483–2491

Ramos MC, Martinez-Casanovas JA (2006) Trends in precipitation concentration and extremes in the Mediterranean Penedés-Anoia Region, NE Spain. Clim Change 74:457–474

Rodrigo FS (2009) Changes in the probability of extreme daily precipitation observed from 1951 to 2002 in the Iberian Peninsula. Int J Climatol 30:1512–1525. doi:10.1002/joc.1987

Rodriguez-Puebla C, Encinas AH, Sáenz J (2001) Winter precipitation over the Iberian Peninsula and its relationship to circulation indices. Hydrol Earth Syst Sci 5:233–244

Ross M, Mitchell RM, Phillips ID (2006) Warm season extreme precipitation events at Sunderland 1940–2000. Weather 61(8):227–232

Rotunno R, Ferretti R (2001) Mechanisms of intense Alpine rainfall. J Atmos Sci 58:1732–1749

Santos JA, Corte-Real J, Ulbrich U, Palutikof J (2007) European winter precipitation extremes and large–scale circulation: a coupled model and its scenarios. Theor Appl Climatol 87:85–102

Schneidereit M, Schär C (2000) Idealised numerical experiments of Alpine flow regimes and southside precipitation events. Meteorol Atmos Phys 72:233–250

Seibert P, Frank A, Formayer H (2007) Synoptic and regional patterns of heavy precipitation in Austria. Theor Appl Climatol 87:139–15

Sénési SP, Bougeault P, Cheze J-L, Cosentino P, Thepenier R (1996) The Vaison-La-Romaine flash flood: mesoscale analysis and predictability issues. Weather Forecast 11:417–442

Sevruk B (1997) Regional dependency of precipitation–altitude relationship in the Swiss Alps. Clim Change 36:355–369

Siwek K (2010) Występowanie na Lubelszczyźnie dobowych sum opadu atmosferycznego ≥100 mm i ich uwarunkowania cyrkulacyjne (1951–2000) (The occurrence of daily precipitation ≥100 mm in Lublin Voivodship and its synoptic background (1951–2000)). Ann UMCS Sec B 65(2):117–125

Steinacker R (1981) Analysis of the temperature and wind field in the Alpine region. Geophys Astrophys Fluid Dyn 17:51–62

Stěpánek P (2010) An Clim-software for time series analysis. http://www.climahom.eu/AnClim.html. Accessed 2 Dec 2011

Stěpánek P, Zahradniček P, Brázdil R, Tolasz R (2012) Metodologiekontroly a homogenizacečasovýchřad v klimatologii (Methodology of data quality control and homogenization of time series in climatology). Česky Hydrometeorologický Ustav, Praha

Strangeways I (2007) Precipitation. Theory, measurements and distribution. Cambridge University Press, New York, p 290

Tibaldi S, Buzzi A, Speranza A (1990) Orographic cyclogenesis. In: Newton CW, Holopainen EO (eds) Extratropical cyclones: the Erik Palmen memorial volume. American Meteorological Society, Washington, DC, pp 107–127

Tolika K, Anagnostopoulou CHR, Maheras P, Kutiel H (2007) Extreme precipitation related to circulation types for four case studies over the Eastern Mediterranean. Adv Geosci 12:87–93

Toreti A, Xoplaki E, Maraun D, Kuglitsch FG, Wanner H, Luterbacher J (2010) Characterisation of extreme winter precipitation in Mediterranean coastal sites and associated anomalous atmospheric circulation patterns. Nat Hazards Earth Syst Sci 10:1037–1050

Trigo IF (2006) Climatology and inter-annual variability of storm-tracks in the EuroAtlantic sector: a comparison between ERA40 and NCEP/NCAR reanalyses. Clim Dyn 26:127–143

Trömel S, Schönwiese CD (2007) Probability change of extreme precipitation observed from 1901 to 2000 in Germany. Theor Appl Climatol 87:29–39

Turato B, Reale O, Siccardi F (2004) Water vapor sources of the October 2000 Piedmont flood. J Hydrometeorol 5:693–712

Twardosz R (1998) Relations entre les précipitations et les indices de circulation atmosphérique à Cracovie durant la dernière période séculaire. Publ Assoc Int Climatol 11:359–366

Twardosz R (1999) Precipitation variability and tendency in Kraków for the period 1850–1997 related to circulation patterns. Acta Geophys Pol 47(1):111–133

Twardosz R (2005) Dobowy przebieg opadów atmosferycznych w ujęciu synoptycznym I probalilistycznym na przykładzie Krakowa (1886–2002) (Daily course of precipitation from synoptic and probabilistic perspectives in Cracow as an example (1886–2002)). Poligrafia Inspektoratu Towarzystwa Salezjańskiego, Kraków

Unkašević M, Radinović D (2000) Statistical analysis of daily maximum and monthly precipitation at Belgrade. Theor Appl Climatol 66:241–249

Ustrnul Z, Czekierda D (2001) Circulation background of the atmospheric precipitation in Central Europe (based on the Polish example). Meteorol Z 10(2):103–111

Ustrnul Z, Czekierda D (2009) Atlas ekstremalnych zjawisk meteorologicznych oraz sytuacji synoptycznych w Polsce (The Atlas of extreme meteorological phenomena and synoptic situations in Poland). IMGW, Warszawa

Valencia JL, Tarquis AM, Saá-Requejo A, Gascó JM (2012) Changes of extreme rainfall indices at Ebro River Basin. Nat Hazards Earth Syst Sci 12:2127–2137

Van Bebber WJ (1891) Die Zugstrassen der barometrischer Minima. Meteorol Z 8:361–366

Von Rudloff H (1967) Die Schwankungen und Pendelungen des Klimas in Europa seitdemBeginn der regelmässigen Instrumenten-Beobachtungen (1670). Vieweg, Braunschweig

Wang GL, You LZ (2004) Delayed impact of the North Atlantic oscillation on biosphere productivity in Asia. Geophys Res Lett 31:L12210. doi:10.1029/2004GL019766

Watson PA, Gunes M, Potter BA, Sathiaseelan V, Leitas J (1982) Development of a climatic map of rainfall attenuation for Europe, Final Report, European Space Agency, ESTEC CONTR. No. 4162/79/NL. Postgratuate School of Electronics and Electronic Engineering, University of Bradford, Bradford

Weltzin JF, McPherson GR (eds) (2003) Changing precipitation regimes and terrestrial ecosystems: a North American perspective. University of Arizona Press, Tuscon

Weltzin JF, Loik ME, Schwinning S, Williams DG, Fay PA, Haddas BM, Harte J, Huxman TE, Knapp AK, Lin G, Pockman WT, Shaw R, Small EE, Smith MD, Smith SD, Tissue DT, Zak JC (2003) Assessing the response of terrestrial ecosystems to potential changes in precipitation. Bioscience 53(10):941–952

Wheeler D (1997) North-east England and Yorkshire. In: Wheeler D, Mayes J (eds) Regional climates of the British Isles. Routledge, London

Wibig J (1999) Precipitation in Europe in relation to circulation patterns at the 500 hPa level. Int J Climatol 19:253–269

Wibig J (2001) Influence of atmospheric circulation on the spatialdifferentiation of temperature and precipitation anomalies in Europe. Rozprawy Habilitacyjne Uniwersytetu Łódzkiego, Łódź

Wibig J, Fortuniak K (1998) The extreme precipitation conditions in Łódź in the period 1931–1995. Acta Univ Lodziensis Folia Geogr Phys 3:241–249

Wijngaard JB, Klein Tank AMG, Können GP (2003) Homogenity of 20th century European daily temperature and precipitation series. Int J Climatol 23:679–692

Wilby RL (1993) The influence of variable weather patterns on river water quantity and quality regimes. J Climatol 13:447–459

Wilby RL (1998) Modelling low-frequency rainfall events using airflow indices, weather patterns and frontal frequencies. J Hydrol 212-213:380–392

Yarnal B (2000) Synoptic climatology in environmental analysis. A primer. Belhaven Press, London

Yu RC, Zhou TJ (2004) Impacts of winter-NAO on March cooling trends over subtropical Eurasia continent in the recent half century. Geophys Res Lett 31:L12204. doi:10.1029/2004GL019814

Zolina O, Kapala A, Simmer C, Gulev SK (2004) Analysis of extreme precipitation over Europe from different reanalyses: a comparative assessment. Global Planet Change 44:129–161

Zolina O, Simmer C, Kapala A, Gulev S (2005) On the robustness of the estimates of centennial-scale variability in heavy precipitation from station data over Europe. Geophys Res Lett 32:L14707. doi:10.1029/2005GL023231

Zorita E, Kharin V, von Storch H (1992) The atmospheric circulation and sea surface temperature in the North Atlantic area in winter: their interaction and relevance for Iberian precipitation. J Climate 5:1097–1108

Chapter 2
Definitions and Indices of Precipitation Extremes

Abstract Definitions of extreme climatological and meteorological events may depend on the objective of the study, the type of data, or the researcher's subjective opinion. This chapter discusses definitions of extreme climatic events and presents indices of precipitation extremes found in the literature. The spatial and seasonal variability in a range of these indices were analysed, and the results were used to select criteria for identifying extreme precipitation in Europe adopted in this study. These criteria were then applied in subsequent chapters. During the study period, the maximum daily precipitation totals in Europe ranged from 37.6 to 520 mm. An overwhelming majority of stations (89 %) recorded daily precipitation greater than 50 mm less than once per year. Because there is great variability between precipitation regimens in Europe, this study adopted a statistical definition of extreme precipitation events. The events were identified separately for each of the weather station and in each of the months using an empirical distribution of the daily precipitation totals. Precipitation totals exceeding or equal to the 95th percentiles of daily precipitation were selected as extreme. The 95th percentile (95P) was calculated from days with daily totals ≥1 mm during the period 1961–1990. The resulting spatial and seasonal variability of the threshold values corresponding to the 95P daily precipitation across Europe is similar to the spatial and seasons variability of the monthly totals.

Keywords Extreme events • Precipitation indices • Precipitation records

2.1 Definition of Extreme Weather Events

There is no single definition of an extreme event that would be shared by all climatologists. Instead, the meaning of the term seems to depend on the objective of a given research project, type of available data, selected methodology, and the researcher's subjective assessment (Karagiannidis et al. 2012). According to the *Glossary of Meteorology* (Glickman 2000), the term "extremum" means the highest or, in some cases, the lowest, value of a climate element observed during a given interval (a month, season, year, or several years). If such value is the highest extremum of the entire observation period, then it is defined as the absolute extremum.

E. Łupikasza, *The Climatology of Air-Mass and Frontal Extreme Precipitation*, Springer Atmospheric Sciences, DOI 10.1007/978-3-319-31478-5_2

According to Beniston et al. (2004) and Beniston (2005) extreme events can be classified using various criteria, including the following:

- Frequency of occurrence;
- Intensity;
- Quantity of material damage.

The first of the criteria has been used in a statistical definition of extreme events derived from the theory of extreme values. A study by MacDonald et al. (1992), *Statistics of Extreme Events with Application to Climate*, is among the first to have discussed the application of this theory in climatology. In this approach weather and climate extrema are rare in terms of the statistical distribution of the frequency of a given weather element at a given site (Klein Tank et al. 2009). An extreme event, therefore, fits below the 10th percentile (10P) or above the 90th percentile (90P) of the statistical distribution of cumulative frequency of a given climate element (such as temperature, precipitation, pressure) (Beniston et al. 2007). This is the definition that has been adopted by the Intergovernmental Panel on Climate Change (IPCC 2001, 2007, 2013). It has also become the most frequently used definition in research of extreme climate events. Often, however, researchers tighten their criteria to 5P or 1P, or above 95P or 99P, of the frequency distribution function (Klein Tank et al. 2002; Alexander et al. 2006; Benestad 2006). In these more stringent definitions, events below 10P or above 90P are referred to as moderate extremes (Beniston et al. 2004; Klein Tank et al. 2009).

The definition of intensity tends to depend on the meteorological phenomenon studied. In the case of precipitation it normally is understood as the quantity of precipitation, regardless of type, falling within a unit of time and is typically expressed in millimetres per minute ($mm*min^{-1}$) (Niedźwiedź 2003b). In most cases data on precipitation intensity thus defined cannot be secured because sufficiently long and detailed observations are not available (Łupikasza 2009). For this reason climatological studies that rely on long-term observation records, especially studies into climate change, must resort to simplified indicators of precipitation intensity (Karl et al. 1995; Karl and Knight 1998; Osborn et al. 2000; Brunetti et al. 2001). The World Meteorological Organisation (WMO) has adopted mean daily precipitation as the measure of precipitation intensity in a given period (Klein Tank et al. 2009). There is also an indirect measure of intensity defined as the quantity of phenomena, such as wind speed and quantity of precipitation, that accompany an extreme phenomenon such as a tropical cyclone or a thunderstorm (Zipser et al. 2006; Łupikasza and Bielec-Bąkowska 2012). In this way an event is categorised as extreme if the intensities of the accompanying phenomena are identified to be exceptionally high. Finally, the most catastrophic extreme events, such as tropical cyclones and floods, can also be measured by the overall cost of the related socioeconomic losses (Changnon and Changnon 1998; Easterling et al. 2000; Beniston 2005).

This third criterion, the cost of material damage, is normally applied to hazardous weather events that cause serious socioeconomic losses. There, however, is a degree of complexity and ambiguity in this criterion, as catastrophic effects do not always require rare or intensive extreme events to happen. An example would be the thawing of permafrost in mountains that triggers mudflows and a breakup of rocks. This criterion is most frequently applied in impact studies for the insurance sector (Changnon and Changnon 1998; Beniston 2005; Changnon 2011).

The existing variety of the definitions of extreme weather event means that in the case of precipitation the term extremum refers to a number of characteristics of this weather element (Strangeways 2007).

2.2 Overview of Extreme Precipitation Indices

An increasing interest in extreme events that are considered to be important indicators of climate change has resulted in the development of many indices of climatological and meteorological extremes. In June 1997, the first international research workshop, organised in Asheville, NC (USA), recommended a list of indices for extreme event research (Nicholls and Murray 1999). The objective of the workshop was to align methods of identifying extreme climatic events that would allow comparability of research between various parts of the world. Subsequent studies in this field worldwide have led to modifications and improvements of the proposed methodologies (Folland et al. 1999; Trenberth and Owen 1999; Peterson et al. 2001; Easterling et al. 2003; Klein Tank and Köennen 2003; WMO 2004).

Alexander et al. (2006) proposed to break down indices of extreme climate and weather events into five categories that differed in the identification methods and features of the event itself (intensity, frequency etc.):

- Indices based on location measures (percentiles), which in the case of precipitation also include the most extreme events. Threshold values for extreme events identified using percentiles for specific location are referred to in the literature as relative threshold values (Beniston et al. 2004). They have become the most popular in the study of climate change (Frich et al. 2002; Klein Tank and Köennen 2003; Groisman et al. 2005; Schmidli and Frei 2005; Kyselý 2009; Lupikasza 2010b; Gajić-Čapka and Cindrić 2011; Łupikasza et al. 2011). This group also includes probabilistic indices used in forecasting the likelihood of an extreme event (Limanówka et al. 1993; Cebulak 1994; Miętus et al. 2005; Ustrnul and Czekierda 2009).
- Indices based on absolute values of weather elements, which include the highest and, in the case of certain elements, also the lowest values recorded within a given period of time. Precipitation indices in this category include, for example, the highest daily total and the highest 5-day total in a year, season, or month (Cebulak 1994; Gajić-Čapka and Cindrić 2011). These indices provide information on the extreme ranges that climate element can reach, thus allowing practical responses of people and communities, but also a valuable input into forecasting climate change (Ustrnul and Czekierda 2009).
- Indices based on fixed threshold values of weather elements. Precipitation indices of this type include the number of days with totals greater than 10 mm

(Paszyński and Niedźwiedź 1999: Polska; Klein Tank et al. 2009: Europa), 20 mm (Mizahari 2000: a central part of southern France; Klein Tank et al. 2009: Europe), 25 mm (Karl et al. 1996: USA) 30 mm (Niedźwiedź et al. 2004; Zwoliński 2008; Łupikasza and Bielec-Bąkowska 2012: Polska), 40 mm (Hellström 2005: popular in Sweden). The highest threshold values adopted in the literature are daily totals higher than 40 mm and 50 mm (Karl et al. 1996: USA; Mekis and Hogg 1997: Canada; Hellström 2005: Sweden; Beniston 2005: Switzerland; Fragoso and Gomes 2008; Changnon 1994; Romero et al. 1999: Portugal); 60 mm (Karagiannidis et al. 2012: Europe), and even 100 mm (Beniston 2005: Switzerland; Siwek 2010: around the southeastern Polish city of Lublin).

- Indices of the duration of an extreme event, such as the highest consecutive number of days without precipitation or the highest consecutive number of days with precipitation greater than 1 mm (Griffiths and Bradley 2007).
- Other indices of extreme monthly, seasonal, or annual values of climate elements, including, for example, the contribution of extreme precipitation to annual or seasonal precipitation totals (Gajić-Čapka and Cindrić 2011).

A study of extreme events needs to always take into account their regional nature (IPCC 2007). The vast diversity of climatic conditions not just between continents, but even within individual countries, means that values of certain elements of the weather may be considered extreme in one type of climate and quite common in another. Strangeways (2007) ascertains that because of their enormous variability values associated with extreme weather events are incomparable on the global scale. Therefore, terms such as the "mean global extreme precipitation" should be avoided. The one group of indices already mentioned that does not take into consideration the regional nature of extreme events is that based on fixed threshold values. For this reason this group is not recommended for research on extreme events covering large territories with varied environmental and, consequently, climatological conditions (Alexander et al. 2006). Klein Tank and Köennen (2003) claim that indices based on fixed threshold values can be useful in the impact studies because they refer to extreme events affecting both the society and the natural environment. Using a simple method of fixed threshold values, one can identify events of the same intensity. The method of relative threshold values (percentiles-based method), on the other hand, identifies events of the same frequency (Haylock and Nicholls 2000; Klein Tank and Köennen 2003; Beniston et al. 2004). Indices of extreme climatic events calculated using relative threshold values are comparable between stations because at each of them they take into consideration the same proportion of the probability distribution of a particular climate element (Klein Tank et al. 2009).

In 2007, the Joint Expert Team on Climate Change Detection and Indices (ETCCDI), including among its members researchers active in the WMO's Commission for Climatology or many global research programmes (e.g., WCDMP, World Climate Data and Monitoring Programme; CLIVAR, the Climate Variability and Predictability; WCRP, Programme of the World Climate Research Programme; JCOMM, Joint WMO–IOC Technical Commission for Oceanography and Marine Meteorology), selected a number of indices of extreme climatic events, including precipitation (Table 2.1), recommended by WMO for studies into extreme climatic

Table 2.1 Extreme precipitation indices recommended by ETCCDI (The Joint Expert Team on Climate Change Detection and Indices) for climate change studies

No	Symbol	Name	Definition
1	RX1day	Maximum 1-day precipitation: highest precipitation amount in 1-day period	RX1dayj = max (RRij), where: *RRij* is the daily precipitation amount on day *i* in period *j*
2	RX5day	Maximum 5-day precipitation: highest precipitation amount in 5-day period	RX5dayj = max (RRkj), where: *RRkj* is precipitation amount for the 5 day interval *k* in period *j*, where *k* is defined by the last day
3	SDII	Simple daily intensity index: mean precipitation amount on a wet day	SDIIj = sum (RRwj) / W, where: *RRwj* is the daily precipitation amount on wet day *w* (RR >1 mm) in period *j*, *W* is the number of wet days in period *j*
4	R10mm	Heavy precipitation days: count of days where RR ≥10 mm	Number of days with RR*ij* ≥10 mm, where: *RRij* is the daily precipitation amount on day *i* in period *j*
5	R20mm	Very heavy precipitation days: count of days where RR ≥20 mm	Number of wet days with RRij ≥20 mm, where: *RRij* is the daily precipitation amount on day *i* in period *j*
6	Rnnmm	Count of days where RR ≥ user-defined threshold in mm	Number of days with RRij ≥ nn mm, where: *RRij* is the daily precipitation amount on day *i* in period *j*, *nn* – user-defined threshold
7	CDD	Consecutive dry days: maximum length of dry spell (RR <1 mm)	The largest number of consecutive days with RR*ij* <1 mm, where: *RRij* is the daily precipitation amount on day *i* in period *j*
8	CWD	Consecutive wet days: maximum length of wet spell (RR ≥1 mm)	The largest number of consecutive days with RR*ij* ≥1 mm, where: *RRij* is the daily precipitation amount on day *i* in period *j*
9	R95pTOT	Precipitation due to very wet days (>95th percentile)	R95pTOTj = sum (RRwj), where RRwj > RRwn95, where: *RRwj* is the daily precipitation amount on a wet day *w* (RR >1 mm) in period *j*, *RRwn95* is 95percentile of precipitation on wet days in the base period *n* (1961–1990)
10	R99pTOT	Precipitation due to extremely wet days (>99th percentile)	R99pTOTj = sum (RRwj), where RRwj > RRwn99, where: *RRwj* is the daily precipitation amount on a wet day *w* (RR >1 mm) in period *j*, *RRwn99* is 99percentile of precipitation on wet days in the base period *n* (1961–1990)
11	PRCPTOT	Total precipitation in wet days (>1 mm)	PRCPTOTj = sum (RRwj), where: *RRwj* is the daily precipitation amount on a wet day *w* (RR >1 mm) in period *j*

RR daily precipitation amount
Source: Klein Tank et al. 2009

events (Klein Tank et al. 2009). These recommended extreme precipitation indices, including symbology and definitions, are summarised in Table 2.1.

Niedźwiedź et al. (2004) proposed a list of extreme weather, hydrological, and geomorphological events for Poland. Details of the criteria for the identification of extreme weather events adopted in that study are also provided by Zwoliński (2008). Some of these, especially probabilistic ones, are highly restrictive in being based solely on the highest seasonal or annual values: M–1. Rainfall: (M–1.1.) maximum hourly, (M–1.2.) maximum daily, especially with probability $p < 10\,\%$ (in the characteristic of this rainfall consideration must be given to days with the total ranges of: ≥10 mm, ≥30 mm, ≥50 mm, and ≥100 mm), (M–1.3.) prolonged precipitation ≥100 mm* ≥100mm*day^{-1}, (M–1.4.) maximum monthly rainfall with the probability of $p < 10\,\%$ (especially >200 mm), (M–1.5.) torrential rain (according to Chomicz scale (1951) and the maximum intensity is in mm*min^{-1} and its duration.

2.3 Spatial and Seasonal Variability of Extreme Precipitation in Europe

Precipitation, including the extreme type, varies between the diverse European types of climates in terms of the timing of its concentration, its totals, and frequency. This section reviews spatial and seasonal variability of high precipitation in Europe using selected indices of their totals and frequencies, such as the maximum daily total and the number of days with precipitation ≥10 mm, ≥30 mm, and ≥50 mm between December 1950 and February 2008. The analysis includes not just absolute maximum totals, but also the mean total of the five highest daily totals recorded. The mean daily maximums obtained in this way reduce the impact of any erroneous maximum values that might escape detection and elimination from the precipitation data records in the homogenisation procedures.

2.3.1 The Highest Daily Precipitation in Europe

Maximum precipitation totals in Europe range from 37.6 mm (2 July 1966) at Hoseda Hard in the northeastern part of the European Russia to 520 mm (24 February 1964) at the French high-mountain station of Mont-Aigoual (1567 m a.s.l.). This Russian station is the only one in Europe where the maximum long-term daily precipitation total has not exceeded 40 mm in the research period. The largest group (34 %) of stations located along the vast expanse stretching from Europe's western coast to its eastern boundary, and rarely north of parallel 45°N, has daily maximums between 50 and 90 mm (Fig. 2.1). In this group of stations the highest daily totals (90 mm) were recorded in Prague (Czech Republic) on 12 July 1981, Lublin (Poland) on 6 September 2007, and at Mozdok (Russia) on 7 January 1997.

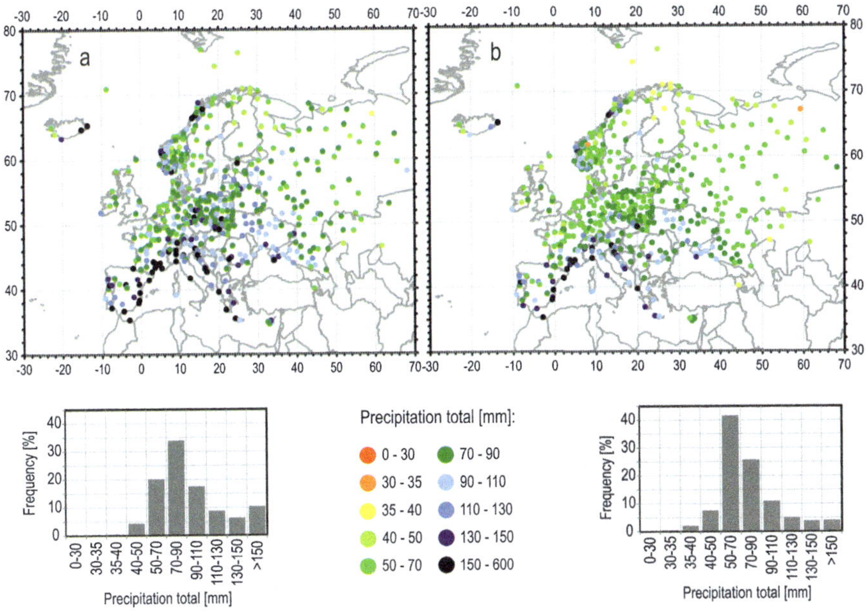

Fig. 2.1 Maximum daily precipitation totals (**a**) and their histograms* and the averages from the five highest daily precipitation totals (**b**) and their histograms for December 1950–February 2008. *Right* closed precipitation intervals

During the study period, the highest daily totals most frequently exceeded 90 mm in the west of the continent. The eastern limit of their occurrence coincided with a line connecting the northern coast of the Scandinavian Peninsula near the meridian 20°E with the western coast of the Caspian Sea near the meridian 45°E (Fig. 2.1). The highest of the absolute daily maximums of more than 150 mm were recorded at 10 % of stations located primarily in the south of the continent: in France at Mont-Aigoual (mentioned above), in Spain's Valencia (178.5 mm, 17 November 1956), in Portugal at Campi (162.4 mm, 25 December 1995), in Croatia at Rijeka (210.3 mm, 31 August 1976), in Italy at Genoa (218.6 mm, 19 September 1953), in Greece at Corfu (239.3 mm, 1 November 2011), and in Cyprus at Amiados (191 mm on 25 December 1995). Other areas with precipitation ≥150 mm included relatively low-altitude stations on the Atlantic coast of the Scandinavian Peninsula, in Iceland, in mountains, where orography was significant in precipitation patterns (Bolliger et al. 2004; Hagen et al. 2000; Milelli et al. 2006; Federico et al. 2008a), and at a few lowland stations, including in Poland (Kielce: 155.2 mm, 24 July 2001) and Germany (Lindenberg: 171.7 mm, 8 August 1978, Dresden: 158 mm, 12 August 2002).

The large number of factors determining the occurrence and amount of precipitation, as well as the complicated and varied spatial relationships linking these factors with precipitation, mean that the timing of individual extreme precipitation events does not always coincide with the normal season of precipitation concentration.

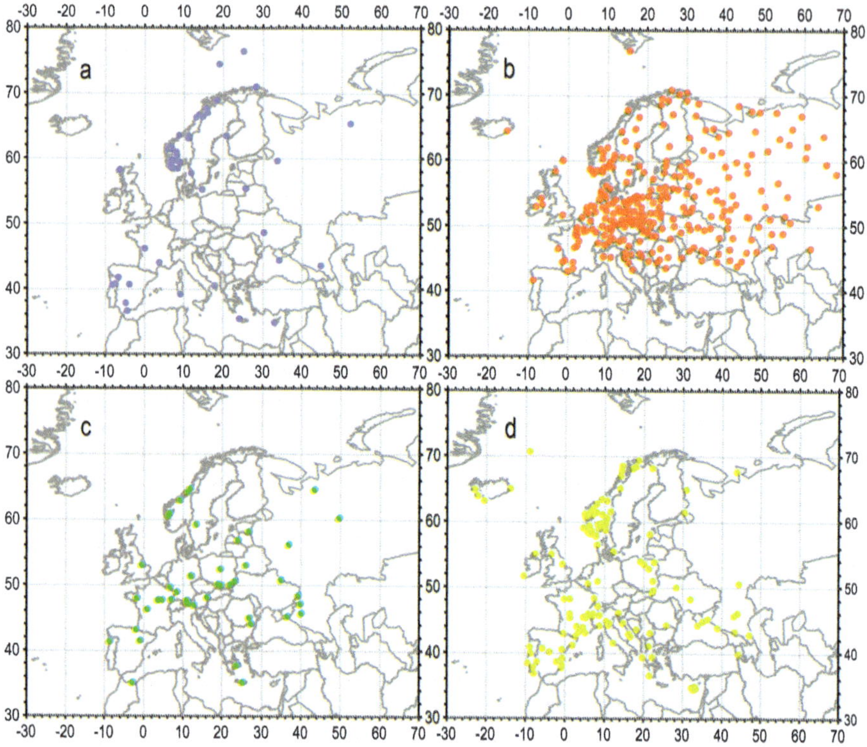

Fig. 2.2 Stations with maximum daily precipitation total in winter (**a**), summer (**b**), spring (**c**), and autumn (**d**) for December 1950–February 2008

Most of the absolute precipitation maximums covered in this section (56 % of stations) occurred in the summer season at stations located just short of the latitude of 43°N. South of the parallel 45°N and at many Norwegian stations (26 % stations in total), the highest daily totals were recorded in autumn (Fig. 2.2).

The remaining 18 % of stations record daily maximum precipitation in spring or winter (equally split at 9 % between them). In spite of a broad scattering of stations with maximums in spring and winter across the continent, it can be noticed that springtime daily maximums rarely occur in Northern and Eastern Europe, whereas winter maximums tend to occur along the western and southwestern coasts of the Scandinavian Peninsula and at isolated stations of Southern Europe, Northern Europe, and in the west of Eastern Europe. Appendix 1 summarises the values of the absolute maximum daily precipitation and their dates of occurrence.

The selection of weather stations for this study that depended on data availability and on the adopted study period does not allow full documentation of extreme precipitation events. Record-breaking daily totals exceeding 300 mm observed at stations that are not covered here include 948.4 mm at Genoa Bolzaneto on 10 September 1970, 480 mm at Crkvice in Montenegro on 28 November 1927 (Martyn 2000), 349 mm at Ciupercenii Vechii in Romania [unclear date] (Furlan 1977),

316.4 mm at Seathwaite in England on 19 November 2009, and 300 mm at Hala Gąsienicowa in Poland on 30 June 1973 (Niedźwiedź 1992). The daily total regarded as the highest ever recorded in the Sudeten Mountains or in Central Europe is 345.1 mm; it occurred on 30 July 1897 at the Czech Nova Louka (Neuwiese) village in Jizera Mountains. On 12 August 2002, the German record daily precipitation of 313 mm was observed at Zinnwald in the Ore Mountains (Niedźwiedź 2003a). Appendix 2 lists other record-breaking daily precipitation totals that have occurred since the beginning of regular measurements (varying from station to station) in Europe. They were found in the literature and in online databases. It is worth also mentioning that the world's greatest daily total, 1825 mm, was observed at Foc Foc station on the island of Reunion on 7–8 January 1966 (Stach 2009, from: WMO 1994).

2.3.2 *Frequency of Daily Precipitation ≥10 mm, ≥30 mm, and ≥50 mm*

The variety of precipitation-determining factors, largely related to the size of the European content, drives the enormous spatial and seasonal variability of its totals (including maximum) and frequency. The extreme precipitation frequency indices covered here and based on fixed threshold values (≥10 mm, ≥30 mm, ≥50 mm) demonstrate how important it is to capture the regional aspect, especially in studies covering the whole continent, for defining extreme precipitation events.

Number of Days with Precipitation ≥10 mm The spatial variability of the frequency of precipitation ≥10 mm in Europe is associated with the time of precipitation concentration over a year. Across vast areas spanning the western European coast and the continent's east, including the lee side (western side) of the Scandinavian Mountains, daily totals ≥10 mm are concentrated in the summer. Their seasonal average varies between 5 and 10 days at nearly 60 % of stations. In southern Europe, where summer is the driest season (Chromow 1977), also days with totals ≥10 mm are a rarity. South of the parallel 40°N their mean frequency of occurrence does not exceed 1 day, and in the eastern Mediterranean days with these totals did not occur at all during the entire study period (Figs. 2.3 and 2.4).

In western and central Mediterranean, these days are most frequent in autumn, and in winter in the eastern part of the Mediterranean. The winter season features the greatest variability of days with precipitation ≥10 mm, which ranges from 0 to 35 days (Table 2.2). This category of precipitation is rare, and its mean frequency peaks at 1 day in an area of Eastern Europe including southern Poland and on the lee side of the Scandinavian Mountains within the Scandinavian Peninsula (Fig. 2.3).

Stations located in the northeastern peripheries of the continent only recorded precipitation ≥10 mm on a few occasions during the 56-year study period (including Zhizhgin Maya: 4 days; Naryan Mar and Hoseda Hard: 9 days each; Uhta and

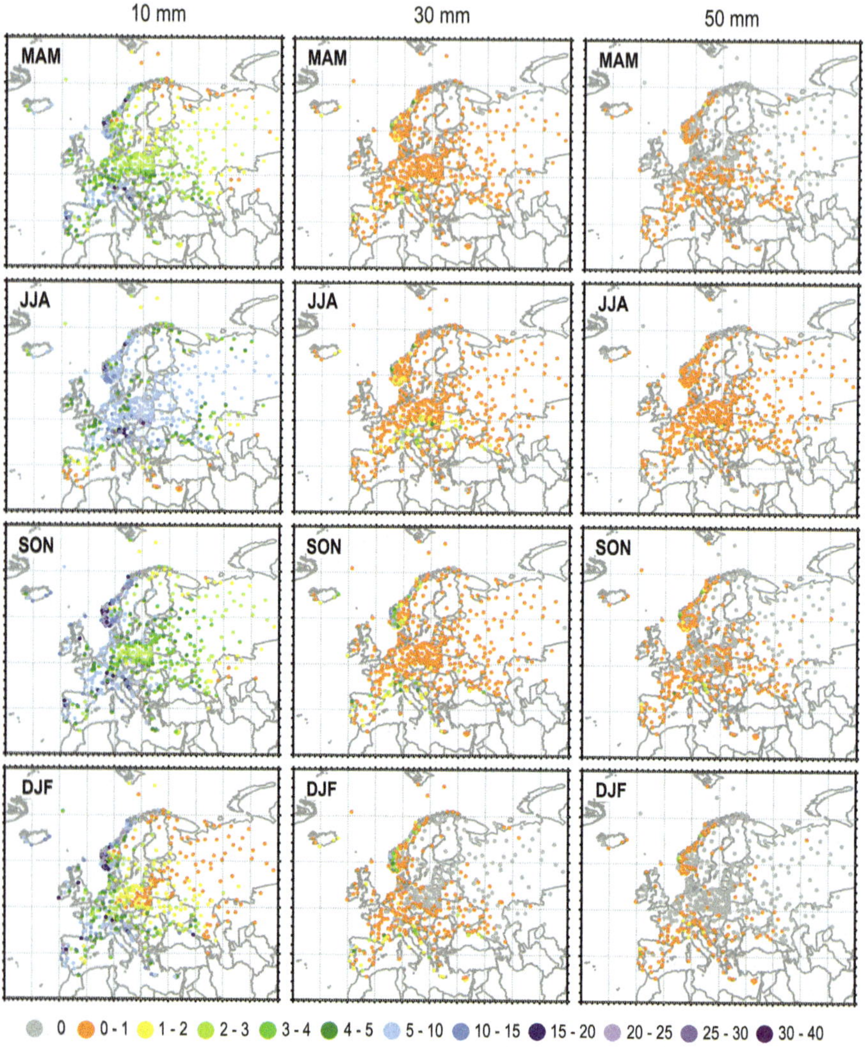

Fig. 2.3 Average number of days with precipitation ≥10 mm, ≥30 mm, and ≥50 mm for December 1950–February 2008

Leushi: 6 days each). In Eastern Europe these numbers increase in the three remaining seasons. In spring, Russian stations record on average 2 such days and 3 days in autumn. On the other hand, areas to the north and northwest of the Caspian Sea (Caspian Depression) have particularly low frequencies of precipitation ≥10 mm in any season, and they never exceed 2 days. This anomaly is explained by the great distance that separates the area from the Atlantic Ocean, Europe's main source of humidity. Indeed, according to a classification proposed by Okołowicz (1969), the area belongs to an extremely arid continental climate type. Westwards from there,

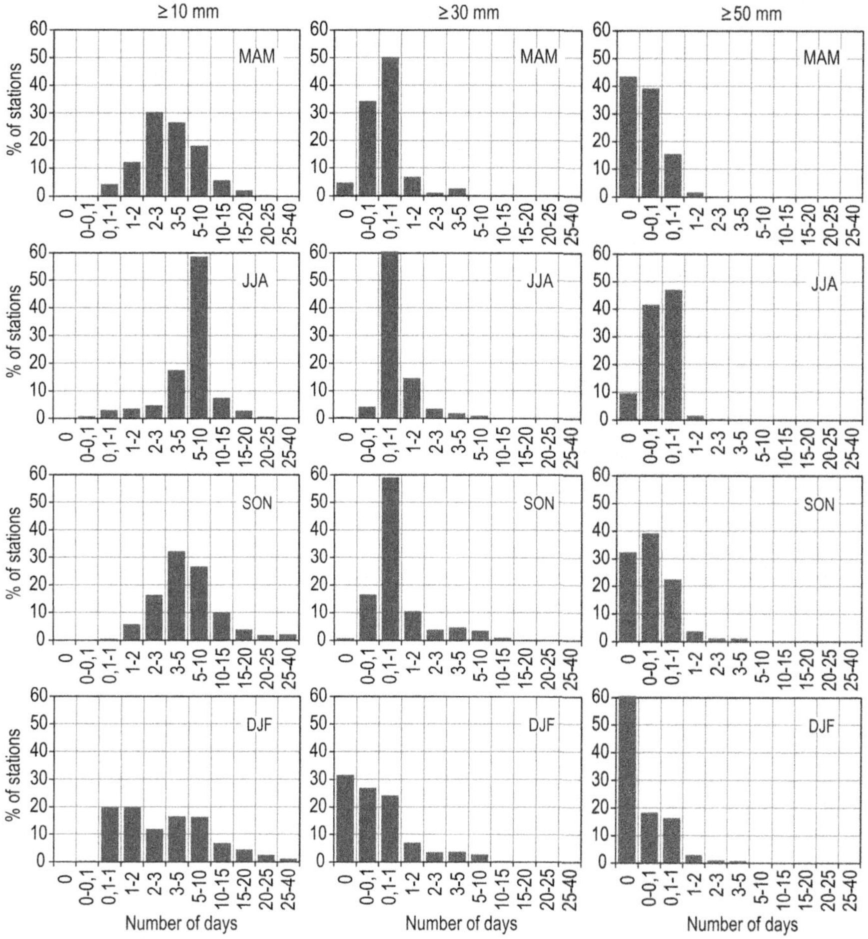

Fig. 2.4 Frequency of stations within the intervals* of extreme precipitation frequency (in days) for December 1950–February 2008. *Right* closed intervals

towards the Atlantic Ocean, daily totals equal or greater than 10 mm are more frequent (Fig. 2.3). At French stations the mean seasonal number is approximately 6 days with the maximum at Oderen (12 days in spring and summer and 18 in winter). The western Scandinavian coast has a considerable number of days with precipitation ≥10 mm all year round, with a maximum in autumn (13 days of seasonal mean). Along the central and southwestern section of this coast, described by Martyn (2000) as the wettest part of Scandinavia, certain stations recorded more than 20 such days per season from winter to spring and more than 30 days in autumn (Takle, Brekke i Sogn, Rørvikvatn ved Vadhim, Hovlandsdal, Osland ved Stongfjorden). The central part of Europe's western coast has the lowest number of days with precipitation ≥10 mm in spring (Figs. 2.3 and 2.4).

Table 2.2 Descriptive statistics for average number of days with precipitation ≥10 mm, ≥30 mm, and ≥50 mm in Europe for December 1950 to February 2008

Index	Season	Average (±SE)	Confidence intervals –95 %	+95 %	Min	Max	SD	ME	Quartiles Lower	Upper
NoD10	MAM	4.4 (±0.16)	4.1	4.7	0.3	20.7	3.5	3.2	2.3	5.1
	JJA	6.5 (±0.16)	6.2	6.8	0.1	26.5	3.6	6.2	4.8	7.5
	SON	6.7 (±0.25)	6.2	7.2	0.6	35.1	5.7	4.5	3.2	7.9
	DJF	5.1 (±0.26)	4.6	5.6	0.0	34.8	5.9	2.8	1.2	6.5
NoD30	MAM	0.4 (±0.03)	0.4	0.5	0.0	4.8	0.8	0.1	0.1	0.4
	JJA	0.8 (±0.04)	0.7	0.9	0.0	7.4	0.8	0.6	0.4	0.9
	SON	1.0 (±0.08)	0.8	1.1	0.0	13.6	1.8	0.3	0.1	0.9
	DJF	0.6 (±0.07)	0.5	0.8	0.0	11.1	1.5	0.1	0.0	0.5
NoD50	MAM	0.1 (±0.01)	0.1	0.1	0.0	2.3	0.2	0.0	0.0	0.1
	JJA	0.0 (±0.01)	0.1	0.2	0.0	2.1	0.3	0.1	0.0	0.2
	SON	0.2 (±0.03)	0.2	0.3	0.0	4.5	0.6	0.0	0.0	0.1
	DJF	0.2 (±0.02)	0.1	0.2	0.0	4.2	0.5	0.0	0.0	0.1

MAM spring, *JJA* summer, *SON* autumn, *DJF* winter, *NoD10, NoD30, NoD50* number of days with precipitation ≥10 mm, ≥30 mm, ≥50 mm; *SE* standard error, *Min* minimum value, *Max* maximum value, *SD* standard deviation, *ME* median

Number of Days with Precipitation ≥30 mm Days with precipitation ≥30 mm are rare in Europe (Table 2.2). In winter, stations located within the large swathes of eastern and central (mainly Poland) parts of the continent and in certain parts of the Scandinavian Peninsula did not record any day with precipitation exceeding 30 mm during the study period. In the remaining portion of Europe (51 % of stations), its frequency is mostly up to 1 day. Only in the southwestern tip of the Scandinavian Peninsula, at certain stations of the Mediterranean, and in the southwestern part of the Iberian Peninsula did the frequency of precipitation ≥30 mm vary between 1 and 4 days on average (14 % of stations), with isolated cases of more (4 % of stations). In winter, three Scandinavian stations stand out with 10 or more days with precipitation ≥30 mm (11 days at Takle and Brekke i Sogn; 10 days at Hovlandsdal).

Daily precipitation ≥30 mm is also rare during other seasons, that is, spring, summer, and autumn. Its average frequency does not exceed 1 day at 89 % of stations in spring, 79 % of stations in summer, and 76 % of stations in autumn (Fig. 2.4). In each of these seasons, there are areas that are much more endowed with this category of precipitation, including the central and southeastern sections of the Scandinavian coast, which record from 1 to 3 days with precipitation ≥30 mm in spring and summer to more than 10 such days at some stations in autumn (Takle: 12 days; Brekke i Sogn: 14 days; Rørvikvatn ved Vadheim, Hovlandsdal and Osland ved Stongfjorden: 11 days). Other areas with more than 1 day with precipitation ≥30 mm include the Alps in spring; Alps and areas stretching eastwards to the western coast of the Black Sea and northwards up to the parallel 50°N in summer; and the northern Mediterranean coast from the Strait of Gibraltar to the Peloponnese and Crete in autumn (Fig. 2.3).

Number of days with precipitation ≥50 mm In Europe, days with precipitation ≥50 mm occur mostly in summer and are notably less frequent than once a year at most (89 %) stations (Fig. 2.4; Table 2.2). Even in summer, this category of precipitation does not occur at all at 10 % of the stations, as well as at 32 % of stations in autumn, 43 % in spring, and 61 % in winter (Figs. 2.3 and 2.4). Mekis and Hogg (1997) and Easterling et al. (1999) independently agree that the average intensity of extreme precipitation events decreases rapidly towards higher latitudes and therefore daily totals with more than 50 mm tend to be rare in these areas.

Seasonal changes in the spatial distribution of high precipitation totals and frequencies in Europe are a result of combined effect of large-scale and local climatic factors. The lower extreme precipitation totals recorded in the eastern part of the continent, compared to its western portions, reflect the former's greater distance from the Atlantic Ocean, Europe's main source of humidity. Air masses undergo transformation as they travel eastwards across the continent, including through dropping of their water vapour content. The highest totals and greatest frequencies of precipitation on the western Scandinavian coast, particularly in autumn, are linked to the effect of the orographic barrier of the Scandinavian Mountains on the humid air masses moving in from the Atlantic Ocean. The continent's heaviest and most frequent extreme precipitation in autumn and winter recorded in the Mediterranean is associated with winter Mediterranean lows developing on the south-shifted polar front at that time of the year (Martyn 2000). The relatively low totals and low frequencies of extreme precipitation in the same area in summertime are an effect of subtropical highs that reach the area with their northern edges (Chromow 1977; Okołowicz 1969; Martyn 2000).

2.4 Criteria for Identifying Daily Precipitation Extremes

The great diversity of precipitation regimens in Europe, as defined by their totals, intensities, and annual course, makes it difficult to select a common European threshold value for extreme precipitation (Karagiannidis et al. 2012). This problem was clearly demonstrated by the frequency distribution of precipitation ≥10 mm, ≥30 mm, and ≥50 mm discussed earlier in Sect. 2.3.2. For this reason, this study has adopted a statistical definition of extreme precipitation event identified separately at each weather station using empirical distribution of the frequency of daily precipitation totals. The most popular criterion of ≥95 percentile (95P) of daily totals was adopted (Klein Tank et al. 2002; Zolina et al. 2005; Alexander et al. 2006; Benestad 2006; Trömel and Schönwiese 2007). There is a range of approaches by which climatologists derive the percentiles in terms of daily totals to be taken into account. For example, Zolina et al. (2004) computed a precipitation threshold corresponding to the 95P taking into account all the days with measurable precipitation (≥0.1 mm) recorded in their study period, although Klein Tank and Köennen (2003), as well as many others (e.g., Gajić-Čapka and Cindrić 2011), only included days with ≥1 mm recorded in the standard 30-year period 1961–1990. Nicholls and Murray (1999) also strongly recommended this latter approach that uses the standard 30-year period for calculating the 95P. As it turns out, the kind of daily totals taken into account (≥0.1 mm versus ≥1 mm) in the calculation of percentiles significantly affects the magnitude of threshold values at some stations and, as a consequence, the number of precipitation events thus derived. This approach is illustrated using data for Poland in Figs. 2.5 and 2.6.

The influence of the baseline period used to derive threshold values on the amount of daily precipitation corresponding to the 95P is lesser, which is illustrated in Fig. 2.6, which presents the same characteristics, but for percentiles derived using a different 30-year baseline period ($95P_{1951-1980}/95P_{1961-1990}/95P_{1971-2000}$). In all three periods, days with precipitation ≥1 mm were used. The values of the percentiles thus obtained were compared by calculating the following differences: $95P_{1951-1980} - 95P_{1961-1990}$ and $95P_{1961-1990} - 95P_{1971-2000}$.

In this study, the 95P value was derived using days with precipitation ≥1 mm in the standard period 1961–1990. It has been demonstrated here that the use of days with precipitation ≥1 mm renders higher the threshold values for extreme precipitation. Also certain databases, including ECA that is crucial for this study, do not contain precipitation <1 mm, although at many stations the accuracy of daily totals equal 1 mm.

An important assumption adopted while determining the threshold values was that precipitation extremes would occur in every month and that their magnitude would be linked to the annual cycle of precipitation totals that is clearly noticeable in Europe. A number of examples of these annual cycles are given by Martyn (2000), including a concentration of precipitation in Southern Europe and on the Crimea Peninsula in autumn and winter, in central Spain, and on the western coast of the Black Sea also in spring.

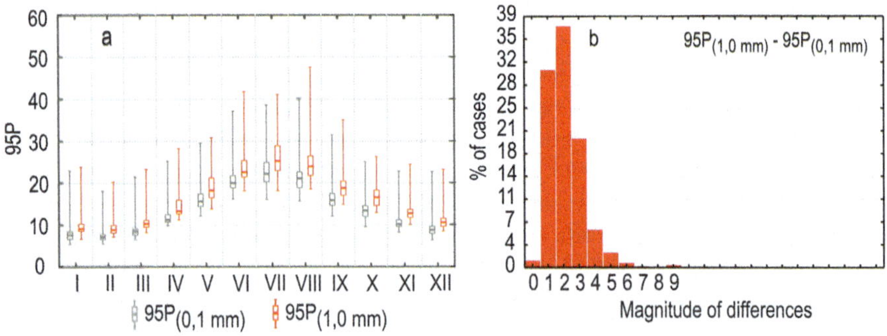

Fig. 2.5 Statistics (median, quartiles, extreme values) of the 95th percentiles (95P) of daily precipitation totals calculated from days with precipitation ≥0.1 mm (95P 0.1 mm) and days with precipitation ≥1,0 mm (95P 1.0 mm) (**a**) and statistical distribution of the differences between these percentiles (95P 1.0 mm–95P 0.1 mm) (**b**) using Polish precipitation data for the 1961–1990 period

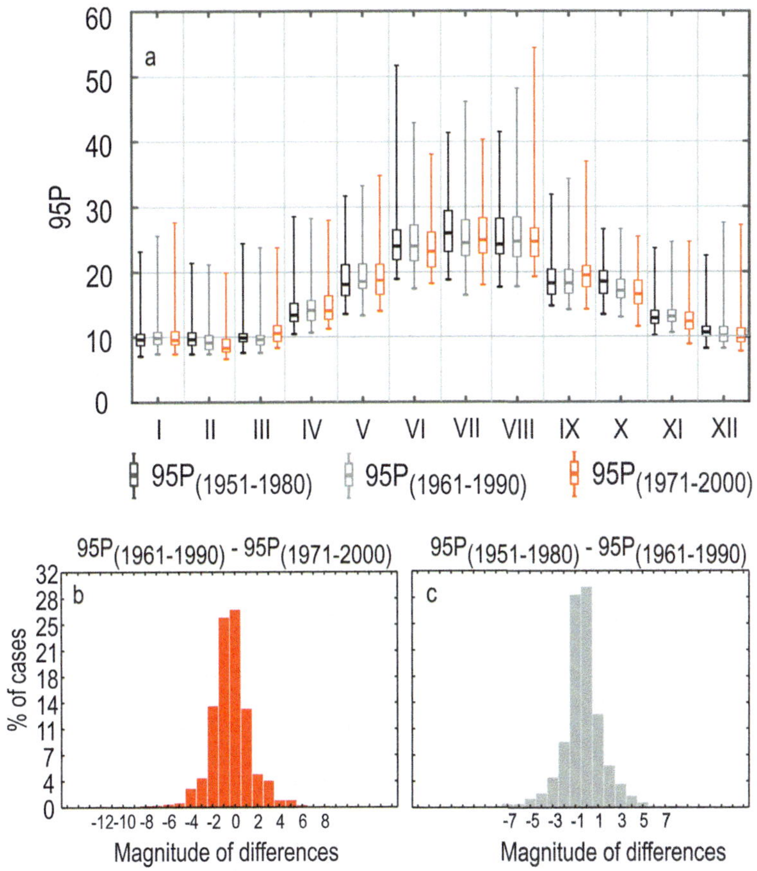

Fig. 2.6 Statistics (median, quartiles, extreme values) of the 95th percentiles (95P) of daily precipitation totals calculated from days with precipitation ≥1.0 mm for different 30-year normal periods (95P 1951–1980, 95P 1961–1990, 95P 1971–2000) (**a**) and statistical distribution of the differences between percentiles for 1961–1990 and 1971–2000 (**b**) and 1951–1980 to 1961–1990 (**c**) using Polish data as an example

Along the western European coast precipitation peaks in autumn and winter, but eastwards from there this peak shifts towards summertime. In the Alps and Carpathians, as well as on inland European lowlands, the annual precipitation maximums occur in spring. Failure to account for this seasonality would lead to extreme events being only identified in seasons characterised by the greatest concentration of precipitation (Zhang et al. 2001). For these reasons, the values of 95P were calculated separately in each month of the year. Precipitation extremes identified using these criteria are less useful for the insurance sector as not all of them actually cause material damage. Indeed, the adopted method is a compromise between extreme precipitation as the object of study and its frequency, thus allowing the use of statistical methods and receiving reliable results (Mizahari 2000; Frei and Schär 2001; Zhang et al. 2001; Beniston 2005; Klein Tank et al. 2009).

Six statistical formulae of calculating percentiles were tested: the weighted mean centred on *Xnp*, the weighted mean centred on $X(n+1)p$, empirical distribution function, empirical distribution function with averaging, empirical distribution function with interpolation, and nearest observation. The weighted mean centred on $X(n+1)p$ was finally selected to determine the value of 95P using the empirical distribution of daily precipitation. In the formula n means the number of cases and p is the value of a percentile divided by 100 (e.g., 50/100 = 5 for the median). This method yielded the highest 95P values of daily precipitation among the six tested. The mean differences between the methods remained within 1 mm (ranging from 0.1 mm in February to 0.6 mm in August) and were the greatest during months characterised by the highest variability of daily precipitation totals at a given station (e.g., in Poland this is August, with the extreme case of 2.6 mm at the high-mountain station Kasprowy Wierch that records the highest precipitation totals in Poland).

Figure 2.7 illustrates an example of empirical statistical distributions of daily totals ≥1 mm for 1961–1990 at stations located in various parts of Europe and in

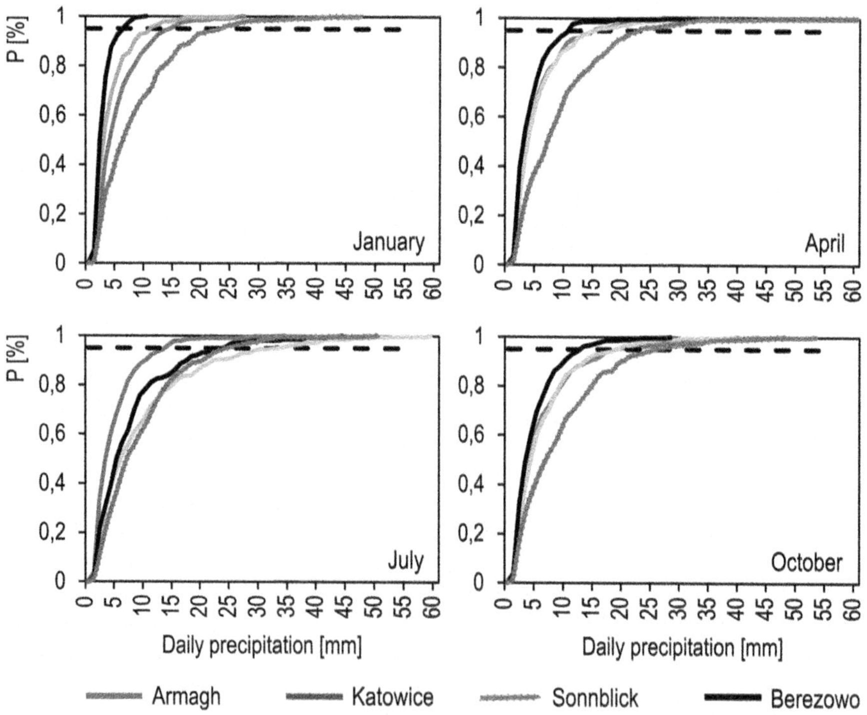

Fig. 2.7 Empirical probability distribution of daily precipitation ≥1 mm in the period 1961–1990 at selected stations in Europe. *P* probability

Table 2.3 Relative threshold values for extreme precipitation corresponding to the 95th percentiles (95P) calculated from days with precipitation ≥1 mm in the period 1961–1990 for selected stations in Europe

	95P of daily precipitation totals (≥1 mm)			
Station	January	April	July	October
Armagh	14.1 mm	14.3 mm	14.0 mm	18.5 mm
Katowice	10.0 mm	14.3 mm	28.9 mm	18.6 mm
Sonnblick	23.0 mm	22.1 mm	23.9 mm	25.3 mm
Berezowo	6.1 mm	8.9 mm	23.7 mm	12.3 mm

selected months representing each season. Table 2.3 summarises values of the 95th percentile calculated using these distributions.

2.5 Spatial and Seasonal Variability in the 95th Percentile Threshold Values of Daily Precipitation

The spatial and seasonal variability of the threshold values corresponding to 95P of daily precipitation in Europe is linked to the spatial and seasonal variability of its monthly totals (Fig. 2.8). The highest average European values of 95P (24 mm) were obtained for the late summer/early autumn (August and September) and the lowest values for winter and spring (more than 16 mm). Nevertheless, because there is a considerable spatial variability of precipitation totals across Europe, these average values must be regarded as generalised and indicative only. Distributions of 95P daily precipitation frequencies (Fig. 2.9), suggest that between December and March the 95P values range from 5 to 10 mm at stations located mainly in Central Europe, Eastern Europe, and on the lee (western) side of the Scandinavian Mountains, whereas they are the highest, up to more than 25 mm, in southern parts of the continent and along the western coast of the Scandinavian Peninsula (Fig. 2.8; Table 2.4).

From April until summer, the value of the 95P continues to increase. For example, in April 95P mostly ranges from 10 to 15 mm (42 % of stations in Western Europe, Central Europe, and in a western portion of Eastern Europe), while between May and August it normally reaches 15–20 mm (from 30 % of stations in August to 42 % of stations in May in Western, Central, and Eastern Europe and on the Scandinavian Peninsula (Figs. 2.8 and 2.9).

In spring, the maximum values of 95P are lower than in other seasons (Table 2.4). These changes in summer, when the number of stations with 95P values ranging from 20 to 30 mm increases (Fig. 2.9), especially in Central Europe (20–25 mm) and Southern Europe (25–30 mm) (Fig. 2.8). Autumn has the greatest range of 95P, which peaks in October (Table 2.2). The values of 95P diminish between September and November at many Central and Eastern European stations. In September and October, 95P typically ranges from 15 to 20 mm. In November, the most frequent

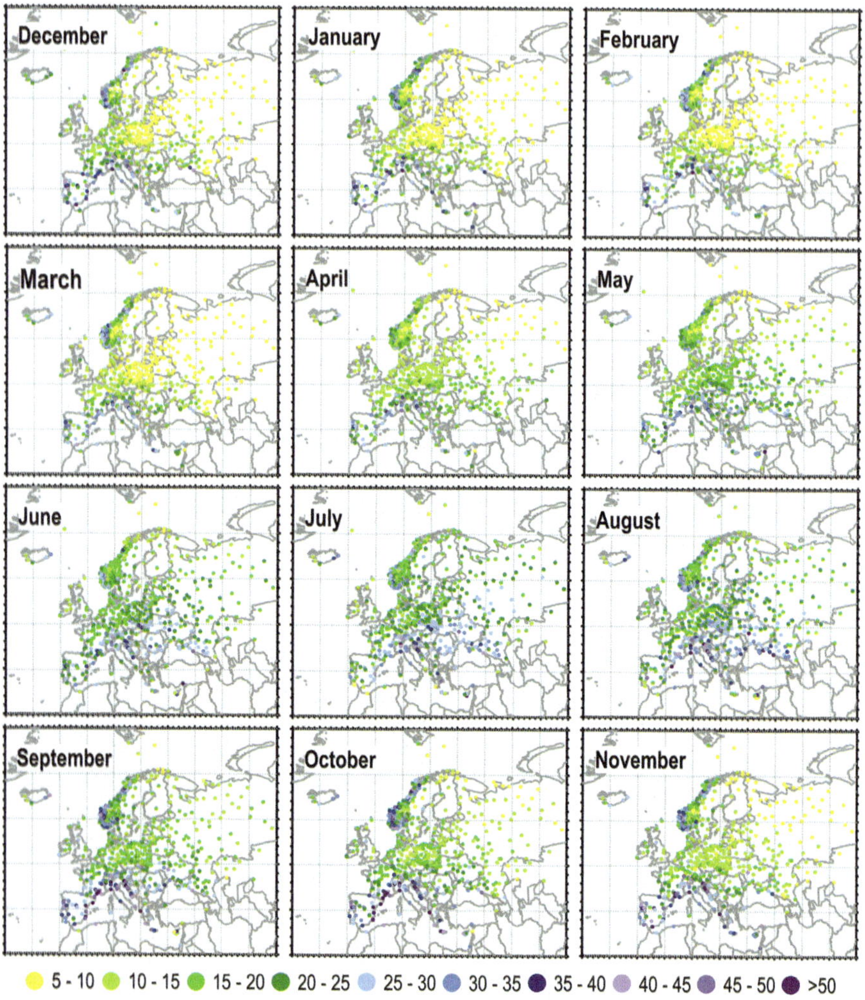

Fig. 2.8 Daily precipitation totals corresponding to the 95th percentile (95P) used as a criteria for extreme precipitation. Percentiles calculated from daily precipitation ≥1 mm in the 1961–1990 period

are threshold values in the range 10–15 mm (36 % of stations). In autumn, the frequency of stations in ranges exceeding 40 mm is higher (on average, 8 % of stations in a month) than in other seasons (i.e., 1 % stations in spring, 3 % in summer, and 4 % in winter). The highest 95P values correspond with stations located in southern Europe (Figs. 2.8 and 2.9). The highest values of 95P all year round are derived for stations located in southern Europe; along the western coast of the Scandinavian Peninsula, especially in its southwestern section; and in mountains. These parts of Europe are also characterised by an increased spatial variability of the 95P values.

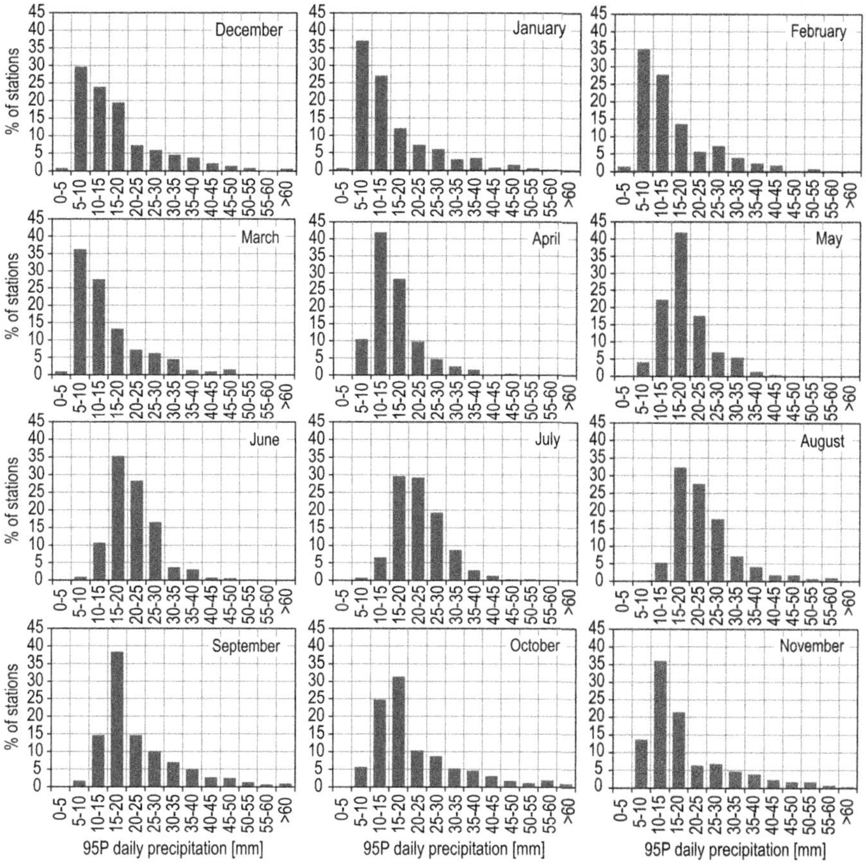

Fig. 2.9 Frequency of stations within the intervals* of daily precipitation corresponding to the 9th percentiles (95P) used as a criteria for extreme precipitation. Percentiles calculated from daily precipitation ≥1 mm in the 1961–1990 period. *Right* closed intervals

Similar conclusions were also reported by Anagnostopoulou and Tolika (2012), who analysed relative threshold values for the purpose of identifying extreme precipitation events in Europe on an annual scale. They explained the high variability of the threshold values in the Mediterranean by the diversity of climatic conditions found in the region. The values of 95P tend to be higher along the tracks of cyclones and in cyclone development areas (Alpert et al. 1990a, 1990b; Maheras et al. 2001; Lionello et al. 2002; Anagnostopoulou and Tolika 2012). Trigo et al. (1999) explained the high values of the 95P in the Iberian Peninsula with the high frequency of a low-pressure system, which determine the local precipitation regimen, and with Atlantic cyclones whose track crosses the central and southern parts of the peninsula on its way into the Mediterranean. Anagnostopoulou and Tolika (2012) add to these factors the effects of altitude and landform.

Table 2.4 Statistical characteristic of precipitation thresholds corresponding to the 95th percentiles used as criteria for extreme precipitation. Percentiles calculated from daily precipitation ≥1 mm in the 1961–1990 period

Month	Average (±SE)	Confidence intervals		ME	Quartiles		Max	Min	Range	SD
		−95 %	+95 %		Lower	Upper				
Jan	16.9 (±0.42)	16.1	17.7	13.3	9.1	22.5	62.4	4.8	57.6	10.4
Feb	16.7 (±0.41)	15.9	17.5	13.6	9.0	21.6	69.0	4.2	64.8	10.2
Mar	16.5 (±0.38)	15.8	17.3	13.4	9.1	21.8	56.6	4.3	52.2	9.6
Apr	16.7 (±0.26)	16.1	17.2	15.0	12.4	19.1	51.9	5.8	46.1	6.5
May	18.7 (±0.24)	18.2	19.2	17.2	14.8	22.1	44.5	6.3	38.2	6.1
Jun	21.3 (±0.25)	20.8	21.8	20.1	17.1	24.8	48.1	7.7	40.4	6.2
Jul	23.1 (±0.28)	22.6	23.7	22.1	18.1	26.7	75.0	6.0	69.0	7.0
Aug	24.0 (±0.33)	23.4	24.7	22.5	18.1	27.1	61.4	8.7	52.7	8.2
Sep	24.0 (±0.43)	23.2	24.9	20.3	16.3	28.6	72.1	8.9	63.1	10.9
Oct	23.4 (±0.51)	22.4	24.4	18.8	14.8	28.5	114.4	7.9	106.5	12.7
Nov	20.5 (±0.46)	19.6	21.4	16.5	12.7	25.5	83.0	5.4	77.6	11.5
Dec	18.6 (±0.44)	17.7	19.5	15.1	9.8	24.0	66.8	4.5	62.3	11.1

SE standard error, *ME* median, *Min* minimum value, *Max* maximum value, *SD* standard deviation

2.6 Conclusions

The considerable degree of variation found in European precipitation regimens, as defined by their totals and concentration seasons, demands the definitions of extreme precipitation events to be regionalised. Seasonal changes in the spatial distribution of high precipitation in Europe are a result of combined effect of large-scale and local climatic factors. The lower extreme precipitation totals recorded in the eastern part of the continent, compared to its western portions, reflect the former's greater distance from the Atlantic Ocean, Europe's main source of humidity. Air masses undergo transformation as they travel eastwards across the continent including through dropping of their water vapour content. The highest totals and greatest frequencies of precipitation on the western Scandinavian coast, particularly in autumn, are linked to the effect of the orographic barrier of the Scandinavian Mountains on the humid air masses moving in from the Atlantic Ocean. The continent's heaviest and most frequent extreme precipitation in autumn and winter recorded in the Mediterranean is associated with winter Mediterranean lows developing on the south-shifted polar front at that time of the year (Martyn 2000). The relatively low totals and low frequencies of extreme precipitation in the same area in summertime are an effect of subtropical highs that reach the area with their northern edges (Chromow 1977; Okołowicz 1969; Martyn 2000). The existing variety of the definitions of extreme weather event means that in the case of precipitation the term extremum refers to a number of characteristics of this weather element. Indices of extreme climate and weather events can be broken down into five categories that differ in the identification methods and features of the event itself (e.g., intensity, frequency). This study adopted the statistical definition of extreme events. An event was identified as extreme if its frequency of occurrence was above the 95th percentile (95P) of the statistical frequency distribution. Threshold values corresponding to the 95P were derived separately for each station and each month using records of precipitation ≥1 mm in the standard period 1961–1990. The spatial and seasonal variability of the daily totals of 95P precipitation (i.e., the threshold values for the identification of days with extreme precipitation) is associated with the variability of monthly precipitation totals. The overall range of variability in the 95P precipitation spans the daily totals of 4.2 mm (January) and 114.4 mm (October). The highest values of 95P were recorded in the Mediterranean (above 40 mm in its western and central parts in autumn) and along the western Scandinavian coast, especially its southwestern section (above 25 mm in winter). Elsewhere, the highest seasonal threshold values of extreme precipitation were identified in summer (15–30 mm) and the lowest in winter (5–10 mm in Central Europe and Eastern Europe; 10–15 mm in Western Europe). The spatial and seasonal variability of 95P daily totals is a result of the combined circulation and altitude effects.

Appendices

Appendix 1

The absolute maxima of daily precipitation in Europe between December 1950 and July 2008

WMO No.[a]	Station	Country	Precipitation mm/day	Date	Origin-based extreme precipitation type
01001	Jan Mayen	Norway	68.5	1979-09-11	Fo
01003	Hornsund (Isfjord)	Norway (Svalbard)	58.3	1994-08-01	Fw
01028	Bjørnøya	Norway	41.5	1988-12-08	Fo
011062	Hopen	Norway	49.1	1986-12-02	A

(continued)

WMO No.[a]	Station	Country	Precipitation mm/day	Date	Origin-based extreme precipitation type
01---	Bjørnsund	Norway	46.0	1986-08-07	Ff
1065	Karasjok	Norway	50.5	1966-07-17	A
1066	Helnes fyr	Norway	41.9	1986-08-08	Ff
01---	Jotkajavre	Norway	43.2	1981-07-06	Fc
1075	Rustefjelbma	Norway	40.7	2001-08-19	Fo
1078	Slettnes fyr	Norway	42.8	2003-01-21	Fw
1092	Makkaur fyr	Norway	62.1	1996-07-15	Fo
01---	Sihccajavri	Norway	53.3	1981-07-06	Fc
1023	Bardufoss	Norway	53.6	1968-02-13	Ff
01---	Bones and Bardu	Norway	66.0	1959-10-06	Fw
1026	Tromsø	Norway	63.5	1964-10-04	Fw
01147	Susendal	Norway	81.7	2005-09-02	A
01---	Dunderlandsdalen	Norway	110.9	1989-12-03	Ff
01---	Lurøy	Norway	181.8	1961-02-14	Ff
01---	Glomfjord	Norway	184.3	1964-01-09	Ff
01---	Sulitjelma	Norway	85.1	2002-01-11	Ff
01152	Bodø VI	Norway	72.7	1988-09-15	Fw
01---	Steigen	Norway	77.0	1971-08-26	Ff
01---	Kråkmo	Norway	171.7	1964-01-09	Ff
01---	Ankenes	Norway	93.2	1959-10-06	Fw
01160	Skrova fyr	Norway	58.2	1975-09-09	Fc
01---	Barkestad	Norway	136.2	1963-09-22	Fw
01270	Værnes	Norway	77.6	1970-09-05	A
01---	Østås and Hegra	Norway	73.4	1987-12-09	Fw
01---	Namdalseid	Norway	78.8	1953-03-25	Ff
01---	Tunnsjø	Norway	61.9	1970-07-11	Fc
01---	Liafoss	Norway	94.5	1982-05-31	Fc
01237	Hitra	Norway	69.6	2003-12-18	Ff
01---	Lien and Selbu	Norway	52.9	1987-12-10	Ff
01241	Ørland III	Norway	62.5	2004-09-22	Fo
01---	Norddal	Norway	71.1	1974-10-26	A
01218	Tafjord	Norway	103.6	1978-09-18	A
01---	Halsafjord II	Norway	124.0	1997-09-14	Fo
01---	Rindal	Norway	77.6	1997-03-31	Ff
01---	Gløtvola – Trøan	Norway	111.5	1985-09-07	Fo
01---	Nord-Odal	Norway	68.8	1996-08-25	Fc
01---	Ørbekkedalen	Norway	74.2	1996-08-25	Fc
01---	Folldal	Norway	49.2	1973-07-21	Fw
01---	Nespåhedmark	Norway	65.0	1956-09-12	Ff
01---	Biri	Norway	79.2	1988-09-03	Fo
01---	Vestre Gausdal	Norway	60.5	1985-09-07	Fo

(continued)

WMO No.[a]	Station	Country	Precipitation mm/day	Date	Origin-based extreme precipitation type
01---	Bøverdal	Norway	45.5	1954-10-14	Ff
01---	Skjåk	Norway	40.3	1960-06-28	Fw
01---	Beito	Norway	75.0	1952-08-03	Fc
01319	Takle	Norway	184.6	1995-10-27	Ff
01---	Brekke and Sogn	Norway	158.4	1995-10-27	Ff
01---	Vik and Sogn III	Norway	72.5	1983-03-09	Ff
01---	Maristova	Norway	60.1	1993-02-04	Ff
01---	Hafslo	Norway	66.6	1986-11-11	Fc
01---	Sogndal – Selseng	Norway	94.2	1995-10-27	Ff
01---	Rørvikvatnved Vadheim	Norway	124.7	1954-09-25	Ff
01---	Lavik	Norway	161.0	1998-11-04	Fc
01---	Hovlandsdal	Norway	170.3	1975-09-23	Ff
01---	Osland ved Stongfjorden	Norway	131.3	1983-10-26	Ff
01---	Botnen and Førde	Norway	131.7	1992-01-11	Ff
01---	Myklebust and Breim	Norway	93.0	1995-10-27	Ff
01---	Briksdal	Norway	104.0	1957-01-09	Fc
01---	Hornindal	Norway	96.2	1957-01-09	Fc
01359	Geilo	Norway	49.5	1986-12-04	Ff
01---	Hiåsen	Norway	77.3	1990-10-30	Fo
01---	Lyngdal and Numedal	Norway	66.0	2000-11-08	Ff
01---	Tunhovd	Norway	86.0	1977-06-26	Ff
01489	Bjørnholt	Norway	98.2	1959-10-28	Ff
01492	Oslo – Blindern	Norway	59.8	1989-08-01	Fo
01---	Kjelsås and Sørkedalen	Norway	71.5	1959-10-28	Ff
01---	Ramnes	Norway	77.3	1962-08-19	Fo
01---	Færder Fyr	Norway	53.5	1975-11-17	A
01---	Hedrum	Norway	108.5	1988-08-23	A
01---	Besstul and Gjerpen	Norway	108.3	1959-10-28	Ff
01---	Rjukan	Norway	61.6	1968-09-04	Fc
01---	Tuddal	Norway	78.2	1996-09-28	Fo
01---	Lifjell	Norway	75.0	1995-01-19	Fc
01---	Rauland	Norway	70.0	2006-01-11	Ff
01---	Drangedal	Norway	77.8	2002-06-11	Fo
01---	Postmyr and Drangedal	Norway	112.5	1990-10-30	Fo
01---	Røldal	Norway	110.8	1962-02-16	Ff
01---	Bulken	Norway	83.8	1983-03-09	Ff
01---	Helleland	Norway	134.3	1997-08-29	Fc

(continued)

WMO No.[a]	Station	Country	Precipitation mm/day	Date	Origin-based extreme precipitation type
1---	Søyland and Gjesdal	Norway	143.0	1983-10-26	Ff
01415	Sola	Norway	92.3	1997-08-29	Fc
01---	Sviland	Norway	129.0	1983-10-26	Ff
01---	Lysebotn	Norway	156.2	1984-01-01	Ff
01424	Sauda	Norway	106.9	2005-11-15	Ff
01403	Utsira Fyr	Norway	81.3	1957-09-13	Fc
01---	Torungen Fyr	Norway	86.1	1986-08-28	Fo
01---	Herefoss	Norway	101.0	1966-12-01	Fo
01---	Mykland	Norway	90.8	1953-08-22	Fo
01452	Kjevik	Norway	99.0	1990-09-05	A
01448	Oksøy Fyr	Norway	108.7	2002-10-23	Fw
01---	Mestad and Oddernes	Norway	151.4	1981-09-19	Fw
01---	Åseral	Norway	103.1	1959-08-15	Ff
01427	Lista Fyr	Norway	117.4	1997-08-29	Fc
01---	Risnes and Fjotland	Norway	85.2	1957-12-21	Ff
01---	Bakke	Norway	111.4	1976-10-13	Ff
01---	Skreådalen	Norway	100.7	1991-11-02	Fo
01---	Øvre Sirdal	Norway	81.1	2000-10-31	Fo
01---	Halden (Ostfold)	Norway	62.0	2001-08-07	A
02836	Sodankyla	Finland	42.5	1992-08-12	Fo
02935	Jyvaskyla	Finland	67.6	2004-07-28	Ff
02974	Helsinki	Finland	79.3	1993-07-24	Ff
02080	Karesuando	Sweden	78.0	2004-09-23	Fo
02120	Kvikkjokk	Sweden	50.4	1972-06-13	Fw
02127	Stensele	Sweden	63.0	1966-07-27	Fc
02196	Haparanda	Sweden	47.0	1986-08-06	Ff
02229	Oestersund	Sweden	54.0	2000-07-19	Fw
02288	Holmogadd	Sweden	106.0	2004-02-04	Ff
02361	Harnosand	Sweden	78.4	1993-07-25	Fo
02410	Malung	Sweden	60.1	2004-08-04	Fw
02418	Karlstad	Sweden	68.4	1952-05-05	Fo
02433	Falun	Sweden	64.8	1951-08-10	Fo
02483	Stockholm	Sweden	59.8	1961-07-21	A
02512	Goteborg	Sweden	99.9	1976-12-24	A
02562	Linkoeping	Sweden	72.7	1972-07-13	A
02584	Gotska sandon	Sweden	99.0	2005-07-23	Fo
02640	Vaexjoe	Sweden	95.7	1959-07-13	Fc
03965	Birr	Ireland	59.1	1984-08-02	Fo
03953	Valentia	Ireland	112.5	1980-11-01	Fc
03969	Dublin	Ireland	85.1	1986-08-25	Ff

(continued)

WMO No.[a]	Station	Country	Precipitation mm/day	Date	Origin-based extreme precipitation type
03980	Malin head	Ireland	60.0	1987-10-21	Ff
03005	Lerwick	Great Britain	79.0	2004-08-19	Ff
03026	Stornoway	Great Britain	65.0	1989-01-19	Fs
03075	Wick	Great Britain	54.9	1964-08-17	Ff
03162	Eskdalemuir	Great Britain	95.0	1977-10-30	Ff
03377	Waddington	Great Britain	101.0	1953-04-27	Fo
03---	Hull	Great Britain	70.4	1968-09-11	Ff
03---	Oxford	Great Britain	87.9	1968-07-10	Ff
03---	Armagh	Great Britain	78.3	1970-08-15	Fo
04013	Stykkisholmur	Iceland	61.2	1958-11-18	Ff
04030	Reykjavik	Iceland	49.2	1959-09-26	Ff
04048	Vestmannaeyjar	Iceland	145.9	1979-11-16	A
04097	Dalatangi	Iceland	200.0	1983-10-03	Fo
04---	Teigarhorn	Iceland	155.7	1994-07-30	Fo
06051	Vestervig	Denmark	55.5	1968-09-01	Fc
06088	Nordby	Denmark	56.1	1988-07-14	Fc
06132	Tranebjerg	Denmark	92.3	1981-08-21	Ff
06186	Koebenhavn	Denmark	55.7	1967-09-23	Ff
06193	Hammer odde Fyr	Denmark	117.0	1995-02-16	Fo
06---	Broderup	Denmark	69.5	1981-08-21	Ff
06---	GrnbAllingskovg	Denmark	76.5	1959-08-01	Fo
06590	Luxembourg	Luxembourg	64.2	1990-08-31	A
06235	De Kooy	Holland	65.1	1980-08-03	Ff
06240	Schiphol	Holland	68.3	1972-08-03	Fc
06260	De Bilt	Holland	66.2	1952-07-04	Fs
06270	Leeuwarden	Holland	62.0	1958-07-03	Fo
06280	Eelde	Holland	69.3	1996-08-28	Fo
06290	Twenthe	Holland	61.3	1989-08-27	Ff
06310	Vlissingen	Holland	80.9	2005-07-04	Ff
06380	Maastricht	Holland	83.0	1966-06-19	A
06645	Basel Binningen	Switzerland	85.0	1991-07-26	A
06660	Zuerich	Switzerland	103.1	1968-09-21	Ff
06680	Saentis	Switzerland	160.0	1991-07-13	Fc
06700	Geneve	Switzerland	92.6	2002-11-14	Ff
06719	Gd. St. Bernhard	Switzerland	159.6	1996-11-12	Fs
06720	Sion	Switzerland	81.3	1971-11-21	Ff
06770	Lugano	Switzerland	176.8	1994-09-12	A
07---	Amiens	France	67.6	1971-06-09	Fs
07---	Montcornet	France	72.8	1988-07-23	Ff
07055	Beauvais-Tille	France	64.7	1953-07-02	Ff
07---	Vouziers	France	73.8	1995-08-07	Fc

(continued)

WMO No.[a]	Station	Country	Precipitation mm/day	Date	Origin-based extreme precipitation type
07070	Reims	France	78.0	2006-07-05	Ff
07130	Rennes	France	82.6	1971-05-15	Fs
07139	Alencon	France	67.2	2002-09-03	Ff
07143	Chartres	France	59.2	1981-10-25	Fo
07145	Trappes	France	91.2	2001-07-06	Fo
07148	Bretigny-Sur-Orge	France	92.0	1997-08-05	Fc
07150	Paris	France	104.2	2001-07-06	Fo
07190	Strasbourg	France	71.0	2006-10-03	Ff
07255	Bourges	France	75.3	1968-08-29	Ff
07265	Auxerre	France	65.3	1973-05-28	Fc
07---	Chatillon-Coligny	France	68.0	1979-08-27	A
07276	Chatillon-Sur-Seine	France	81.2	1988-05-08	A
07---	Oderen	France	145.0	2000-05-30	Ff
07354	Deols	France	84.0	2002-06-05	Ff
07---	L-ile-d-yeu	France	94.1	1951-08-06	Fc
07---	Montmorillon	France	81.1	1964-05-24	Fc
07---	Lezay	France	72.6	1978-01-23	Fw
07---	La souterraine	France	96.3	1961-09-29	Fc
07480	Lyon	France	89.2	1958-09-30	Ff
07---	Saint-Germain-Les-Belles	France	106.8	1960-10-03	Ff
07---	Faverges-de-la-Tour	France	120.0	1999-09-25	Fc
07500?	Lege-cap-Ferret	France	68.9	1955-06-18	Fc
07510	Bordeaux	France	87.6	1992-08-08	Fc
07---	Catus	France	112.8	2001-07-04	Ff
07---	Le Massegros	France	173.8	2000-09-28	Fc
07560	Mont-Aigoual	France	520.0	1964-02-24	Ff
07579	Orange	France	219.2	2002-09-08	Fc
07---	Espoey	France	94.6	1957-11-09	Fo
07621	Tarbes	France	79.7	1978-06-10	Fo
07630	Toulouse	France	82.7	1977-07-07	Fo
07641	Sete	France	151.2	1985-10-25	A
07645	Nimes	France	266.8	1990-10-12	A
07650	Marseille	France	161.3	1973-10-02	Fw
07747	Perpignan	France	222.0	1999-11-12	Ff
08027	San Sebastian	Spain	167.7	1997-05-31	Ff
08160	Zaragoza	Spain	140.0	2006-04-17	A
08180	Barcelona	Spain	187.0	1953-09-25	Ff
08202	Salamanca	Spain	59.0	1955-11-03	Ff
08215	Navacerrada	Spain	150.0	1996-01-21	Fo
08222	Madrid	Spain	87.0	1972-09-21	A

(continued)

WMO No.[a]	Station	Country	Precipitation mm/day	Date	Origin-based extreme precipitation type
08238	Tortosa	Spain	176.5	1965-10-19	Fw
08280	Albacete	Spain	147.0	1996-09-11	Fo
08285	Valencia	Spain	262.6	1956-11-17	Ff
08330	Badajoz Talavera	Spain	119.1	1997-11-05	Ff
08338	Melilla	Spain	180.0	1976-05-01	Ff
08359	Alicante	Spain	220.2	1982-10-19	A
08410	Cordoba	Spain	120.0	1962-12-27	Fw
08433	Torrevieja	Spain	240.0	1989-09-04	A
08482	Malaga	Spain	151.0	1969-02-22	Ff
08282	Tavira	Portugal	186.0	1983-10-29	Fc
08562	Beja	Portugal	111.3	1997-11-06	A
08535	Lisboa	Portugal	95.6	1983-11-19	Fw
08546	Porto	Portugal	101.2	1953-11-28	Fc
08549	Coimbra	Portugal	88.0	2006-10-24	Ff
08---	Campi	Portugal	180.0	1995-12-25	Fc
08---	Aguiar da Beira	Portugal	148.0	1979-02-05	Ff
08575	Braganca	Portugal	72.0	2000-12-07	Ff
08---	Barcelos	Portugal	115.0	1969-05-09	A
08---	Ponte de Lima	Portugal	137.0	1958-06-26	Fc
10015	Helgoland	Germany	64.9	2006-08-28	Fo
10020	List auf Sylt	Germany	82.5	1981-06-29	Ff
10035	Schleswig	Germany	85.9	1989-08-07	Fc
10147	Hamburg	Germany	68.2	1994-08-18	Fw
10162	Schwering	Germany	118.5	1969-06-24	Fc
10170	Rostock	Germany	61.6	1960-06-29	A
10224	Bremen	Germany	78.5	1964-08-12	Fs
10270	Neuruppin	Germany	83.1	1993-06-12	Fc
10338	Hannover	Germany	76.1	2002-07-17	Ff
10361	Magdeburg	Germany	67.1	1955-06-09	Fc
10379	Potsdam	Germany	105.7	1978-08-08	Fo
10384	Berlin	Germany	119.5	1978-08-08	Fo
10393	Lindenberg	Germany	171.7	1978-08-08	Fo
10427	Kahler Asten	Germany	102.5	1998-09-14	A
10436	Kassel	Germany	82.3	1992-07-31	Fw
10466	Halle	Germany	77.1	1994-04-12	Fo
10469	Leipzig	Germany	77.1	1994-04-12	Fo
10488	Dresden	Germany	158.0	2002-08-12	Ff
10499	Gorlitz	Germany	76.2	1964-08-10	Ff
10501	Aachen	Germany	66.6	1956-09-26	Ff
10554	Erfurt	Germany	75.0	2006-08-10	Fc

(continued)

WMO No.[a]	Station	Country	Precipitation mm/day	Date	Origin-based extreme precipitation type
10---	Jena	Germany	110.0	1997-07-18	A
10578	Fichtelberg	Germany	137.8	2002-08-12	Ff
10609	Trier	Germany	57.8	1970-05-10	Ff
10637	Frankfurt	Germany	109.7	1981-08-09	Fd
10655	Wurzburg	Germany	71.1	1995-08-07	Fc
10675	Bamberg	Germany	75.3	1975-06-24	Fc
10685	Hof	Germany	67.7	1992-07-21	Fc
10708	Saarbrucken	Germany	73.1	1968-08-08	Fs
10727	Karlsruhe	Germany	88.8	1978-05-22	Fo
10738	Stuttgart	Germany	68.3	1971-06-06	Fc
10763	Nurnberg	Germany	90.6	1955-08-01	A
10852	Augsburg	Germany	71.0	1965-06-10	Fo
10865	Muenchen	Germany	101.7	1979-06-17	A
10946	Kempten	Germany	86.1	1971-06-07	Fc
10961	Zugspitze	Germany	133.9	1999-05-21	A
10962	Hohenpeissenberg	Germany	138.5	1999-05-21	A
11012	Kremsmuenster	Austria	110.0	1955-07-22	Fs
11035	Wien	Austria	92.8	1951-05-10	Fc
11120	Innsbruck	Austria	92.2	1999-05-21	Fw
11146	Sonnblick	Austria	102.2	1962-05-14	Ff
11150	Salzburg	Austria	135.0	1954-07-08	A
11290	Graz	Austria	87.0	2005-08-21	Ff
11028	Praha	The Czech Republic	90.0	1981-07-19	A
11464	Milesovka	The Czech Republic	87.0	1970-08-02	A
11---	Oravska Lesna	Slovakia	163.2	1970-07-18	Ff
11858	Hurbanovo	Slovakia	81.8	1992-07-12	Ff
11934	Poprad Tatry	Slovakia	79.3	1996-08-28	Fc
11968	Kosice	Slovakia	110.5	1963-08-13	Fc
12100	Kołobrzeg	Poland	85.2	1996-07-09	Ff
12105	Koszalin	Poland	101.3	1991-08-18	Fo
12115	Ustka	Poland	94.2	1988-07-24	Ff
12120	Łeba	Poland	141.0	1988-07-24	Ff
12125	Lębork	Poland	83.7	1980-07-10	Fo
12135	Hel	Poland	76.4	1980-07-10	Fo
12160	Elbląg	Poland	83.8	1992-09-06	Fo
12195	Suwałki	Poland	64.2	1956-10-06	Ff
12200	ŚwinoujŚcie	Poland	58.7	1969-08-29	Fo
12205	Szczecin	Poland	74.3	1978-08-08	Fo
12210	Resko	Poland	74.8	1996-06-08	Fd

(continued)

WMO No.[a]	Station	Country	Precipitation mm/day	Date	Origin-based extreme precipitation type
12215	Szczecinek	Poland	84.1	1980-06-15	Ff
12235	Chojnice	Poland	84.6	1974-07-19	Ff
12250	Toruń	Poland	101.6	1980-06-15	Ff
12272	Olsztyn	Poland	98.9	1992-09-06	Fo
12280	Mikołajki	Poland	78.4	1957-08-17	Fo
12285	Ostrołęka	Poland	94.9	1995-09-03	Fo
12295	Białystok	Poland	90.6	1985-06-26	Fo
12300	Gorzów	Poland	77.4	1977-08-08	Fc
12310	Słubice	Poland	132.5	1978-08-08	Fo
12330	Poznań	Poland	85.7	1996-07-08	Ff
12345	Koło	Poland	79.1	1956-08-23	Fo
12360	Płock	Poland	83.8	1962-05-14	Ff
12375	Warszawa	Poland	69.6	2002-08-05	Fc
12385	Siedlce	Poland	81.0	2006-08-11	Fs
12400	Zielona G.	Poland	80.0	1961-06-04	Fw
12415	Legnica	Poland	85.9	2001-07-20	A
12424	Wrocław	Poland	74.4	2001-07-20	A
12435	Kalisz	Poland	86.8	1985-08-08	Fw
12455	Wieluń	Poland	78.7	1997-07-06	Fs
12465	Łódź	Poland	99.8	1980-06-15	Ff
12495	Lublin	Poland	90.0	2007-09-06	Ff
12497	Włodawa	Poland	88.9	2006-08-11	Fs
12500	Jelenia G.	Poland	119.3	2001-07-20	A
12510	Śnieżka	Poland	149.7	1977-07-31	Fo
12520	Kłodzko	Poland	84.1	2001-07-20	A
12530	Opole	Poland	99.0	1998-07-27	Fs
12540	Racibórz	Poland	92.9	1997-07-07	Fs
12550	Częstochowa	Poland	83.2	1988-06-08	Fw
12560	Katowice	Poland	81.6	1972-04-22	Fw
12566	Kraków	Poland	87.4	1985-05-17	Fc
12570	Kielce	Poland	155.2	2001-07-24	Ff
12575	Tarnów	Poland	110.8	1970-07-18	Ff
12580	Rzeszów	Poland	62.2	1987-05-22	Fw
12585	Sandomierz	Poland	75.6	2000-07-29	Fo
12595	ZamoŚć	Poland	89.9	1980-05-31	Fw
12600	Bielsko B	Poland	147.4	1972-08-21	Ff
12625	Zakopane	Poland	138.7	1973-06-30	Fc
12650	Kasprowy Wierch	Poland	232.0	1973-06-30	Fc
12690	Lesko	Poland	80.7	1996-09-06	Ff
12695	PrzemyŚl	Poland	83.5	1996-09-06	Ff
12942	Pecs Pogany	Hungary	97.0	1972-07-12	Ff

(continued)

WMO No.[a]	Station	Country	Precipitation mm/day	Date	Origin-based extreme precipitation type
13584	Prilep	Macedonia	120.2	1979-11-19	Fc
13274	Beograd	Serbia Montenegro	94.0	1994-06-14	Fo
13388	Nis	Serbia Montenegro	76.6	1954-11-05	Ff
13008	Kredarica	Slovenia	195.0	2007-09-18	Fc
13017	Ljubljana	Slovenia	124.3	1971-08-28	Fc
14652	Sarajevo	Bosnia and Herzegovina	118.5	2003-10-23	Ff
14113	Rijeka	Croatia	210.3	1976-08-31	A
14236	Zagreb	Croatia	95.8	1989-07-03	Fw
14159	Osijek	Croatia	101.2	1968-06-09	Fc
14218	Zavizan	Croatia	186.0	2002-11-20	A
14223	Gospic	Croatia	120.9	1969-08-23	Ff
14333	Split Marjan	Croatia	131.6	1975-08-24	Ff
14335	Hvar	Croatia	159.0	2005-10-03	Fc
14439	Lastovo	Croatia	169.8	2005-11-07	Ff
15120	Cluj Napoca	Romania	81.6	1969-07-29	A
15200	Arad	Romania	71.0	1998-07-28	Fc
15280	Varfu Omul	Romania	102.4	1979-06-22	Fo
15340	Tg Jiu	Romania	131.8	1998-07-16	Ff
15350	Buzau	Romania	90.5	1976-05-22	Fo
15410	Drobeta Turnu Severin	Romania	224.0	1999-07-12	A
15420	Bucuresti	Romania	126.0	2005-09-20	Ff
15460	Calarasi	Romania	84.0	1967-05-25	Fw
15490	Turnu Magurele	Romania	132.4	1970-07-05	A
16080	Milan	Italy	161.2	1979-08-18	Ff
16090	Verona	Italy	198.0	1984-08-09	Ff
16---	Mantova	Italy	115.0	2004-09-16	Ff
16120	Genoa	Italy	218.6	1953-09-19	Fc
16138	Ferrara	Italy	89.6	1958-11-13	Ff
166140	Bologna	Italy	114.6	1990-10-05	Fw
16239	Roma (Ciampino)	Italy	213.0	1951-09-25	Fc
16320	Brindisi	Italy	127.4	1956-02-03	Fo
16560	Cagliari	Italy	109.6	1974-02-17	Fo
16641	Corfu	Greece	239.3	2000-11-01	A
16648	Larissa	Greece	141.1	1978-09-14	Ff
16716	Hellinikon	Greece	142.0	1998-03-26	Fo
16734	Methoni	Greece	208.2	1963-10-10	Fc
16746	Souda	Greece	185.4	1981-01-09	Fo
16754	Heraklion	Greece	107.5	1963-04-22	A

(continued)

WMO No.[a]	Station	Country	Precipitation mm/day	Date	Origin-based extreme precipitation type
17---	Amiandos	Cyprus	191.0	1968-12-25	Ff
17609	Larnaca	Cyprus	99.1	1984-11-04	A
17---	Limassol	Cyprus	78.8	2000-11-28	Fc
17607	Nicosia	Cyprus	130.0	1951-09-11	A
17---	Polis	Cyprus	82.5	1968-11-26	Ff
22113	Murmansk	Russia	56.5	1977-08-01	Fw
22165	Kanin nos	Russia	84.9	1959-07-12	Ff
22217	Kandalaksa	Russia	48.3	1966-06-27	Fs
22271	Sojna	Russia	72.0	2005-09-28	Fw
22282	Mys-Mikulkin	Russia	48.6	1966-07-01	Ff
22292	Indiga	Russia	58.2	1984-08-25	A
22408	Ukhta Kalevala	Russia	47.7	1989-09-04	A
22438	Zhizhgin Mayak	Russia	64.0	2003-08-29	Fo
22520	Kem	Russia	48.0	2007-08-26	Ff
22550	Archangelsk	Russia	60.8	1991-08-22	Fw
22563	Pinega	Russia	81.0	2005-05-22	A
22583	Kojnas	Russia	55.8	1995-07-23	Ff
22602	Reboly	Russia	77.8	1997-07-28	Fw
22641	Onega	Russia	57.6	1985-08-02	A
22802	Sortavala	Russia	59.0	1997-09-11	Fo
22820	Petrozawodsk	Russia	67.6	2003-08-06	Ff
22837	Vytegra	Russia	94.6	1997-07-09	Fc
22845	Kargopol	Russia	72.9	1998-08-27	Fo
22887	Kotlas	Russia	78.4	1983-07-09	Fc
22996	Objacevo	Russia	60.0	1955-05-27	Ff
23205	Narjan mar	Russia	81.6	2004-06-25	Ff
23219	Hoseda hard	Russia	37.6	1966-07-02	Fo
23405	Ust tzilma	Russia	82.0	2007-12-10	Fo
23418	Petsjora	Russia	54.0	1998-06-17	Fw
23606	Uhta	Russia	47.5	1996-06-24	Fo
23631	Berezovo	Russia	70.3	1955-06-24	A
23711	Troitzko	Russia	66.1	1986-07-02	A
23724	Nioaksimvol	Russia	51.5	1991-07-09	Fw
23804	Syktyvar	Russia	73.8	1993-07-26	Ff
23921	Ivdel	Russia	72.2	1976-08-12	A
26038	Tallinn	Estonia	179.1	1995-06-08	A
26214	Vilsandi	Estonia	78.4	2000-09-04	Fo
26231	Pyarnu	Estonia	60.6	1960-07-30	Fw
26242	Tartu	Estonia	73.0	1980-05-19	A
26249	Voru	Estonia	75.6	1993-07-24	Ff

(continued)

WMO No.[a]	Station	Country	Precipitation mm/day	Date	Origin-based extreme precipitation type
26314	Ventspils	Latvia	113.0	1996-07-09	Ff
26422	Riga	Latvia	114.0	1996-05-21	Fo
26544	Daugavpils	Latvia	102.1	1993-06-27	A
26509	Klaipeda	Lithuania	73.9	1988-07-28	Ff
26518	Lazdijai	Lithuania	79.8	1972-08-04	Ff
26524	Siauliai	Lithuania	63.2	2001-06-23	Fo
26531	Birzai	Lithuania	84.7	1999-06-23	Fw
26629	Kaunas	Lithuania	82.9	2005-08-09	Ff
26633	Utena	Lithuania	134.1	1996-02-24	Ff
26634	Ukmarge	Lithuania	50.4	1972-08-04	Ff
26730	Vilnius	Lithuania	85.1	2005-08-09	Ff
26554	Verkhnedvinsk	Belarus	71.5	1973-07-26	Ff
26666	Vitebsk	Belarus	72.0	2006-08-23	Fo
26941	Baranovici	Belarus	71.9	1982-05-10	Fw
26961	Bobrujsk	Belarus	110.0	1986-08-21	Ff
26059	Kingisepp	Russia	63.0	2004-06-30	Fo
26063	St. Petersburg	Russia	69.0	2002-07-16	Fc
26094	Tihvin	Russia	84.0	2001-12-19	Ff
26067	Nikolayevskoye	Russia	115.0	2006-06-06	Ff
26258	Pskow	Russia	103.1	2003-08-06	Ff
26275	Staraja-Russa	Russia	72.2	1963-07-29	A
26389	Ostaskov	Russia	91.7	1977-07-29	Ff
26477	Velikie Lukie	Russia	105.8	1987-08-07	Ff
26702	Kaliningrad	Russia	118.3	2005-08-10	Ff
26781	Smolensk	Russia	87.5	1986-07-10	Fw
26976	Krasnaja-Gora	Russia	83.2	1983-08-12	A
27037	Vologda	Russia	57.3	1976-08-04	Ff
27051	Tot-Ma	Russia	81.0	2007-07-01	Fo
27066	Nikolsk	Russia	57.7	1963-06-19	Fw
27272	Sar-ja	Russia	80.4	1971-07-21	Fw
27333	Kostroma	Russia	80.0	1997-06-28	Ff
27417	Klin	Russia	84.0	1969-04-24	Fo
27509	Mozajsk	Russia	62.3	1967-08-31	Ff
27595	Kazan	Russia	64.0	1999-07-01	A
27612	Moskou	Russia	62.5	1970-06-14	Ff
27648	Elatma	Russia	74.7	1990-07-25	Fc
27719	Tula	Russia	89.9	1999-08-11	Fw
27823	Paveletz	Russia	76.3	1977-07-03	Fo
27857	Zametchino	Russia	89.9	1993-07-15	Fc
27930	Lipeck	Russia	77.2	1982-07-03	Fw
27947	Tambov	Russia	72.0	1998-08-24	Fc

(continued)

WMO No.[a]	Station	Country	Precipitation mm/day	Date	Origin-based extreme precipitation type
28064	Leusi	Russia	62.3	1970-07-30	A
28275	Tobolsk	Russia	102.4	1976-08-13	A
28411	Izevsk	Russia	80.0	1984-08-05	A
28722	Ufa Dema	Russia	58.2	1961-07-24	Fc
28900	Samara	Russia	59.6	1997-06-20	Ff
28952	Kustanaj	Russia	83.5	2000-07-06	Ff
33008	Brest Zonalnaya	Belarus	85.8	1974-07-19	Ff
33038	Vasiliewiczi	Belarus	81.9	1958-06-30	Fo
33041	Gomel	Belarus	74.4	1971-06-13	Fc
33088	Sarny	Ukraine	118.0	1999-07-26	Fs
33275	Sumy	Ukraine	109.0	1994-04-24	A
33300	Kharkov	Ukraine	72.2	1967-08-16	Fo
33317	Shepetivka	Ukraine	111.5	1969-07-13	Fw
33325	Zhitomir	Ukraine	119.9	1996-10-02	A
33345	Kyiv	Ukraine	88.2	1974-07-01	Fw
33377	Lubny	Ukraine	107.5	1995-08-04	A
33393	Lvov	Ukraine	86.3	1966-06-26	Fs
33466	Imeni-Starcenko	Ukraine	110.0	1997-06-06	A
33506	Poltava	Ukraine	96.0	2007-06-02	Fc
33523	Lugansk	Ukraine	147.0	2006-03-03	Ff
33524	Debalcevo	Ukraine	88.9	1977-08-15	Ff
33562	Vinnica	Ukraine	112.0	1972-07-18	A
33587	Uman	Ukraine	103.1	1999-01-29	Fw
33631	Uzhgorod	Ukraine	68.9	1980-08-13	Fc
33658	Chernovtsy	Ukraine	111.2	1991-07-27	Fw
33711	Kirovograd	Ukraine	99.5	1971-09-20	Ff
33837	Odessa	Ukraine	106.3	2002-08-15	Fo
33846	Nikolaev	Ukraine	144.0	1955-06-30	Fc
33889	Izmail	Ukraine	98.4	1997-07-27	Fo
33910	Geniczesk	Ukraine	114.4	1961-07-26	Fc
33915	Askaniia nova	Ukraine	103.0	2005-07-01	Ff
33946	Simpheropol	Ukraine	119.2	2004-08-11	Fc
33---	Opasnoe	Ukraine	102.1	2002-09-16	Ff
333976	Feodosia	Ukraine	132.3	1991-09-02	Fw
33983	Kercz	Ukraine	100.1	1973-05-15	Fw
33990	Yalta	Ukraine	140.0	1996-01-07	A
33999	Ai-petri	Ukraine	186.0	1968-09-06	A
34009	Kursk	Russia	98.3	1953-06-18	Fo
34122	Voronezh	Russia	95.1	1988-06-26	Fd
34139	Kamennaja stiep	Russia	57.3	2006-06-09	Ff
34163	Oktjabrskij gorodok	Russia	111.4	1976-08-01	A

(continued)

WMO No.[a]	Station	Country	Precipitation mm/day	Date	Origin-based extreme precipitation type
34172	Saratow	Russia	81.0	1985-06-27	A
34262	Rudnya	Russia	67.0	1971-09-14	Ff
34336	Boguchar	Russia	82.4	1963-07-11	A
34391	Aleksandrow	Russia	73.3	1976-08-01	A
34545	Morozovsk	Russia	80.2	1973-06-21	A
34579	Verhnij-Baskuncak	Russia	76.0	2004-07-25	A
34730	Rostov na Donu	Russia	87.9	1989-05-27	A
34747	Celina	Russia	87.2	1974-08-16	A
34824	Primorsko-Ahtarsk	Russia	100.2	1972-07-02	A
34838	Tikhoretsk	Russia	95.0	1997-05-26	A
34858	Divnoe	Russia	72.5	1995-10-04	Ff
34861	Elista-Stepnoy	Russia	71.0	1974-09-09	A
34866	Jaskul	Russia	64.3	1996-06-06	Fc
34880	Astrakan	Russia	72.2	1999-08-13	A
35108	Uralsk	Russia	79.3	1992-07-25	A
35121	Orenburg	Russia	62.2	1981-08-07	A
35225	Aktiubinsk	Russia	58.6	1984-06-01	A
35406	Kalmykovo	Russia	76.2	1966-08-29	Ff
35416	Uil	Russia	55.8	1977-08-21	Ff
35700	Atyrau	Russia	40.6	1989-06-27	A
35746	Aralsk	Russia	44.8	2003-06-02	Fc
37000	Novorossiysk	Russia	90.1	1975-06-10	A
37018	Tuapse	Russia	122.0	2003-11-20	Ff
37031	Armavir	Russia	89.2	1960-06-25	A
37050	Pjatigorsk	Russia	94.8	1975-06-13	Fc
37061	Budennovsk	Russia	110.0	2002-06-20	Fc
37145	Mozdok	Russia	90.0	1997-01-07	A
37212	Nal-cik	Russia	66.2	1971-09-13	A
37472	Machaczkala	Russia	84.4	1990-09-30	A
37789	Erevenerebuni	Russia	50.6	1974-09-04	A

[a]World Meteorological Organisation (WMO) number for synoptic stations only; explanation of abbreviation of origin-based extreme precipitation types in Table 3.1

Appendix 2

The highest daily precipitation totals in Europe based on different sources and different periods: data from published sources and the Internet

Station	Country/region	Precipitation [mm]	Data	Sources
Genua Bolzaneto	Italy	948.4	10-09-1970	Martyn (2000)
Générargues	France. Languedoc-Roussilon	588.6	8-09-2002	www.euroforecaste.org/newsletter9/gard.pdf
Crkvice	Montenegro	550.0	?	Furlan (1977)
Crkvice	Montenegro	480.0	28-11-1927	Martyn (2000)
Falmenta	Switzerland	420.0	22-06-1933	Zeller et al. (1983)
Falmenta	Switzerland	380.0	10-09-1965	Zeller et al. (1983)
Valencia	Spain	361.0	?-10-1957	Escardó (1970)
Mosogno	Switzerland	359.0	24-09-1924	Zeller et al. (1983)
Mont Matajur	Slovenia. Julian Alps	356.0	?	Furlan (1977)
Ciuperceni Vechii	Romania	349.0	?	Furlan (1977)
Nova Louka (Neuwiese)	The Czech Republic. Sudeten –Izerskie Range	345.1	30-07-1897	Niedźwiedź (2003a)
Camedo	Switzerland	340.0	9-09-1965	Zeller et al. (1983)
Dornbirn	Austria. Voralberg	336.0	31-08-1910	Schüepp and Schirmer (1977)
Falmenta	Switzerland	335.0	10-09-1937	Zeller et al. (1983)
Bilyasar	Azerbaijan	333.8	16-08-1955	Lydolph (1977)
Semmering	Lower Austria	323.0	5-06-1947	Schüepp and Schirmer (1977)
Seathwaite	Great Britain. Cumbria	316.4	19-11-2009	www.metoffice.gov.uk
Zinnwald-Georgenfeld	Germany. Ore Mountains (Erzgebirge)	313.0	12-08-2002	Niedźwiedź (2003a)
Hala Gąsienicowa	Poland. Tatras	300.0	30-06-1973	Niedźwiedź (1992)
Catania/ Fontanarossa	Italy	299.0	25-10-2006	ECA&D
?	Russian. the Northern Caucasus	298.2	?	Lydolph (1977)
Pjarnu	Estonia	298.0	14-01-1999	ECA&D
Izana	Spain. The Canary Islands	296.6	24-11-1968	ECA&D
Sandanski	Bulgaria	296.0	25-04-2007	ECA&D
Mogilev Podolski	Ukraine	292.0	24-08-2008	ECA&D
Shkodra	Albania	291.0	26-09-1952	www.twojapogoda.pl

(continued)

Station	Country/region	Precipitation [mm]	Data	Sources
Witów	Poland. Podhale	285.0	16-07-1934	Cebulak and Niedźwiedź (2000)
Podgajcy	Ukraine	281.6	3-06-1957	Lydolph (1977)
Martinstown	Wielka Brytania. Anglia. South Dorset	279.0	18-07-1955	Manley (1970); www.metoffice.gov.uk
Bohinjska Cesnjica	Slovenia	278.7	18-09-2007	ECA&D
Encumeada – Madera	Portugal	277.0	09-12-1976	ECA&D
Furnas. S. Miguel. Azory	Portugal	276.0	03-10-1974	ECA&D
Leskowiec	Poland. Little Beskids	275.1	18-07-1970	Cebulak and Niedźwiedź (2000)
Amiados	Greece	273.0	03-12-1936	ECA&D
Ust-Hairyuzovo	Russia	271.0	24-10-1946	ECA&D
Hala Kondratowa	Poland. Tatras	269.4	30-06-1973	Cebulak and Niedźwiedź (2000)
Locarno-Monti	Switzerland	268.4	09-09-1965	ECA&D
Lugano	Switzerland	263.0	21-08-1911	Zeller et al. (1983)
Luqa	Malta	259.1	30-03-1976	ECA&D
Warna	Bulgaria	257.8	?-08-?	Furlan (1977)
Hala Gąsienicowa	Poland. Tatras	255.2	18-07-1934	Cebulak and Niedźwiedź (2000)
Passo del San Bernardino	Switzerland	254.0	28-09-1868	Zeller et al. (1983)
Ierapetra	Greece	246.3	24-09-1986	ECA&D
Cloone Lake. Co.	Ireland	243.5	18-09-1993	ECA&D
Sopron	Hungary	243.1	06-10-1992	ECA&D
Leskowiec	Poland. Little Beskids	242.9	08-07-1955	www.twojapogoda.pl
Gibraltar	Gibraltar	241.0	03-12-1989	ECA&D
Śnieżka	Poland. Sudeten – Karkonosze Mountains	239.3	30-07-1897	TN (Czerwiński 1989. Hellmann)
Batumi	Georgia	238.7	12-09-1962	Lydolph (1977)
Sloy Main Adit	Great Britain. Scotland	238.0	17-01-1974	www.metoffice.gov.uk
Ostende	Belgium	238.0	14-07-1984	ECA&D
Stańcowa	Poland. Kotlina Orawska	234.4	18-07-1970	Cebulak and Niedźwiedź (2000)
Lysa Hora	The Czech Republic	233.8	6-07-1997	ECA&D
Kasprowy Wierch	Poland. Tatras	232.0	30-06-1973	Cebulak and Niedźwiedź (2000)
Indre Matre	Norway	230.0	26-11-1940	Johannessen (1970)

(continued)

Station	Country/region	Precipitation [mm]	Data	Sources
Sighetu Marmatiei	Romania	230.0	12-08-1994	ECA&D
Exmoor	Great Britain	230.0	?-08-1952	Manley (1970)
Split Marjan	Croatia	228.5	05-09-1948	ECA&D
Śnieżka	Poland. Sudeten – Karkonosze Mountains	226.0	17-07-1882	TN (Czerwiński 1989)
Hala Gąsienicowa	Poland. Tatras	223.5	8-07-1997	Archiwum IMGW
Opstveit	Norway	223.0	14-11-2005	ECA&D
Bad Reichenhall	Germany	222.0	13-09-1899	Schüepp and Schirmer (1977)
Czerniowcy	Ukraine	222.0	6-06-1965	Lydolph (1977)
Kujawsko-Pomorskie	Poland. Nieszawa	221.7	08-08-2006	www.twojapogoda.pl
Penhas de Suide	Portugal	220.0	14-01-1977	Schüepp and Schirmer (1977)
Świętokrzyskie	Poland. Sienno	218.5	16-07-1934	www.twojapogoda.pl
Kiszyniew	Moldavia	218.0	08-07-1948	http://fhmzbih.gov.ba/
Negotin	Serbia	211.1	09-10-1955	ECA&D
Lluest Wen Reservoir	Great Britain. Wales	211.0	11-11-1929	www.metoffice.gov.uk
Valley	Great Britain	210.0	13-12-1994	ECA&D
DolnoŚląskie	Poland. Wałbrzych	206.5	08-08-2006	www.twojapogoda.pl
Opasnoje	Ukraine	206.3	17-07-1956	Lydolph (1977)
Wałbrzych	Poland. Sudeten	205.6	17-06-1979	Precipitation Yearbook (IMGW)
DolnoŚląskie	Poland. Jakuszyce	204.3	08-08-2006	www.twojapogoda.pl
Bastia	France. Corsica	201.0	?-08-?	Arléry (1970)
Dalatangi	Island	200.0	20.10.1983	ECA&D
Arnis	Germany	199.9	04.03.1950	ECA&D
Fagerheden	Sweden	198.0	28.07.1997	ECA&D
Herrenwies	Germany. Black Forest	196.3	20-05-1906	von Rudloff (1967)
Samnanger	Norway	195.0	?	Alfnes and Førland (2006)
Ostroszowice	Poland. Sudeten	191.8	10-08-1964	Precipitation Yearbook (IMGW)
Horta	Portugal. Azores	191.0	10-01-1983	ECA&D
Komarno	Poland Sudeten – The Kaczawskie Mountains	190.6	29-05-1968	Precipitation Yearbook (IMGW)
Przełęcz Okraj	Poland. Sudeten	190.6	1-08-1977	Precipitation Yearbook (IMGW)

(continued)

Station	Country/region	Precipitation [mm]	Data	Sources
Banie Mazurskie	Poland	189.7	17.08.1957	Precipitation Yearbook (IMGW)
Ostroszowice	Poland. Sudeten	188.7	17-06-1979	Precipitation Yearbook (IMGW)
Perpignan	France	186.0	?-09-?	Arléry (1970)
Ivan Sedlo	Bosnia and Herzegovina	185.0	?-08-1989	
Yelgava	Latvia	185.0	16-09-1994	ECA&D
Panevezhis	Lithuania	179.1	22-07-1995	ECA&D
Słupia Nowa	Poland. Świętokrzyskie Mountains	172.5	21-06-1955	Precipitation Yearbook (IMGW)
Zakopane	Poland	172.3	24.07.2001	www.twojapogoda.pl
Feuerkogel	Austria	171.1	12-08-1959	ECA&D
Komrat	Moldavia	169.9	11-06-1975	ECA&D
Mieroszów	Poland. Sudeten	169.3	17-06-1979	Precipitation Yearbook (IMGW)
Store	Denmark. Jyndevad	167.5	08-07-1931	ECA&D
Sokołowsko	Poland. Sudeten	165.5	17-06-1979	Precipitation Yearbook (IMGW)
Samtredia	Greece	164.2	17-08-1977	ECA&D
Oravska Lesna	Slovakia	163.2	18-07-1970	ECA&D
Świeradów	Poland. DolnoŚląskie voivodship	163.0	14-06-1999	www.twojapogoda.pl
Walim	Poland. DolnoŚląskie voivodship	161.6	31-07-1977	Precipitation Yearbook (IMGW)
?	Latvia	161.2	?	Lydolph (1977)
Kerimaki Ylakuona	Finland	159.4	04-07-1988	ECA&D
Tallymore Forest	Great Britain. Northern Ireland	159.0	31-10-1968	www.metoffice.gov.uk
Sanski Most	Bosnia and Herzegovina	157.8	?-08-1976	http://fhmzbih.gov.ba/
Kielce	Poland. Świętokrzyskie voivodship	155.2	21-08-1972	www.twojapogoda.pl
Mława	Poland. Mazowieckie voivodship	155.1	24-07-1988	www.twojapogoda.pl
Jakuszyce	Poland. Sudeten	154.1	1-08-1977	Precipitation Yearbook (IMGW)
Gaik-Brzezowa	Poland. Raba basin	151.2	18-07-1970	TN

(continued)

Station	Country/region	Precipitation [mm]	Data	Sources
Śnieżka	Poland. Sudeten – Karkonosze Mountains	149.7	31-07-1977	Precipitation Yearbook (IMGW)
Neumn	Bosnia and Herzegovina	148.1	?-?-1984	http://fhmzbih.gov.ba/
Bielsko-Biała (Aleksandrowice)	Poland	147.4	16-07-1934	www.twojapogoda.pl
Nicea	France	147.0	?-11-?	Arléry (1970)
Gouda	Holland	145.5	23-06-1975	ECA&D
?	Russia. Wołga basin	144.4	?	Lydolph (1977)
Barcelona	Spain	143.2	?-02-?	Escardó (1970)
Pomorskie	Poland. Łeba	141.0	09-07-2001	www.twojapogoda.pl
Stolac	Bosnia and Herzegovina	140.0	05-05-1940	Bosna i Hercegovina M E T E O B I H
Caplijna	Bosnia and Herzegovina	138.0	?-?-1984	Bosna i Hercegovina M E T E O B I H
Bochnia	Poland	134.0	3-08-1877	TN. Austrian Yearbook
Mostar	Bosnia and Herzegovina	134.0	?-11-1955	
Słubice	Poland	132.5	22-05-1987	www.twojapogoda.pl
Livno	Bosnia and Herzegovina	132.0	?-12-1959	
?	Lithuania	130.7	?	Lydolph (1977)
Baligród-Mchawa	Poland. the Eastern Carpathian	130.6	26-07-2005	Starkel (2011)
Tarnów	Poland	128.7	15-06-1980	Precipitation Yearbook (IMGW)
Gdańsk	Poland	127.7	15-07-1980	Precipitation Yearbook (IMGW)
Bjelasnica	Bosnia and Herzegovina	122.0	?-10-1961	http://fhmzbih.gov.ba/
Gradacac	Bosnia and Herzegovina	122.0	?-07-1967	http://fhmzbih.gov.ba/
Bihac	Bosnia and Herzegovina	121.6	?-07-1976	http://fhmzbih.gov.ba/
Prilep	Macedonia	120.0	18-11-1979	ECA&D
Goleniowy	Poland. Pilica basin	124.5	18-07-1982	Archive IMGW
?	Moldavia	124.3	?	Lydolph (1977)
Sarajevo-Bjelave	Bosnia and Herzegovina	118.5	23-11-2003	
?	Russia. Karelia	116.1	?	Lydolph (1977)
?	Estonia	112.4	?	Lydolph (1977)
?	Belarus	110.2	?	Lydolph (1977)
Bugojno	Bosnia and Herzegovina	105.4	?-09-1983	http://fhmzbih.gov.ba/

(continued)

Station	Country/region	Precipitation [mm]	Data	Sources
Wisła – Malinka	Poland. Silesian Beskids	104.3	19-01-1974	Precipitation Yearbook (IMGW)
Krosno	Poland. Podkarpackie voivodship	102.8	06-09-1992	www.twojapogoda.pl
Bydgoszcz	Poland	102.5	15-06-1980	Precipitation Yearbook (IMGW)
Toruń	Poland	101.6	15-06-1980	Precipitation Yearbook (IMGW)
Koszalin	Poland	101.3	18-08-1991	Archive IMGW

Sources:

Alfnes E, Førland EJ (2006) Trends in extreme precipitation and return values in Norway 1900–2004. Metereol No Rep Climate 2:1–40.
Arléry R (1970) The climate of France. Belgium, The Netherlands, and Luxembourg. In: Wallen CC (ed) Climates of Northern and Western Europe. World Survey of Climatology, Vol 6. Elsevier Scientific, Amsterdam, pp 135–193.
Cebulak E, Niedźwiedź T (2000) Zagrożenie powodziowe dorzecza górnej Wisły przez wysokie opady atmosferyczne. In: Maciejewski M (ed) Ryzyko w Gospodarce Wodnej. Monografie Komitetu Gospodarki Wodnej PAN 17:55–70.
Escardó AL (1970) The climate of the Iberian Peninsula. In: Wallen (ed) Climates of Northern and Western Europe. World Survey of Climatology, Vol 6. Elsevier Scientific, Amsterdam, pp 195–239.
Furlan D (1977) The climate of Southeast Europe. In: Wallen (ed) Climates of Northern and Western Europe. World Survey of Climatology, Vol 6. Elsevier Scientific, Amsterdam, pp 185–235.
Johannessen TW (1970) The climate of Scandinavia. In: In: Wallen (ed) Climates of Northern and Western Europe. World Survey of Climatology, Vol 6. Elsevier Scientific, Amsterdam, pp 23–79.
Kundzewicz ZW (2005) Intense precipitation and high river flows in Europe: observations and projections. Acta Geophys Pol 53 (4):385–400.
Lydolph PE (1977) Climates of the Soviet Union. In:World Survey of Climatology, Vol 7. Elsevier Scientific, Amsterdam, p 443.
Manley G (1970) The climate of the British Isles. In: In: Wallen (ed) Climates of Northern and Western Europe. World Survey of Climatology, Vol 6. Elsevier Scientific, Amsterdam, pp 81–133.
Schüepp M, Schirmer H (1977) Climates of Central Europe. In: In: Wallen (ed) Climates of Northern and Western Europe. World Survey of Climatology, Vol 6. Elsevier Scientific, Amsterdam, pp 3–73.
Starkel L (2011) ZłożonoŚć czasowa i przestrzenna opadów ekstremalnych: ich efekty geomorfologiczne i drogi przeciwdziałania im. Landform Analysis 15:65–80.
Zeller J, Geiger H, Röthlisberger G (1983) Starkniederschläge des schweizerischen Alpen und Alpenrandgebietes. Intensitäten und Häufigkeiten. Ausgewertete Periode 1901–1970. Eidgenössische Anstalt für das forstliche Versuchswesen. Birmensdorf.

References

Alexander LV, Zhang X, Peterson TC, Caesar J, Gleason B, Klein Tank AMG, Haylock M, Collins D, Trewin B, Rahimzadeh F, Tagipour A, Rupa Kumar K, Revadekar J, Griffiths G, Vincent L, Stephenson DB, Burn J, Aguilar E, Brunet M, Taylor M, New M, Zhai P, Rusticucci M, Vazquez-Aguirre JL (2006) Global observed changes in daily climate extremes of temperature and precipitation. J Geophys Res 111:D05109. doi:10.1029/2005JD006290

Alpert P, Neeman BU, Shay-El Y (1990a) Climatological analysis of Mediterranean cyclones using ECMWF data. Tellus 42A:65–77

Alpert P, Neeman BU, Shay-El Y (1990b) Intervariability of cyclone tracks in the Mediterranean. J Climate 3:1474–1478

Anagnostopoulou CH, Tolika K (2012) Extreme precipitation in Europe: statistical threshold selection based on climatological criteria. Theor Appl Climatol 107:479–489

Benestad RE (2006) Can we expect more extreme precipitation on the monthly time scale? J Climate 19:630–637

Beniston M (2005) An overview of extreme climatic events in Switzerland: implications for assessing economic damages and costs. In: Haurie A, Viguier L (eds) The coupling of climate and economic dynamics. Springer, The Netherlands

Beniston M, Stephenson DB, Christensen OB, AT Ferro ATCh, Frei Ch, Goyette S, Halsnaes K, Holt T, Jylhä K, Koffi B, Palutikof J, Schöll R, Tido Semmler T, Woth K (2004) Future extreme events in European climate: an exploration of regional climate model projections. Joint Paper PRUDENCE WP5

Beniston M, Stephenson DB, Christensen OB, Ferro CAT, Frei C, Halnaes GK, Holt T, Julhä K, Koffi B, Palutikof J, Schöll R, Semmler T, Woth K (2007) Future extreme events in European climate: an exploration of regional climate model projections. Clim Change 81:71–95

Bolliger M, Binder P, Roesli HP, Wanner H (2004) Satellite- and radar-based investigations of heavy precipitation systems on the southern side of the European Alps. Meteorol Atmos Phys 87:219–234

Brunetti M, Colacino M, Maugeri M, Nanni T (2001) Trends in the daily intensity of precipitation in Italy from 1951 to 1996. Int J Climatol 21(3):299–316

Cebulak E (1994) ZmiennoŚć maksymalnych opadów dobowych w Dorzeczu Górnej Wisły (Variability of precipitation maximums in upper Vistula basin). WiadomoŚci IMGW 16(38):83–91

Changnon D (1994) Regional and temporal variations in heavy precipitation in South Carolina. Int J Climatol 14:165–177

Changnon SA (2011) Temporal distribution of weather catastrophes in the USA. Clim Change 106:129–140

Changnon D, Changnon SA (1998) Evaluation of weather catastrophe data for use in climate change investigations. Clim Change 38:435–445

Chomicz K (1951) Ulewy i deszcze nawalne w Polsce (Downpours and showers in Poland). WiadomoŚci Służby Hydrologiczno-Meteorologicznej 2(3):5–88

Chromow SP (1977) Meteorologia i klimatologia (Meteorology and Climatology). PWN, Warszawa

Easterling DR, Diaz HF, Douglas AV, Hogg W, Kunkel KE, Rogers JC, Wilkinson JF (1999) Long-term observations for monitoring extremes in the Americas. Clim Change 42:285–308

Easterling DR, Evans JL, Groisman PYA, Karl TR, Kunkel KE, Ambenje P (2000) Observed variability and trends in extreme climate events: a brief review. Bull Am Meteorol Soc 81(3):417–425

Easterling DR, Alexander LV, Mokssit A, Detemmerman V (2003) CCl/CLIVAR workshop to develop priority climate indices. Bull Am Meteorol Soc 84:1403–1407

Federico S, Avolio E, Bellecci C, Lavagnini A, Colacino M, Walko RL (2008) Numerical analysis of an intense rainstorm occurred in southern Italy. Nat Hazards Earth Syst Sci 8:19–35

Folland CK, Horton B, Scholfield P (eds) (1999) Report of WMO working group on climate change detection task group on climate change indices, Bracknell, 1–3 September 1998, WMO–TD No 930, WCDMP No 37

Fragoso M, Gomes PT (2008) Classification of daily abundant rainfall patterns and associated-large-scale atmospheric circulation types in Southern Portugal. Int J Climatol 28:537–544

Frei C, Schär C (2001) Detection probability of trends in rare events: theory and application to heavy precipitations in the Alpine region. J Climate 4:1568–1584

Frich P, Alexander LV, Della-Marta P, Gleason B, Haylock M, Tank AMGK, Peterson T (2002) Observed coherent changes in climatic extremes during the second half of the twentieth century. Clim Res 19:193–212

Furlan D (1977) The climate of southeast Europe. In: Wallen CC (ed) Climates of Central and Southern Europe – World Survey of Climatology 6. Elsevier, Amsterdam, pp 185–235

Gajić-Čapka M, Cindrić K (2011) Secular trends in indices of precipitation extremes in Croatia 1901–2008. Geofizika 28:293–312

Glickman TS (ed) (2000) Glossary of meteorology. American Meteorological Society, Boston

Griffiths ML, Bradley RS (2007) Variations of twentieth-century temperature and precipitation extreme indicators in the northeast United States. J Climate 20:5401–5417

Groisman PY, Knight RW, Easterling DR, Karl TR, Hegerl GC, Razuvaev VN (2005) Trends in intense precipitation in the climate record. J Climate 18:1326–1350

Hagen M, Schiesser H-H, Dorninger M (2000) Monitoring of mesoscale precipitation systems in the Alps and the Northern Alpine foreland by radar and rain gauges. Meteorol Atmos Phys 72:87–100

Haylock M, Nicholls N (2000) Trends in extreme rainfall indices for an updated high quality data set for Australia, 1910–1998. Int J Climatol 20:1533–1541. doi:10.1002/1097-0088(20001115)20

Hellström C (2005) Atmospheric conditions during extreme and non-extreme precipitation events in Sweden. Int J Climatol 25:631–648

IPCC (2001) Climate change 2001: the scientific basis. In: Houghton JT, Ding Y, Griggs DJ, Noguer M, van der Linden PJ, Dai X, Maskell K, Johnson CA (eds) Contribution of working group I to the third assessment report of the intergovernmental panel on climate change, 2001st edn. Cambridge University Press, Cambridge, UK

IPCC (2007) Climate change 2007: the physical science basis. In: Solomon S, Qin D, Manning M, Chen Z, Marquis M, Averyt KB, Tignor M, Miller HL (eds) Contribution of working group I to the fourth assessment report of the intergovernmental panel on climate change. Cambridge University Press, Cambridge, UK

IPCC (2013) Climate change 2013: the physical science basis. In: Stocker TF, Qin D, Plattner G-K, Tignor M, Allen SK, Boschung J, Nauels A, Xia Y, Bex V, Midgley PM (eds) Contribution of working group I to the fifth assessment report of the intergovernmental panel on climate change. Cambridge University Press, Cambridge, UK

Karagiannidis AF, Karacostas T, Maheras P, Makrogiannis T (2012) Climatological aspects of extreme precipitation in Europe, related to mid-latitude cyclonic systems. Theor Appl Climatol 107:165–174

Karl TR, Knight RW (1998) Secular trends of precipitation amount frequency and intensity in the United States. Bull Am Meteorol Soc 79(2):231–241

Karl TR, Knight RW, Plummer N (1995) Trends in high-frequency climate variability in the twentieth century. Nature (Lond) 377:217–220

Karl TR, Knight RW, Easterling DR, Quayle RG (1996) Indices of climate change for the United States. Bull Am Meteorol Soc 77:279–292

Klein Tank AMG et al (2002) Daily dataset of 20th century surface air temperature and precipitation series for the European Climate Assessment. Int J Climatol 22:1441–1453

Klein Tank AMG, Köennen GP (2003) Trends in indices of daily temperature and precipitation extremes in Europe, 1946–99. J Climate 16:3665–3680

Klein Tank AMG, Zwiers FW, Zhang X (2009) Guidelines on analysis of extremes in a changing climate in support of informed decisions for adaptation. Climate Data and Monitoring WCDMP-No 72:52

Kyselý J (2009) Trends in heavy precipitation in the Czech Republic over 1961–2005. Int J Climatol 29:1745–1758

Limanówka D, Niedźwiedź T, Ustrnul Z (1993) Prawdopodobieństwo występowania wybranych ekstremalnych charakterystyk termicznych w Polsce (Probability of the occurrence of selected characteristics of air temperature in Poland). Zeszyty IGiPZ PAN 18:31–48

Lionello P, Dalan F, Elvini E (2002) Cyclones in the Mediterranean region: the present and the doubled CO_2 climate scenarios. Clim Res 22:147–159

Lupikasza E (2010) Spatial and temporal variability of extreme precipitation in Poland in the period 1951–2006. Int J Climatol 30:991–1007. doi:10.1002/joc.1950

Łupikasza E (2009) Zmiany intensywnoŚci opadów w Hornsundzie (Spitsbergen) w okresie 1978–2008. Problemy Klimatologii Polarnej 19:169–188

Łupikasza E, Bielec-Bąkowska Z (2012) Dependence of severe storm occurrence on circulation types in Poland. Badania Fizjograficzne, Geografia Fizyczna A63:255–279

Łupikasza E, Hänsel S, Matschullat J (2011) Regional and seasonal variability of extreme precipitation trends in southern Poland and central-eastern Germany 1951–2006. Int J Climatol 15:2249–2271

MacDonald G, Abarbane H, Koonin S, Levine H, Rothaus O (1992) Statistics of extreme events with application to climate. U.S. Department of Energy, Washington, DC

Maheras P, Flocas AH, Patrikas I, Chr A (2001) A 40 year objective climatology of surface cyclones in the Mediterranean region: spatial and temporal distribution. Int J Climatol 21:109–130

Martyn D (2000) Klimaty kuli ziemskiej. Wydawnictwo Naukowe PWN, Warszawa

Mekis E, Hogg WD (1997) Rehabilitation and analysis of Canadian daily precipitation time series. In: Preprints, 10th conference in applied climatology, Reno, AMS Boston, 20–23 October 1997

Miętus M, Filipiak J, Owczarek M, Jakusik E (2005) ZmiennoŚć warunków opadowych w rejonie polskiego wybrzeża Morza Bałtyckiego w Świetle kwantylowej klasyfikacji opadowej (Variability of precipitation on the polish coast of the Baltic Sea in the light of the quantile classification of precipitation), Materiały Badawcze, Seria: Meteorologia 37, IMGW, Warszawa

Milelli M, Llasat MC, Ducrocq V (2006) The cases of June 2000, November 2002 and September 2002 as examples of Mediterranean floods. Nat Hazards Earth Syst Sci 6:271–284

Mizahari F (2000) Heavy daily precipitation distribution in east-central France and west European meteorological patterns. Theor Appl Climatol 66:199–210

Nicholls N, Murray W (1999) Workshop on indices and indicators for climate extremes: Asheville, NC, USA, 3–6 June 1997 breakout group B: precipitation. Clim Change 42:23–29

Niedźwiedź T (1992) Climate of the Tatra Mountins. Mt Res Dev 12(2):131–146

Niedźwiedź T (2003a) Słownik meteorologiczny (Meteorological dictionary). Polskie Towarzystwo Geofizyczne. IMGW, Warszawa

Niedźwiedź T (2003b) Extreme precipitation events on the northern side of the Tatra mountains. Geogr Pol 76(2):15–23

Niedźwiedź T, Michalczyk Z, Starkel L, Ustrnul Z (2004) Criteria for extreme meteorological and hydrological events. in Poland (ocena zdarzeń oraz prognozowanie ich skutków dla Środowiska życia człowieka). Paper presented at the conference Extreme meteorological and hydrological events in Poland (assessment and forecast of their consequences for human life environment), Sosnowiec, 13 December 2004

Okołowicz W (1969) Klimatologia Ogólna (General climatology). PWN, Warszawa

Osborn TJ, Hulme M, Jones PD, Basnett TA (2000) Observed trends in the daily intensity of United Kingdom precipitation. Int J Climatol 20:347–364

Paszyński J, Niedźwiedź T (1999) Klimat. In: Starkel L (ed) Geografia Polski. Środowisko Przyrodnicze. PWN, Warszawa

Peterson TC, Folland C, Gruza G, Hogg W, Mokssit A, Plummer N (2001) Report on the activities of the Working Group on Climate Change Detection and Related Rapporteurs 1998–2001.

World Meteorological Organisation Rep WMO-TD No 1071, WCDMP-No 47, Geneva, Switzerland

Romero R, Sumner G, Ramis C, Genovés A (1999) A classification of the atmospheric circulation patterns producing significant daily rainfall in the Spanish Mediterranean area. Int J Climatol 19:765–785

Schmidli J, Frei C (2005) Trends of heavy precipitation and wet and dry spells in Switzerland during the 20th century. Int J Climatol 25:753–771

Siwek K (2010) Występowanie na Lubelszczyźnie dobowych sum opadu atmosferycznego ≥100 mm i ich uwarunkowania cyrkulacyjne (1951–2000) (The occurrence of daily precipitation ≥100 mm in Lublin Voivodship and its synoptic background (1951–2000)). Ann UMCS Sec B 65(2):117–125

Stach A (2009) Analiza struktury przestrzennej i czasoprzestrzennej maksymalnych opadów dobowych w Polsce w latach 1956–1980 (Analysis spatial and spatio-temporal structure of maximum daily precipitation in Poland in the period 1956–1980), vol 85, Seria Geografia. Adam Mickiewicz University Press, Poznań, p 323

Strangeways I (2007) Precipitation. Theory, measurements and distribution. Cambridge University Press, New York

Trenberth KE, Owen TW (1999) Workshop on indices and indicators for Climate Extremes, Asheville, NC, USA, 3–6 June 1997. Clim Change 42:9–21

Trigo I, Davies T, Bigg G (1999) Objective climatology of cyclones in the Mediterranean region. J Climate 12:1685–1696

Trömel S, Schönwiese CD (2007) Probability change of extreme precipitation observed from 1901 to 2000 in Germany. Theor Appl Climatol 87:29–39

Ustrnul Z, Czekierda D (2009) Atlas ekstremalnych zjawisk meteorologicznych oraz sytuacji synoptycznych w Polsce (The Atlas of extreme meteorological phenomena and synoptic situations in Poland). IMGW, Warszawa

WMO (2004) Sixth WMO long-term plan 2004–2011. WMO No 962, Geneva, Switzerland

Zhang X, Hogg WD, Bonsal BR (2001) A cautionary note on the use of seasonally varying thresholds to assess temperature extremes. Clim Change 50:505–507

Zipser EJ, Cecil DJ, Liu CH, Nesbitt SW, Yorty DP (2006) Where are the most intense thunderstorms on Earth? Bull Am Meteor Soc 87:1057–1071

Zolina O, Kapala A, Simmer C, Gulev SK (2004) Analysis of extreme precipitation over Europe from different reanalyses: a comparative assessment. Glob Planet Change 44:129–161

Zolina O, Simmer C, Kapala A, Gulev S (2005) On the robustness of the estimates of centennial-scale variability in heavy precipitation from station data over Europe. Geophys Res Lett 32:L14707. doi:10.1029/2005GL023231

Zwoliński Z (2008) Wybrane zjawiska ekstremalne pojezierzy polskich (Selected extreme phenomena of Polish Lake District). Landform Anal 8:98–106

Chapter 3
Air-Mass and Frontal Extreme Precipitation

Abstract The occurrence as well as the daily and annual patterns of precipitation are inseparably linked with cloud formation processes. Upward air movement, alongside its sufficient humidity, is a precondition of precipitation. This chapter uses existing studies to discuss mechanisms leading to the formation of clouds. These mechanisms are then used to identify seven origin-based extreme precipitation types, that is, air-mass precipitation and a breakdown of frontal precipitation depending on the front. The chapter discusses the spatial extent of extreme air-mass and frontal precipitation in general and then focusses on the spatial and seasonal variabilities in all the origin-based types of precipitation in Europe.

The spatial extent of frontal precipitation events (defined as the number of stations involved on a single day) is much greater than that of air-mass events. The frequency of occurrence of various origin-based types of precipitation in Europe follows discernible spatial and seasonal variabilities, which are driven by the varying pace of cyclones life and by ground relief. This latter factor also affects cyclogenesis, thus having an influence on the spatial variability of origin-based types of extreme precipitation. A clear majority of extreme precipitation in Europe is linked to weather fronts. In each season, the average proportion of frontal precipitation in the overall number of days with extreme precipitation was several times greater than that of air-mass precipitation.

Keywords Air-mass precipitation • Frontal precipitation • Weather fronts • Cold front • Warm front • Occlusion

The occurrence of precipitation, as well as its daily and annual patterns, is inseparably linked with cloud formation processes (Okołowicz 1969). The earliest research into these processes, which started around 1940, targeted cloud droplet formation as part of a larger discipline of cloud physics (Strangeways 2007). In 1957, B.J. Mason noticed the importance of movements within the atmosphere, whether at a scale of 1 kilometre (km) or more than 1000 km, which was named cloud dynamics. Precipitation develops as a result of the growth of cloud droplets from their original cloud-building size to one a thousand times larger when they are too heavy to be lifted by upward movements of air (Aguado and Burt 1999). Mechanisms of the droplet growth inside the cloud are among the fundamental aspects of meteorology and have been covered in numerous academic studies and textbooks (Trewartha

E. Łupikasza, *The Climatology of Air-Mass and Frontal Extreme Precipitation*, Springer Atmospheric Sciences, DOI 10.1007/978-3-319-31478-5_3

1968; Okołowicz 1969; Chromow 1977; Barry and Chorley 1998; Kożuchowski 1998; Aguado and Burt 1999; Strangeways 2007). This string of studies begins with Bergeron's classic book of 1935 about the development of precipitation wherein the author observes that for precipitation to develop there must be a simultaneous occurrence of ice crystals and a considerable number of supercooled water droplets in the part of the cloud where the temperature falls below −10 °C. Although the physics of precipitation development is outside the scope of this volume, it is still worth noting a fundamental study by Mason of 1955 entitled "The physics of natural precipitation processes," which provides in-depth coverage of physical fundamentals of precipitation processes.

The formation of precipitation is therefore a product of both large-scale dynamic processes that lead to the development of clouds and smaller-scale physical processes inside the cloud. This study focuses on the forces that initiate these dynamic processes (which lead to the development of clouds through adiabatic cooling) and uses them as a basis to categorise extreme precipitation into origin-based types. This chapter provides a short discussion of these forces.

Upward air movement, alongside a sufficiently humid air-mass (Trewartha 1968), is a precondition for the development of precipitation. It leads to the development of clouds and then to precipitation. According to Strangeways (2007), clouds, depending on their type, may develop as a result of either advective or convective movement or of the activity of an atmospheric front. Advection of warm air over a cooled ground may lead to the formation of a thin stratus cloud cover. Precipitation from such a stratiform cloud is typically little. An opposite situation with cool air moving in over warm ground can lead to an increase in atmosphere instability and conditions favourable for the development of convection.

Convection, defined as a vertical movement in the atmosphere caused by buoyancy forces (free convection) or by dynamic forces (forced convection) (Niedźwiedź 2003), leads to the development of *Cumulus* and *Cumulonimbus* clouds. Depending on their type these clouds either give no precipitation (*Cu humilis*, *mediocris*) or, in the conditions of deep convection, lead to abundant but short-lasting precipitation (*Cu congestus* and *Cumulonimbus*) (Strangeways 2007). Precipitation that develops as a result of free convection caused by a difference in temperature, that is, density between the moving particles of air and their environment (Niedźwiedź 2003), is typical of temperate continental climates and normally occurs in summer over land and in the afternoon (Martyn 2000), lasting generally between 30 min and 1 h (Barry and Chorley 1998). Forced convection occurs on atmospheric fronts or over high-altitude areas that form a barrier to oncoming air-masses. In the latter cases orographic clouds can develop. The type of clouds depends on the atmospheric stability. Well-developed tall cumuliform orographic clouds can yield torrential precipitation, sometimes flood inducing, wherever there are mountains (Strangeways 2007). According to Barry and Chorley (1998), convective precipitation develops either ahead of a cold front in the warm air sector, sometimes as a squall line, or behind a warm front, but always it is parallel to the front line. Convection also occurs at cyclone centres, in troughs, as well as on cold fronts, warm fronts (especially lifted ones), and stationary fronts (Raddatz and Hanesiak 2008). Convection plays an enormous role in the development of extreme, so-called convective precipitation, which can lead to flooding in and outside of mountainous areas (Doswell 1993; Wheeler 1988, 1991, 1995; Llasat et al. 1996; Hagen et al. 2000; Hand et al. 2004; Federico et al. 2008a, b).

Depending on the nature of the atmospheric movement, that is, its vertical component, on weather fronts different types of clouds develop and a different type of precipitation is produced in terms of its duration and intensity. Precipitation associated with the passage of fronts is referred to as frontal precipitation. Its nature, that is, duration, intensity, and, ultimately, amount, depend on the type of the atmospheric front and of the accompanying clouds (Blüthgen 1966).

Clouds associated with the warm front and described by Chromow (1977) as "overrunning clouds" develop through the adiabatic cooling of the air that rises over a cold air-mass wedged underneath it. The type of clouds generating during this process and, as a consequence, the location of the precipitation zone and its nature depend on the altitude to which the warmer air-mass ascends, which, in turn, depends on its thermal and humidity properties (Trewartha 1968; Strangeways 2007). If the warm air is unstable, then its lift may be violent and may lead to intensive precipitation and thunderstorm phenomena (Trewartha 1968). Warm fronts are not always linked with low-pressure systems, as they can also develop along the coast or mountain ranges (Schneider 1996). Warm front-related precipitation typically occurs ahead of the front and covers large areas up to hundreds of thousands of square kilometres and is normally continuous, long lasting, and has medium intensity (Chromow 1977; Schneider 1996; Martyn 2000). Its intensity increases in proportion to the thickness of the clouds associated with the warm front and peaks just ahead of the front line. Various recent studies have demonstrated that the intensity of the warm-front precipitation varies greatly within the frontal area. Small areas of intensive precipitation that develop within vast frontal precipitation bands are organised in the form of rainbands, or banded precipitation (Schneider 1996). Continuous precipitation accounts for the highest proportion of overall precipitation in moderate latitudes (Chromow 1977) and is more intensive in the cold half of the year (Okołowicz 1969; Martyn 2000).

Mechanisms driving cloud development in the cold front zone are far more dynamic than these of the warm front because the front surface is much steeper. The convective air movement on the cold front is an effect of an upward displacement of warm air by cold air wedging in at the surface level (Strangeways 2007). Cold-front precipitation normally takes the form of showers, especially in summer (Okołowicz 1969; Chromow 1977; Martyn 2000). In general, the intensity of cold-front precipitation increases with temperature and with air humidity in the warm sector of the low-pressure system (warm air-mass on the front) (Trewartha 1968). The nature of precipitation associated with the cold front changes along its line from the most intensive and persistent at the northern end, near the centre of the system, to light or potentially nonexistent at its far southern tip (Schneider 1996). According to W. Chromow (1977, p. 238): "The intensity of convective precipitation associated with fronts strongly fluctuates. Even within a single rainfall event precipitation amount can vary by up to 50 mm at a distance of merely 1–2 km."

At the onset of the final stage of the low-pressure system, the faster-moving cold front catches up and merges with the slower warm front and an occluded front develops. An occlusion lifts warm air higher up the atmosphere and, as a result, the bulk of the low-pressure system at the ground surface is occupied by a cold air-mass. Precipitation accompanying an occlusion displays features typical of both cold- and warm-front precipitation, which means that continuous precipitation of moderate intensity from a *nimbostratus* cloud can turn into heavy precipitation (sometimes thunderstorm precipitation) which, according to Schneider (1996), is the most frequent type along a raised warm front or in the prefrontal surge zone. Satellite images suggest that occluded fronts can also develop by an instant occlusion, as a comma cloud that develops on a low-pressure trough in a polar air-mass merges with clouds developed on the polar front (Djurić 1994). Figure 3.1 illustrates a general model of a low-pressure system and vertical sections through that model along AB and CD lines (Fig. 3.1c), including a schematic distribution of precipitation.

Hobbs (1981) used satellite and radar images to provide the first accurate and detailed distribution of rainbands in frontal systems (Fig. 3.2). He identified six mesoscale rainbands:

Type 1. Warm-frontal bands occur within the leading portion of the frontal system, where warm advection occurs through a deep layer. They have orientations similar to that of the warm front. These bands are typically about 50 km wide. They may be located ahead of the warm front (type 1a), coincide with a surface warm front (type 1b), or have an orientation similar to that of a warm front.

Type 2. Warm-sector bands are in the warm sector and are oriented parallel to the surface cold front. They are typically up to about 50 km wide.

Type 3. Wide cold-frontal bands are oriented parallel to the cold front and either straddle or are behind the surface cold front. In the case of occlusions they are associated with the cold front aloft. They are about 50 km wide.

Type 4. The narrow cold-frontal band differs markedly from the other types of bands. It is very narrow (about 5 km) and coincides with the position of the cold front at the surface.

Type 5. Prefrontal, cold-surge bands are associated with the surges of cold air ahead of the cold front or are essentially the same type of features as the wide cold-frontal bands.

Type 6. Prefrontal bands are lines of convective clouds that form well behind and parallel to the cold front.

This frontal system structure is representative of all extratropical cyclones. Naturally, these rainbands do not always occur in all lows. A detailed characteristic of all these rainbands can be found in House and Hobbs (1982).

Synoptic charts include not just warm, cold, and occluded fronts, but also stationary fronts, which, according to Aguado and Burt (1999), have structures similar to the structure of the cold front. Movement, or the lack of it, is their identification criterion. Stationary fronts are either do not move, or move very slowly, or in an irregular manner (Schneider 1996). Djurić (1994) added one more condition,

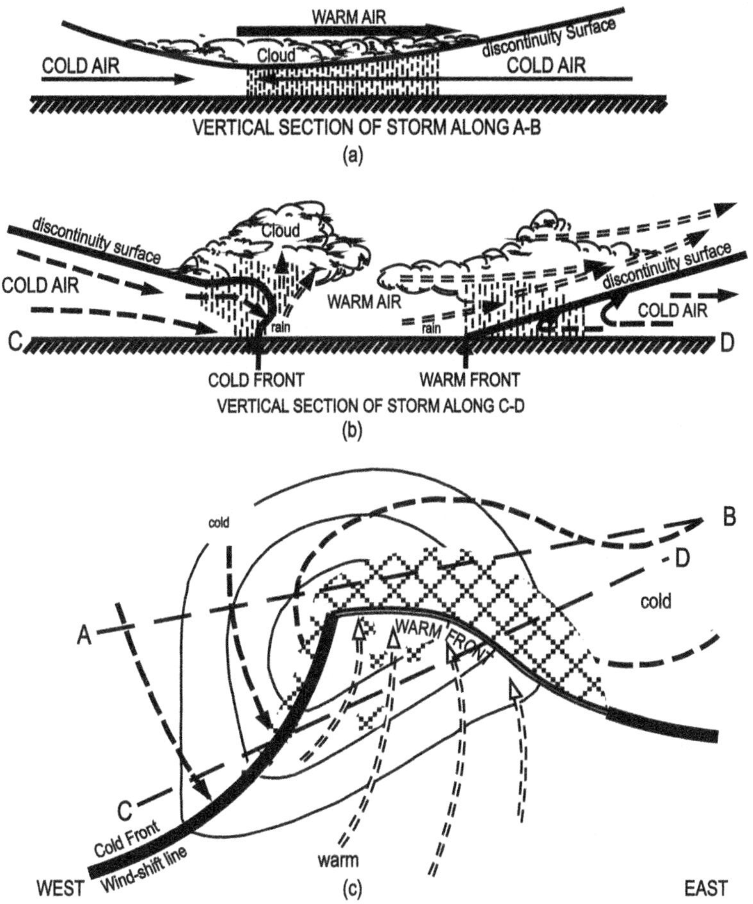

Fig. 3.1 The cyclone model. Ground plan (**c**) and vertical sections (**a, b**) of a fully developed wave cyclone (Revised from Trewartha 1968)

namely, that the isobars do not cross the stationary front line, or, if they do, the distance between them is greater than 500 km.

Discontinuity lines, referred to as squall lines, are productive of heavy precipitation. They typically develop in the warm sector of a cyclone ahead of the cold front and have features similar to that of a cold front (Niedźwiedź 2003). Schneider (1996) defines them as convective systems arranged in a line. Once the line is formed, it immediately produces a narrow band of intensive precipitation. As the band develops it expands while the precipitation becomes less intensive and more even. The strongest convection and most intensive precipitation occur in the easternmost part of the squall line (Schneider 1996).

These mechanisms, which initiate upward movements of the air leading to the development of clouds and finally to precipitation, are not mutually exclusive.

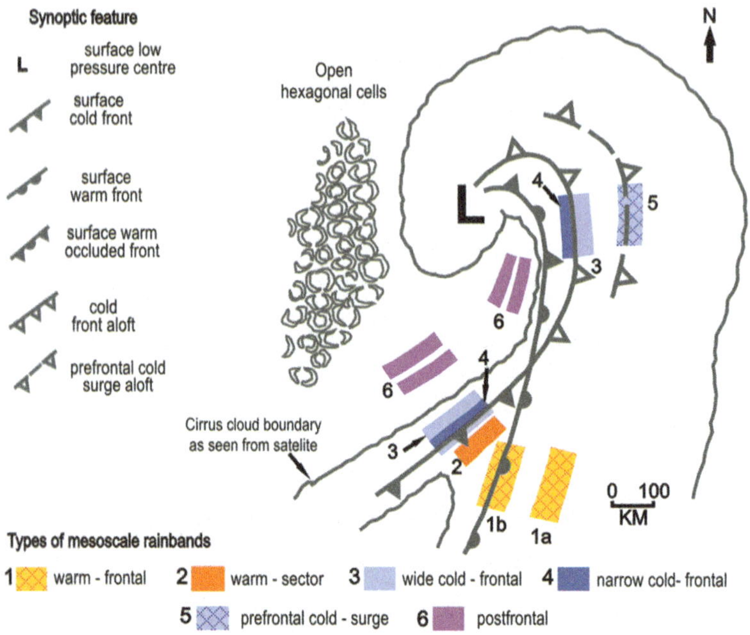

Fig. 3.2 Schematic depiction of the types of rainbands (*numbers 1–6*) observed in extratropical cyclones (Revised from Hobbs 1981)

Specific precipitation events may therefore be a result of more than one of these dynamic processes (Trewartha 1968; Little et al. 2008).

3.1 Classification of Origin-Based Extreme Precipitation Types

Considering two basic mechanisms that lead to precipitation formation, free convection and weather fronts, the latter of which can readily be found on synoptic charts, seven types of extreme precipitation origin (ExPT) were distinguished (Table 3.1). Chomicz (1971), in a discussion of the classification and the structure of precipitation, claims that precipitation can be broken down into origin-based types linked to synoptic situations and to corresponding cloud configurations. Thus, extreme precipitation types (ExPTs) identified here by associating precipitation with weather fronts, which are components of the atmospheric circulation, are origin based.

The air-mass type (type A) represents precipitation associated with free convection in a homogenous air-mass. All other frontal types were identified taking into account all types of fronts marked on synoptic charts, such as associated with the passage of a warm front (type Fw), with the passage of a cold front (type Fc), with an occluded front (type Fo) with a stationary front (type Fs) and associated with a

Table 3.1 Origin-based extreme precipitation types

Origin-based type of extreme precipitation	Symbol	Description
Air-mass (convective according to Okołowicz 1969)	A	Precipitation developing in a homogeneous air mass. No front has been observed to pass through the extreme precipitation area during the observation day. Also, during the whole precipitation day, the nearest fronts were at least 300 km away from the extreme precipitation area.
Associated with the passage of several different fronts	Ff	Frontal precipitation associated with the passage of at least two different fronts through the extreme precipitation area during the precipitation day.
Associated with the passage of cold front	Fc	Frontal precipitation associated with the passage of a cold front through the extreme precipitation area during the precipitation day.
Associated with the passage of warm front	Fw	Frontal precipitation associated with the passage of a warm front through the extreme precipitation area or with a front line less than 300 km away (Djurić 1994) from the extreme precipitation area during the precipitation day.
Associated with an occluded front	Fo	Frontal precipitation associated with the passage of an occluded front through the extreme precipitation area or with a stagnant occluded front in the extreme precipitation area during the precipitation day.
Associated with a stationary front	Fs	Frontal precipitation associated with a stationary front in the extreme precipitation area during the precipitation day.
Associated with a discontinuity line	Fd	Precipitation associated only with an occurrence of a discontinuity line in the extreme precipitation area during the precipitation day (no fronts).

discontinuity line (type Fd). An additional type was identified that was associated with the passage of several different fronts (type Ff).

Extreme precipitation events (≥95 P) selected from the database were assigned to one of the seven origin-based precipitation types using their descriptions from Table 3.1. Figures 3.3, 3.4, 3.5, 3.6, 3.7, 3.8, and 3.9 provide a sample of synoptic charts used to identify the origin of the extreme precipitation events. For each day in the study period (December 1950–February 2008), a chart of extreme precipitation occurrence was drawn. Each of more than 20,500 charts thus produced was compared with corresponding synoptic charts matching the timing of the precipitation day at each station. For each precipitation day, synoptic charts were available at least at two observation times. The entire classification process was manual and as such it was subjective in nature. Mätlik and Post (2008) used synoptic charts in a similar manual procedure of attribution of extreme precipitation (greater than 50 mm) to weather fronts and synoptic situations in Estonia, in 1951–2006. An application of an objective method to produce an origin-based classification of precipitation based on weather fronts would not be possible because of the unavailability of a reliable objective, that is, automatic, method of locating weather fronts and defining their dynamics during the day. This difficulty is linked with the fact that suitable data covering a broad range of meteorological elements at a resolution of

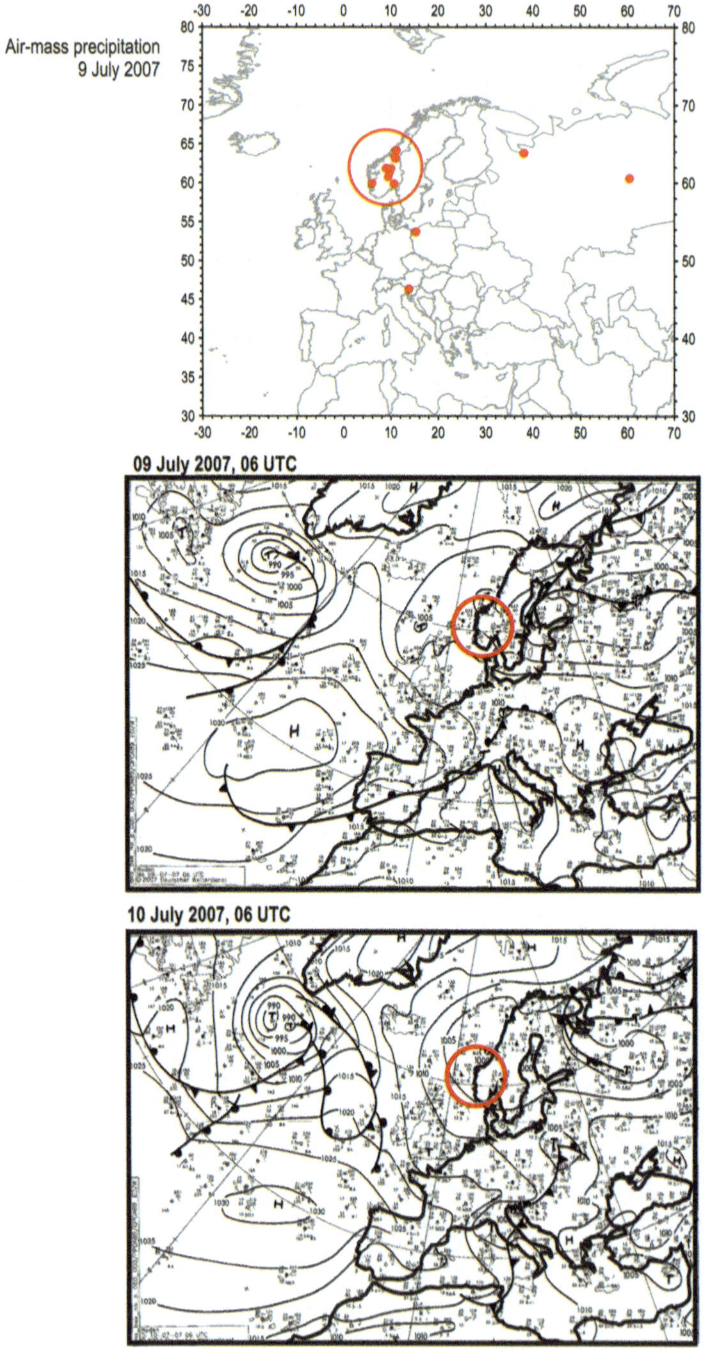

Fig. 3.3 Distribution of extreme precipitation in Europe on 9 July 2007 and synoptic maps used to recognize its origin at station within the *red circle*; air-mass extreme precipitation

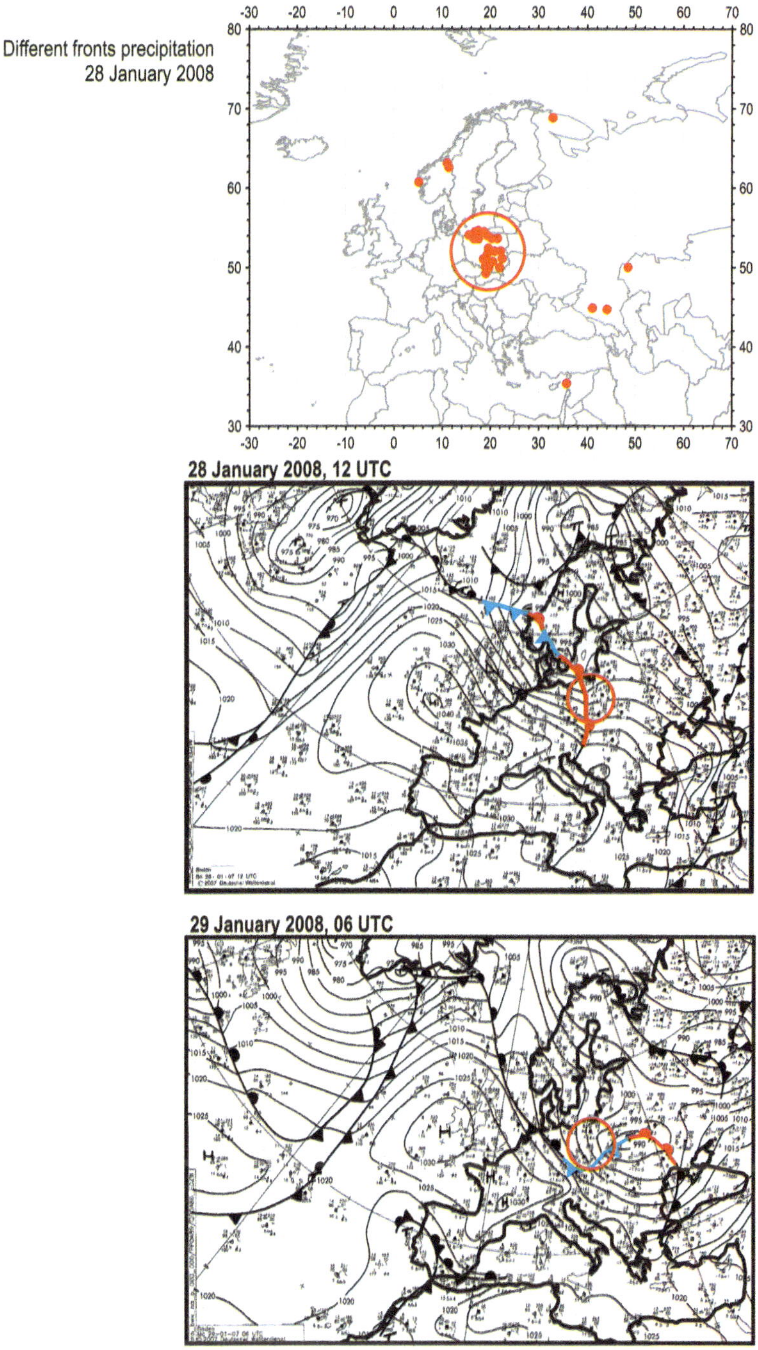

Fig. 3.4 Distribution of extreme precipitation in Europe on 28 January 2008 and synoptic maps used to recognize its genetic type at station within the *red circle*; extreme precipitation associated with the passage of different fronts

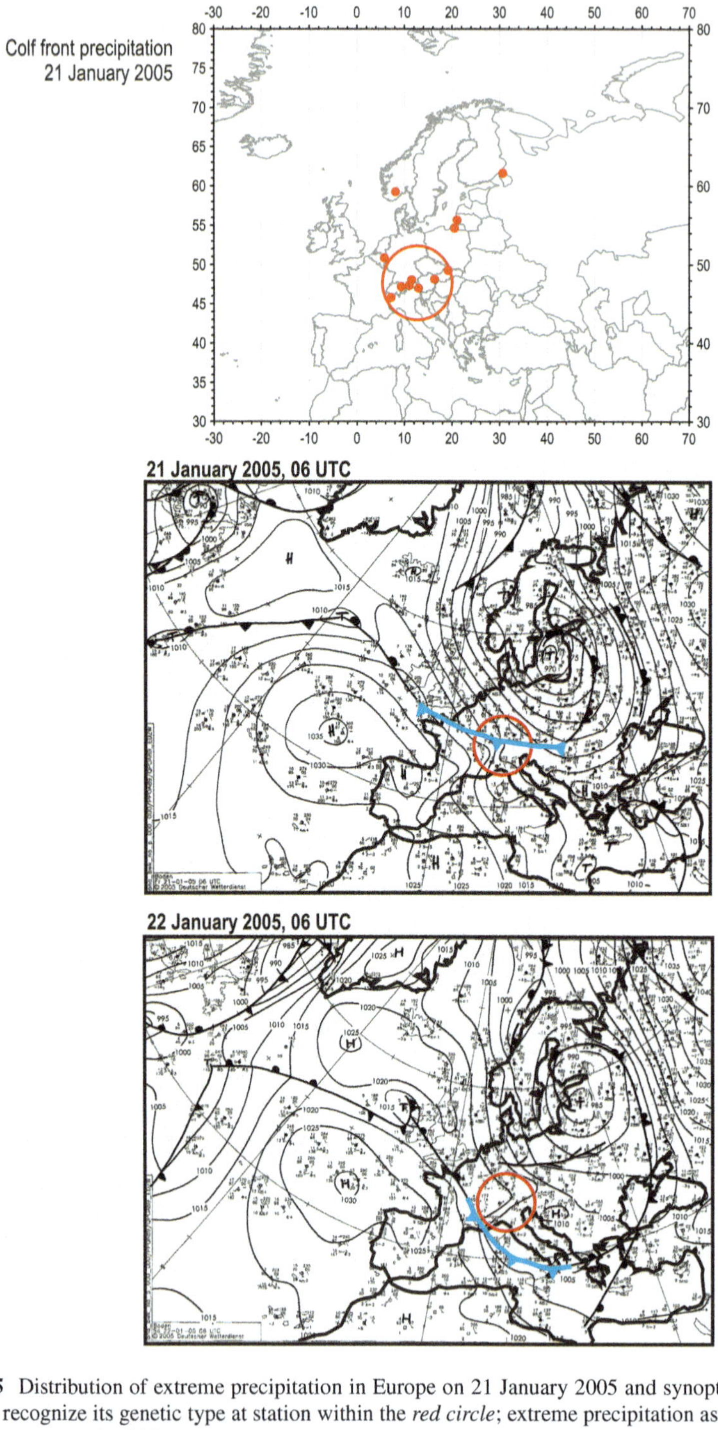

Fig. 3.5 Distribution of extreme precipitation in Europe on 21 January 2005 and synoptic maps used to recognize its genetic type at station within the *red circle*; extreme precipitation associated with the passage of cold front

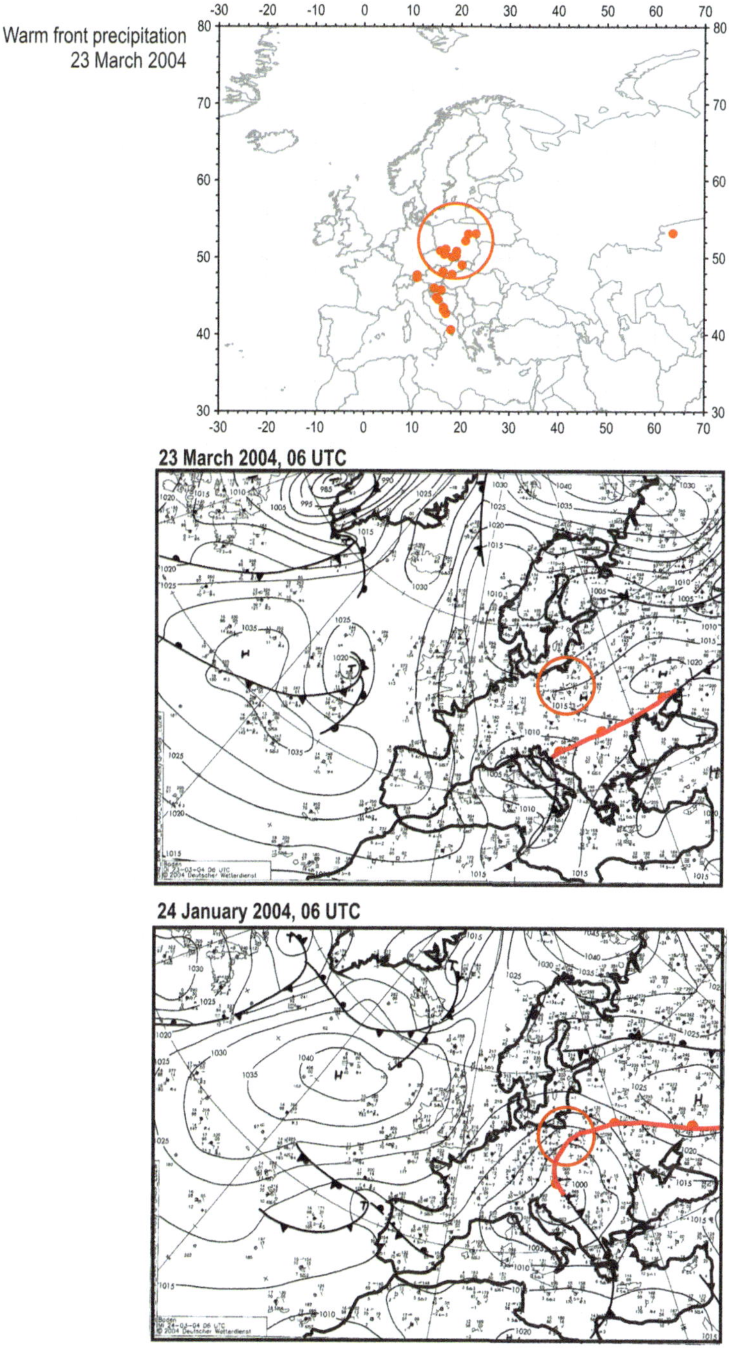

Fig. 3.6 Distribution of extreme precipitation in Europe on 23 March 2004 and synoptic maps used to recognize its genetic type at station within the *red circle*; extreme precipitation associated with the passage of warm front

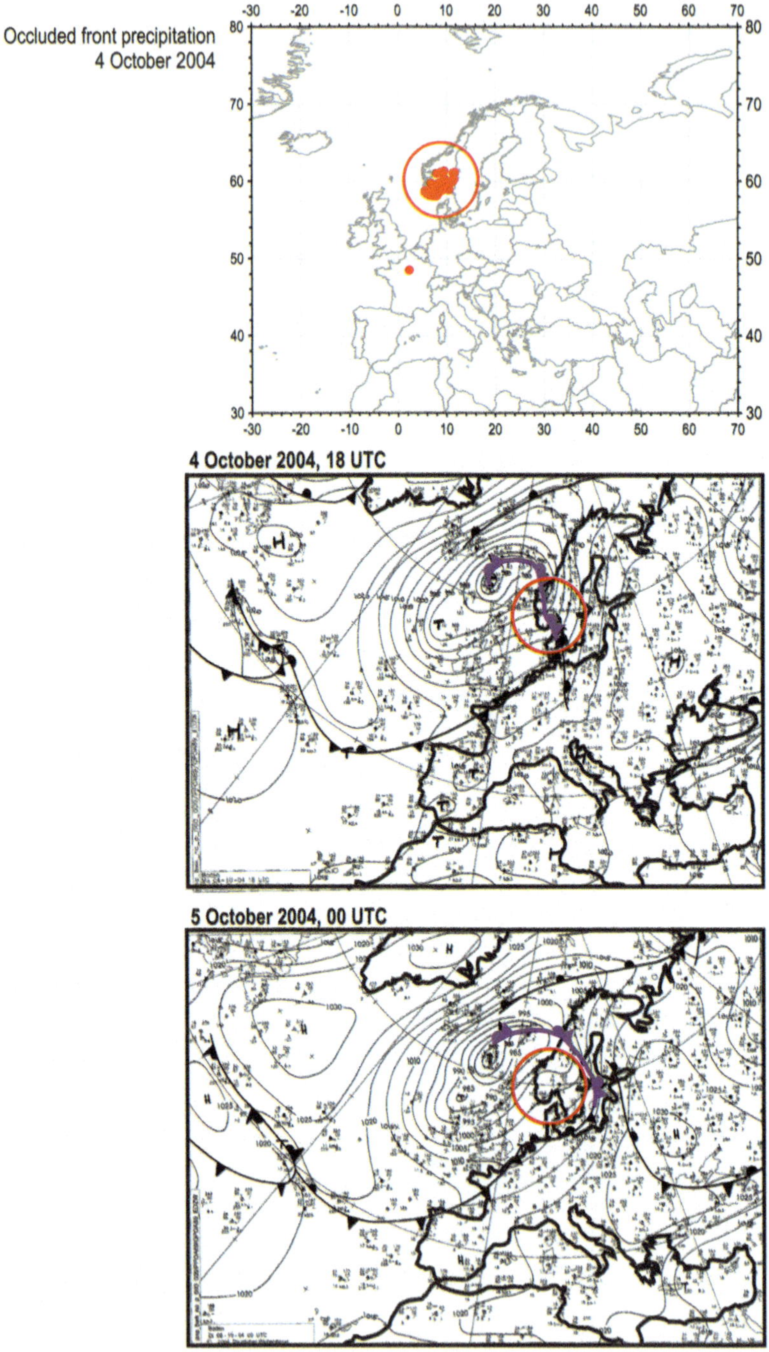

Fig. 3.7 Distribution of extreme precipitation in Europe on 04 October 2004 and synoptic maps used to recognize its genetic type at station within the *red circle*; extreme precipitation associated with occluded front

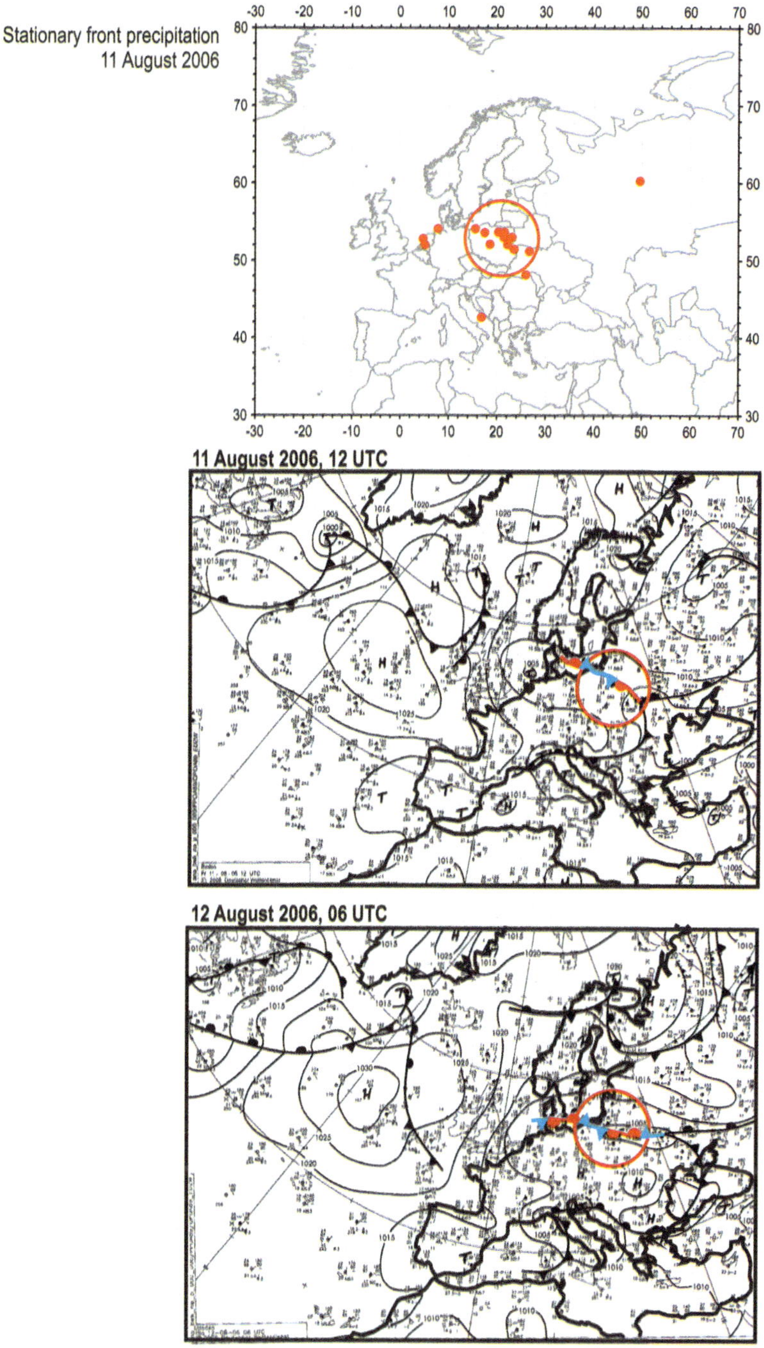

Fig. 3.8 Distribution of extreme precipitation in Europe on 11 August 2006 and synoptic maps used to recognize its genetic type at station within the *red circle*; extreme precipitation associated with stationary front

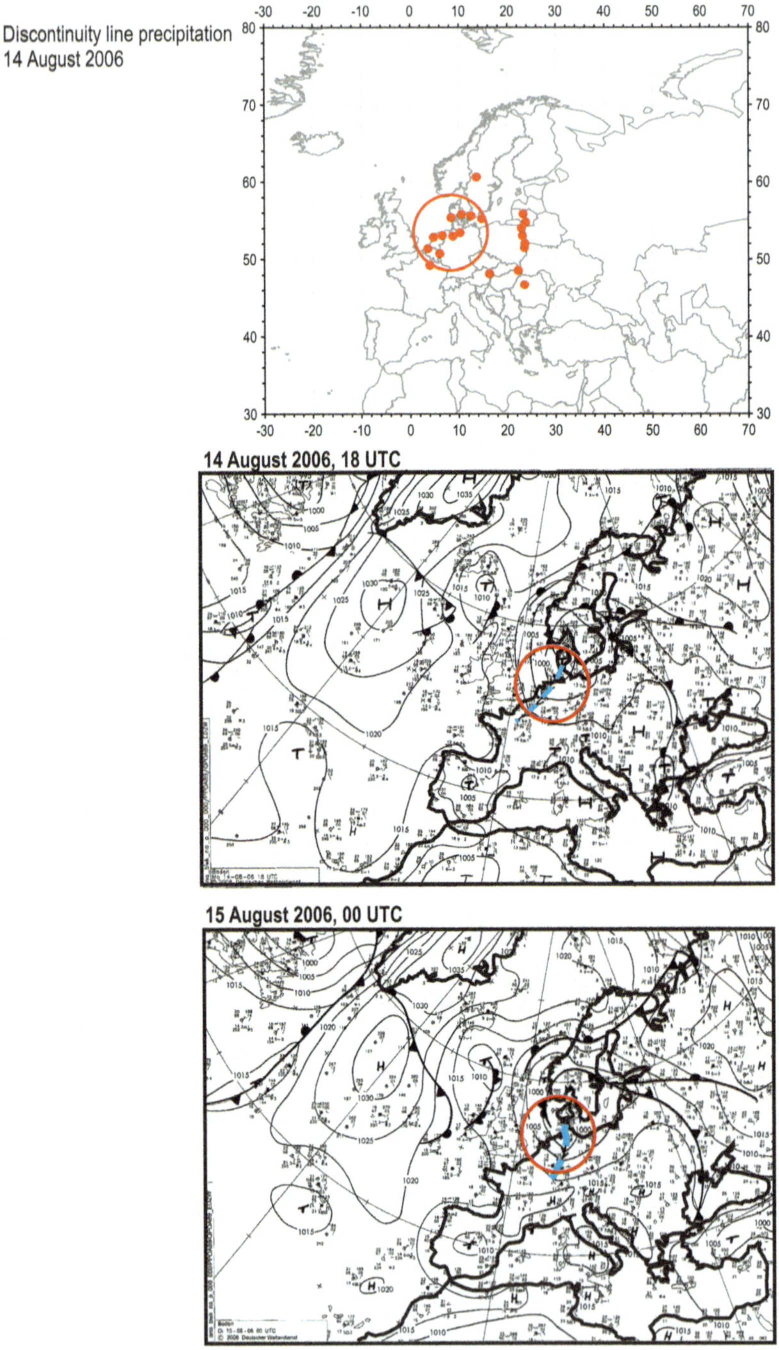

Fig. 3.9 Distribution of extreme precipitation in Europe on 14 August 2006 and synoptic maps used to recognize its genetic type at station within the *red circle*; extreme precipitation associated with discontinuity line

individual hours are not available and that during the passage of a front various meteorological elements often behave differently than is generally understood. Barry and Carleton (2001) claim that a coherent and well-defined procedure to locate atmospheric fronts does not exist. They also say that fronts are not always located in areas of strong temperature gradients. Raddatz and Hanesiak (2008) also employed a manual procedure of identifying origin-based types of heavy precipitation (>10 mm*24h^{-1}) for summer season events in the Canadian prairie between 2000 and 2004. Similarly to this study, the authors also observed that the attribution of precipitation to synoptic elements, that is, fronts, went mostly smoothly (Raddatz and Hanesiak 2008). Indeed, the occurrence of extreme precipitation is linked with a clearly defined synoptic situation.

Another source of the subjectivity in this method stems from the types of fronts and the way they are marked on synoptic charts, which still requires the involvement of a highly skilled and experienced meteorologist. The high degree of subjectivity involved in the identification of weather fronts on synoptic charts was criticised by Berry et al. (2011a), who chose to apply an automatic front location method in a study discussing climatic aspects of fronts. While criticising the use of synoptic charts, highly subjective in their view, in analysing weather fronts these researchers still concluded that both manual and automatic front analyses produced highly compatible results.

The least subjective aspect of this approach is the breakdown of precipitation into main origin-based types, that is, air-mass (type A) and frontal (type F), provided that the identification of weather fronts on the synoptic charts is correct. Because the detection of cold fronts, with their typically good visibility in the pattern of meteorological elements, is easier than that of warm fronts (Schneider 1996), a high degree of certainty may also be assumed for the classification of precipitation associated with the passage of a cold front (type Fc). The greatest degree of uncertainty is linked with the classification of precipitation associated with the discontinuity line. The frequency of this precipitation may be underestimated because not all squall lines (discontinuity lines) are marked on synoptic charts.

Orographic precipitation depends on one single precondition for its development and that is the existence of an orographic barrier, which always causes precipitation when precipitation-bearing air-masses arrive from a specific direction.

In the case of air-mass precipitation mountains force convective movement of air, but an orographic barrier also influences the amount of frontal precipitation. For this reason the proposed synoptic-based classification of extreme precipitation does not include a separate orographic type. Mechanisms involved in the development of orographic precipitation are complicated. Hand et al. (2004), in his classification system of selected origin-based extreme precipitation events, used synoptic charts and various kinds of information on other weather elements to identify a range of origin-based precipitation types over elevated areas, including frontal, convective, and orographic. In the orographic type a mountain barrier was the only factor initiating this kind of precipitation. Although Hand et al. (2004) were able to identify this factor by using various types of information about weather elements other than just synoptic charts, this study relies on synoptic charts only, making it impossible to reliably identify the leading mechanism of an extreme precipitation formation in mountainous areas.

3.2 Territorial Extent of Air-Mass and Frontal Extreme Precipitation

Spatial extent of precipitation is among fundamental features that help discern air-mass precipitation from frontal precipitation. Air-mass events, linked in moderate latitudes to locally occurring free convection, are much smaller in size than events associated with travelling cyclones with their systems of weather fronts (Kane et al. 1987; Karagiannidis et al. 2012).

The spatial extent of extreme precipitation events including a comparison between the extents of air-mass and frontal precipitation was analysed using a simple index defined as the number of stations that recorded extreme precipitation on the same day. The procedure was applied to all extreme precipitation events and, separately, to their major types, air-mass (type A) and frontal (type F). The results were presented as histograms illustrating the number of days with extreme precipitation in spatial ranges, that is, numbers of stations (Figs. 3.10 and 3.11). This frequency was expressed in percentages of all days of relevant seasons during the study period (e.g., 5244 days in spring). The influence of the changing numbers of stations involved in the study on the histograms was assessed by comparing histograms produced for the entire period with one for the period 1961–1990, when the

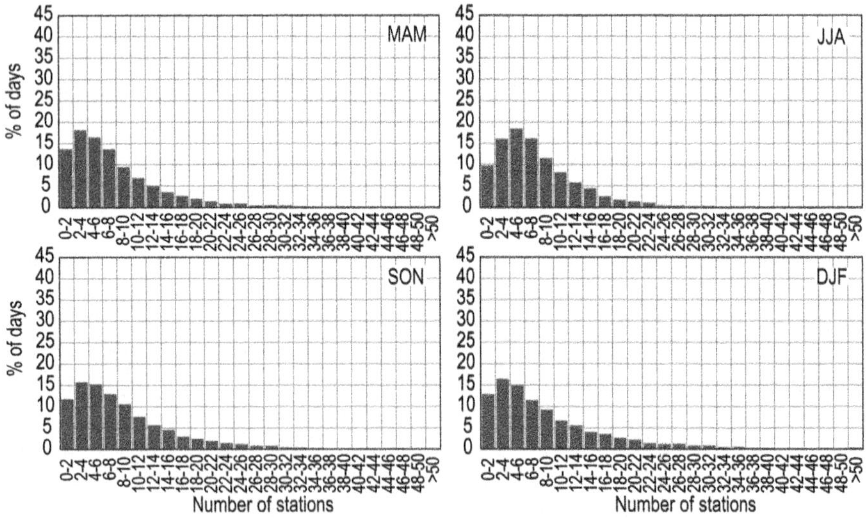

Fig. 3.10 Frequency of days within the intervals of extreme precipitation spatial extent for December 1950–February 2008. Right closed intervals. Spatial extent expressed as the number of stations with extreme precipitation during the same day

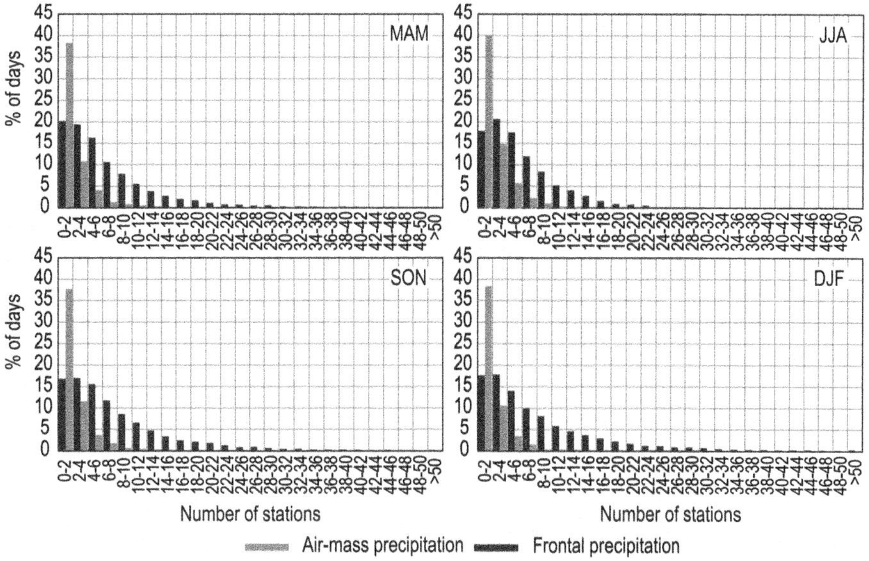

Fig. 3.11 Frequency of days within the intervals of spatial extent of air-mass and frontal extreme precipitation for December 1950–February 2008. Right closed intervals. Spatial extent means the number of stations with extreme precipitation during the same day

variation in the number of stations was insignificant and stayed within 4 %. The overall influence was found to be insignificant. The index of spatial extent does not take into account the spatial continuity of a precipitation event, that is, the distance between stations that recorded an ExP on the same day, which may have a greater impact on the results of spatial extent analysis of locally occurring air-mass events than of frontal precipitation that tends to cover large areas. A degree of inaccuracy in the results of the analysis of the spatial extent of extreme precipitation may also stem from uneven distribution of weather stations in the study area. Indeed, the number of stations recording extreme precipitation on the same day in Eastern Europe would have probably been greater if the density of the station network had been similar to that in Western Europe. It also must be mentioned that this chapter analyses the spatial extent covered by precipitation generated by entire frontal systems.

Extreme Precipitation (ExP) In all seasons extreme precipitation events covered typically between 2 and 6 stations, with the summer frequency of 4–6 stations being slightly higher then that in other seasons (Fig. 3.10). In Europe, winter extreme precipitation are of the largest spatial extent. On 10 % of winter days these events were recorded on the same day at 20 stations or more, compared to 8 % in autumn, 6 % in spring, and 4 % in summer. There was an even greater seasonal difference in the spatially largest events covering more than 60 stations on the same day.

These events were rare overall, but 17 of them were recorded in winter compared to just 5 in autumn (0.1 % of the total of days), 3 in spring (0.06 %), and only 1 in summer (0.02 %).

Air-Mass Extreme Precipitation (Type A) Throughout the year, the spatial extent of air-mass precipitation was smaller than that of frontal events (Fig. 3.11). According to Mason (1955), torrential air-mass precipitation typically covers areas up to several kilometres in size. In each season, smaller-scale events (up to 2 stations) are clearly more frequent than larger ones (more than two stations) (Fig. 3.11). There is little variability in the frequency of small-area ExPs, as it ranges from 38 % of days in autumn and spring to 40 % days in summer. There were also very few, that is, seven, air-mass events that extended over more than 20 stations. The largest such event covered 38 stations and occurred in autumn on 16 October 1974. The largest extreme precipitation events in other seasons included a springtime event that covered 25 stations on 23 April 1980, a summer event recorded at 24 stations on 31 August 1995, and an event observed at 22 stations in winter on 28 December 1979.

Frontal Extreme Precipitation (Type F) Distribution of frontal precipitation frequency is also relatively even throughout the year (Fig. 3.11). Seasonal variability in the frequency of type F for corresponding spatial ranges peaks at 4% of days. The spatially largest frontal extreme precipitation was recorded in winter on 12 February 1962 and covered 79 stations. In other seasons, the maximum coverage events included 71 stations in spring on 23 March 1986, 50 stations in summer on 1 June 1973, and 65 stations in autumn on 23 November 1984.

3.3 Spatial and Seasonal Variability in the Frequency of Air-Mass and Frontal Extreme Precipitation

The percentages of origin-based precipitation types with reference to the seasonal numbers of days with extreme events were calculated to recognize seasonal and spatial variability in the occurrence of the extreme precipitation types (ExPT). The results were then transferred onto charts of the frequency distribution of each ExPT in Europe, and corresponding histograms were produced for each season (Figs. 3.12, 3.13, 3.14, 3.15, 3.16, 3.17, 3.18, and 3.19). Descriptive statistics of these events are summarised in Tables 3.2 and 3.3.

3.3.1 Air-Mass Precipitation (Type A)

The average contribution of air-mass precipitation to the total number of days with extreme precipitation (ExP) varied from 14 % in winter to 25 % in summer. In summer, the frequency of extreme air-mass precipitation is characterised by the greatest

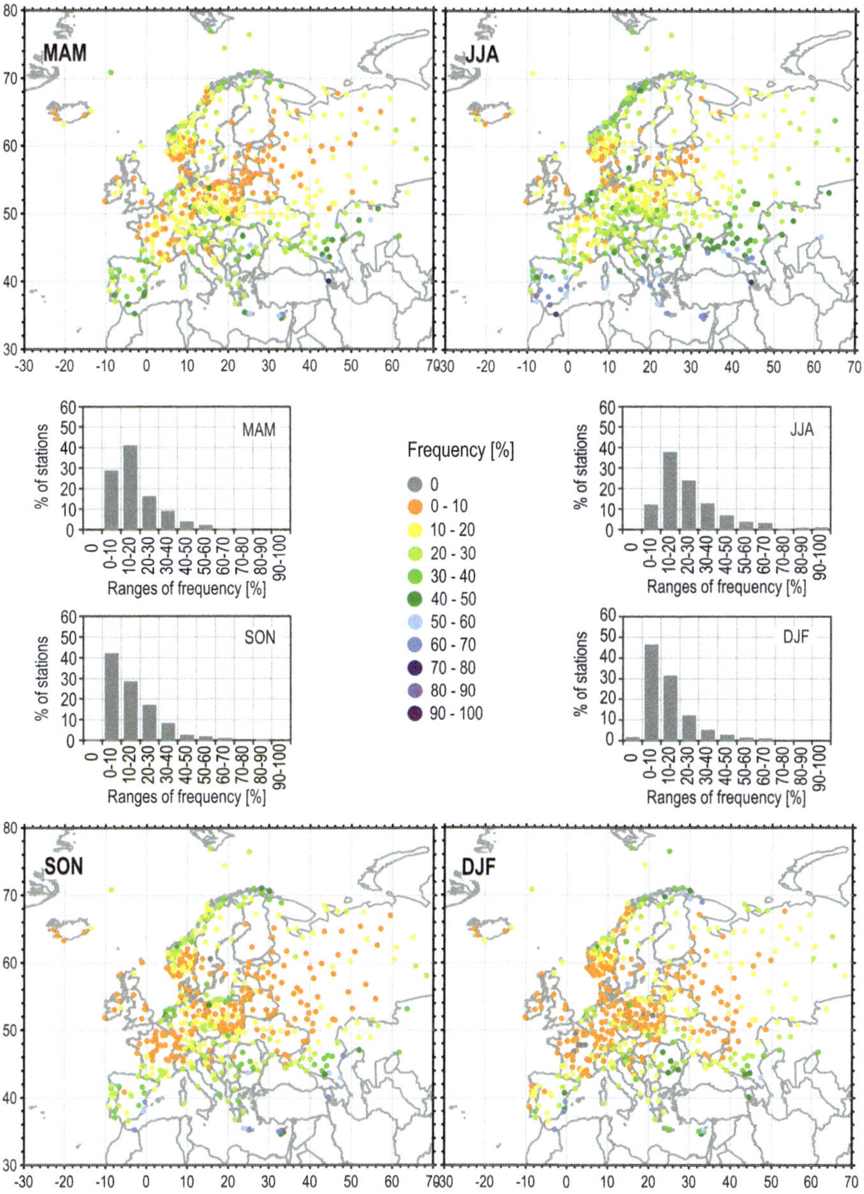

Fig. 3.12 Frequency of air-mass extreme precipitation (type A): spatial distribution and histograms for December 1950–February 2008. Right closed intervals

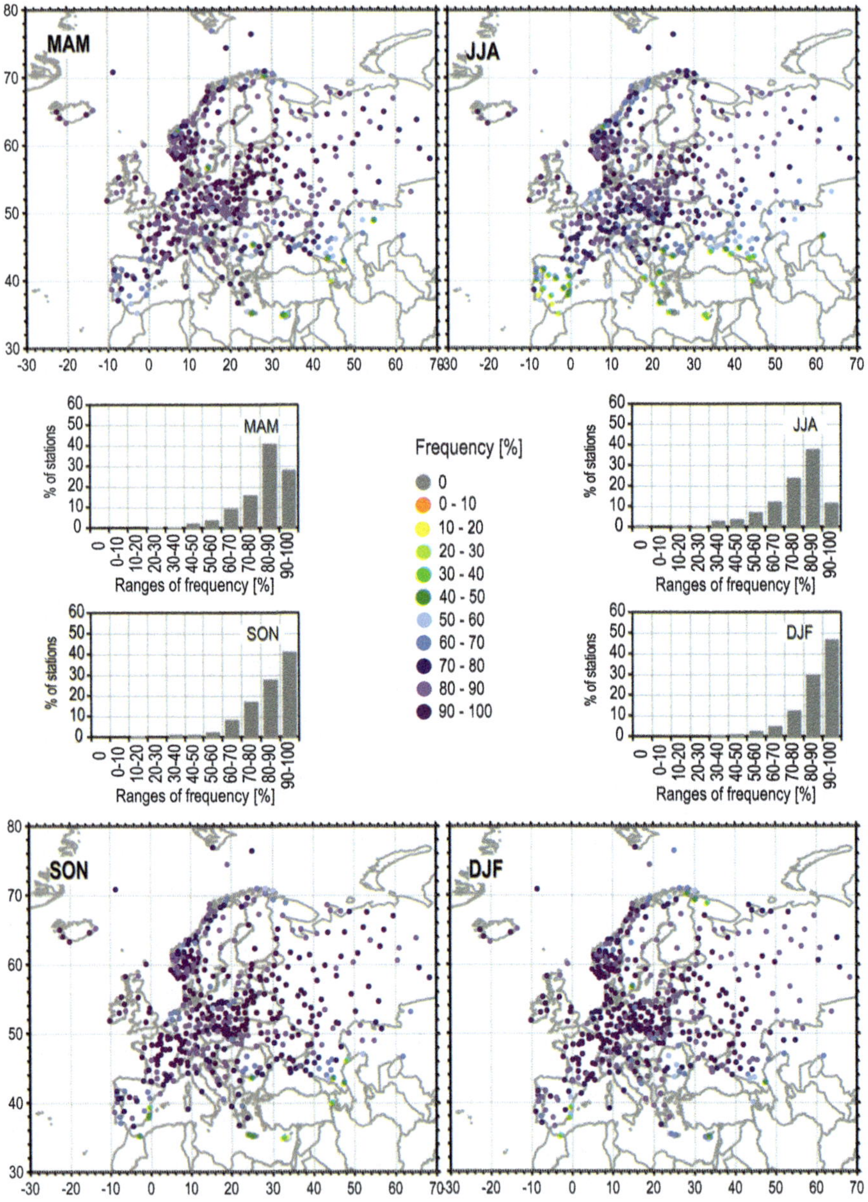

Fig. 3.13 Frequency of frontal extreme precipitation (type F): spatial distribution and histograms for December 1950–February 2008. Right closed intervals

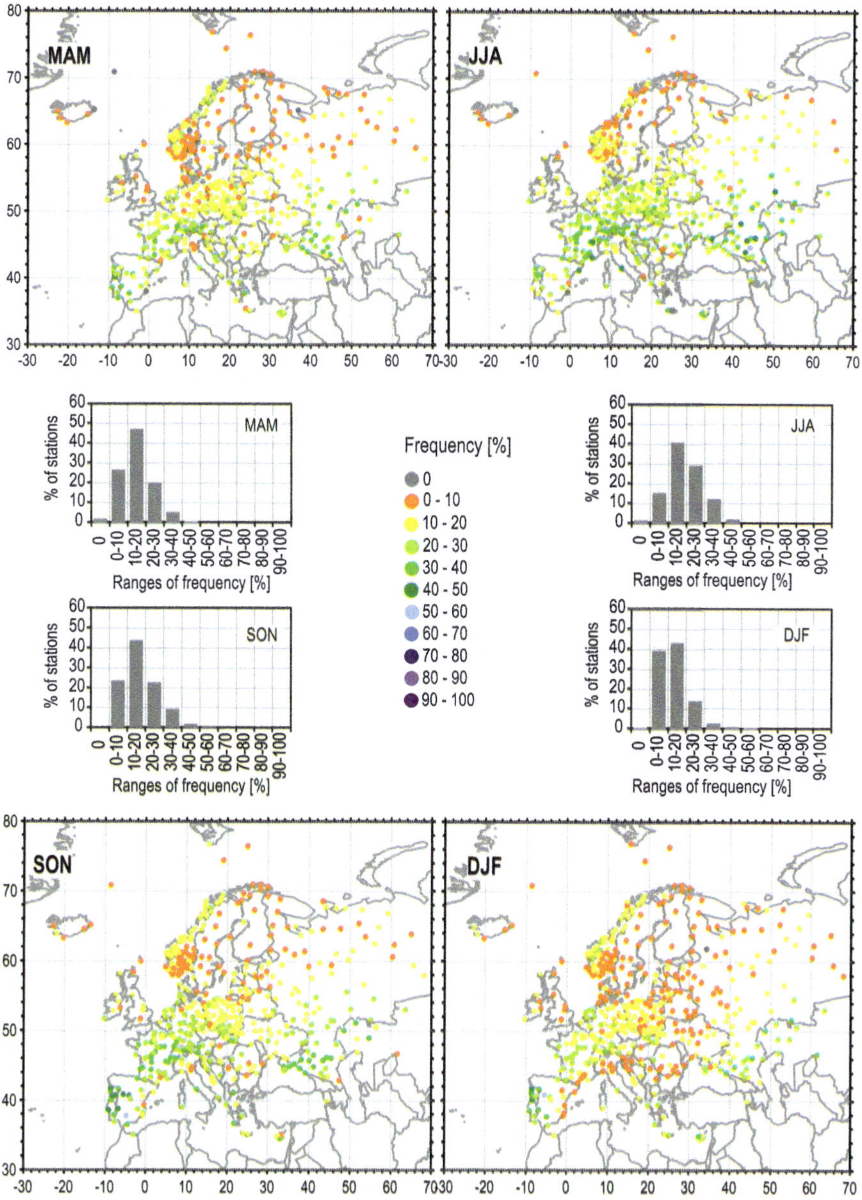

Fig. 3.14 Frequency of cold front extreme precipitation (type Fc): spatial distribution and histograms for December 1950–February 2008. Right closed intervals

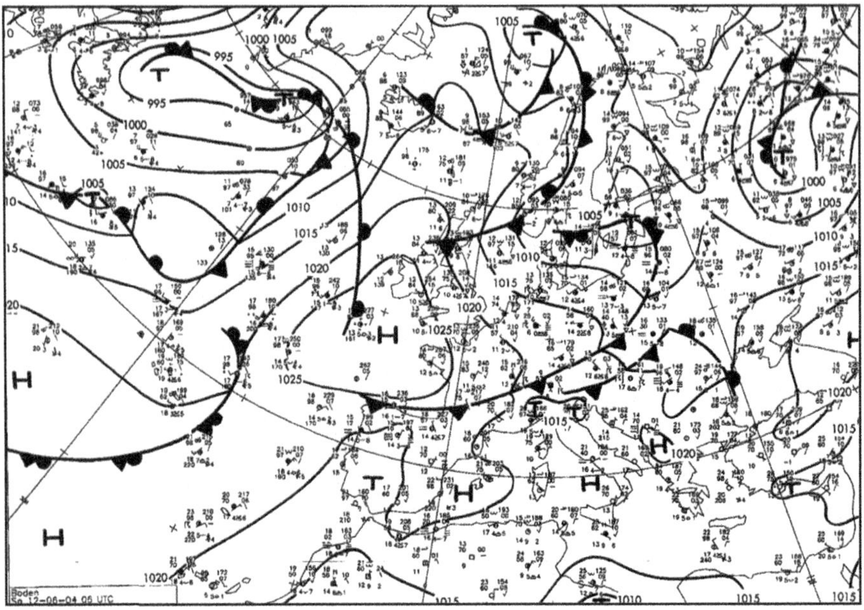

Fig. 3.15 Cold front over the Alps. Synoptic map from 12 June 2004, 06. UTC (*Source:* (http://www.wetter3.de))

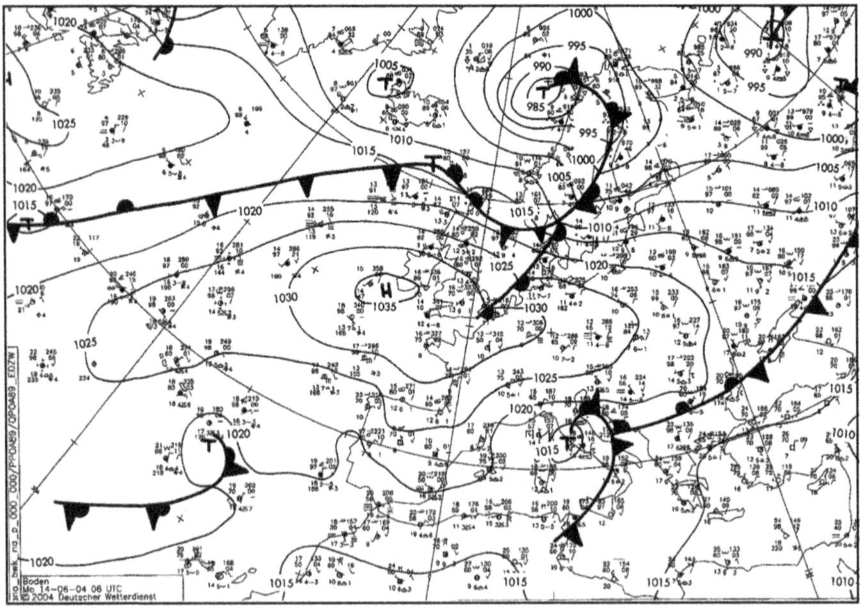

Fig. 3.16 Cold front flowing in over the vicinity of the Balck and Caspian Seas. Synoptic map from 14 June 2006, 06. UTC (*Source:* http://www.wetter3.de)

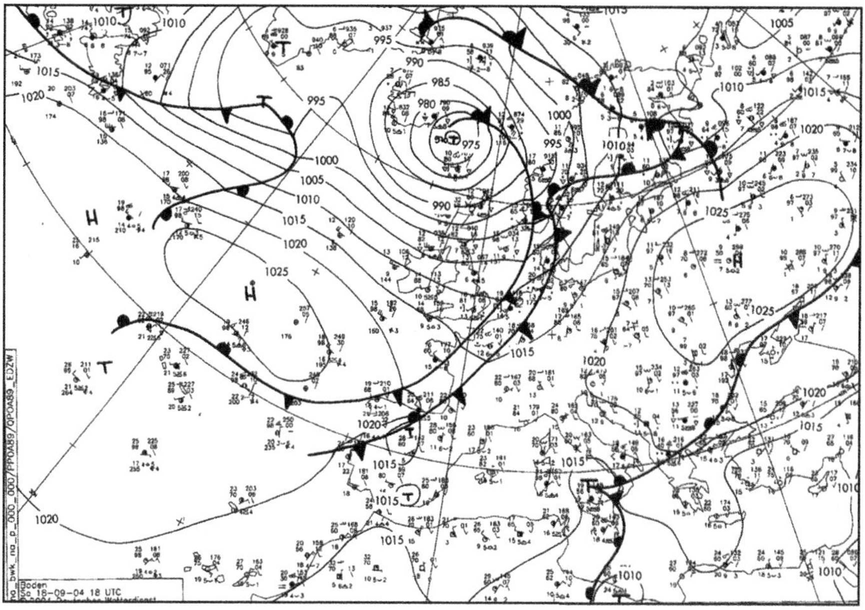

Fig. 3.17 Cold fronts flowing in over the Iberian Peninsula. Synoptic map from 14 June 2006, 18 UTC (*Source:* http://www.wetter3.de)

variability, as measured by standard deviation and range of variability (Max-Min), both of which are greater than in other seasons (Table 3.2). At 50 % of stations, air-mass precipitation accounted for between 13.5 % (bottom quartile) and 31.3 % (top quartile) of all extreme precipitation events occurring in this season.

In summer, type A precipitation is the most frequent in southern Europe (Fig. 3.12). Indeed, more than 50 % of extreme precipitation develops in homogenous air masses at a majority of stations located in the western and southern part of southern Europe (from the Iberian Peninsula to the Balkan Peninsula) south of the parallel 42°N, as well as a majority of stations in Eastern Europe south of the parallel 45°N. In summer, southern Europe is under the influence of a subtropical high-pressure zone (Okołowicz 1969; Chromow 1977; Martyn 2000). The combination of the area high temperatures and its complicated landform favours this part of the continent for torrential air-mass rainfall (Font and Tullot 1983; Ramis 1995; Riosalido 1990). The Iberian Peninsula is under a permanent influence of the Azores high, and the afternoon rainfall that develops at the time is a result of free convection. Two other factors that can cause the development of this type of rainfall are (1) convergence zones developing on contact between land and maritime surfaces (breeze effect) and (2) a thermal low that typically also occurs over the peninsula (Alonso et al. 1994; Romeo et al. 1998). Northwards from this area the frequency of air-mass precipitation declines (Fig. 3.12). This type of extreme precipitation accounts for 20–30 % of ExP at 24 % of weather stations located in mountainous and upland areas of the Central Europe, in western European coast (especially along

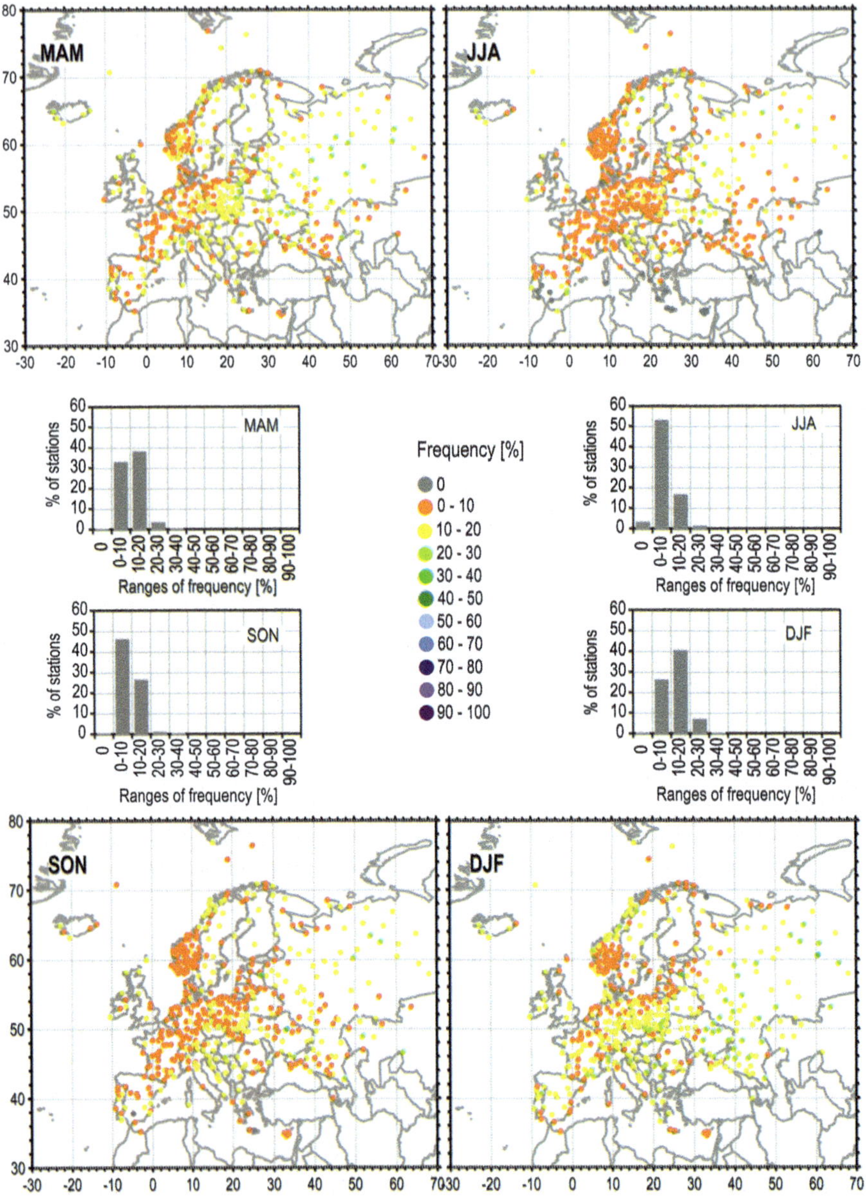

Fig. 3.18 Frequency of warm front extreme precipitation (type Fw): spatial distribution and histograms for December 1950–February 2008. Right closed intervals

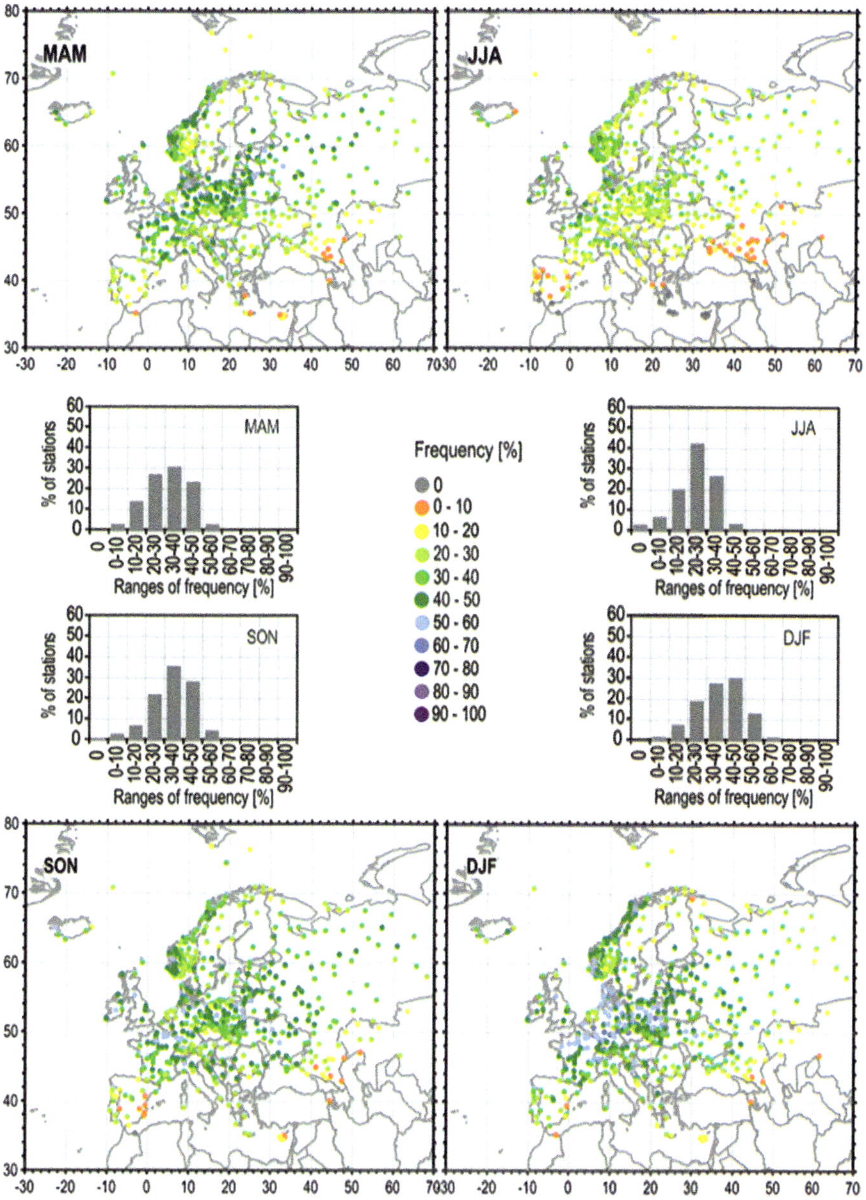

Fig. 3.19 Frequency of extreme precipitation associated with the passage of different fronts (type Ff): spatial distribution and histograms for December 1950–February 2008. Right closed intervals

Table 3.2 Descriptive statistics of the frequency (%) of air mass (type A) and frontal (type F) extreme precipitation in Europe for December 1950–February 2008

ExPT	Season	Average (±SE)	Confidence intervals −95 %	+95 %	Min	Max	SD	CV	ME	Quartiles Lower	Upper
Typ A	MAM	17.6 (±0.5)	16.6	18.6	1.1	77.5	11.8	66.8	13.8	9.6	22.9
	JJA	24.9 (±0.7)	23.5	26.3	3.4	100	16.4	65.9	20.2	13.5	31.3
	SON	16.2 (±0.6)	15.1	17.3	0.0	85.2	12.6	77.5	11.9	7.2	22.4
	DJF	13.9 (±0.5)	12.9	14.9	0.0	62.5	11.7	83.8	10.5	5.8	18.3
Typ F	MAM	82.3 (±0.5)	81.3	83.4	22.5	98.9	11.8	14.3	86.2	76.9	90.4
	JJA	75.1 (±0.7)	73.7	76.5	0.0	96.6	16.4	21.9	79.8	68.8	86.5
	SON	83.8 (±0.6)	82.7	84.9	14.8	100.0	12.6	15.0	88.1	77.6	92.8
	DJF	86.1 (±0.5)	85.1	87.1	37.5	100.0	11.7	13.6	89.5	81.7	94.2

ExPT extreme precipitation type, *MAM* spring, *JJA* summer, *SON* autumn, *DJF* winter, *SE* standard error, *Min* minimum, *Max* maximum, *SD* standard deviation, *CV* variability coefficient, *ME* median

the Bay of Biscay, southeastern North Sea coast, and the coast of the Scandinavian Peninsula) and in the east of the continent.

Along the European lowlands from the central section of the western European coast, protected from the open ocean by the British Isles, eastwards to Poland and on to the Central Russian Upland and the East European Lowland, the proportion of type A in extreme precipitation total falls to 10–20 %. Mätlik and Post (2008) obtained a similar result of about 12 % of convective precipitation in the total extreme rainfall in Estonia. The frequency of extreme precipitation of the air-mass type falls very low (no more than 10 % ExP) around the Gulf of Riga and the Gulf of Finland, at certain stations on the Jutland Peninsula, at the southern tip of the Norwegian portion of the Scandinavian Peninsula, and at certain stations in France, the UK, and in Iceland. The frequency of type A ExP decreases northwards. Overall, at most of the stations the summer frequencies are greater than in other seasons.

In other seasons, air-mass precipitation follows the main features of its summer occurrence patterns. In general, throughout the year, air-mass precipitation accounts for a larger proportion of extreme precipitation at stations located in the south of the continent, especially between the Black Sea and the Caspian Sea (the Caucasus and the Caspian Depression), than anywhere else in Europe. In these areas extreme air-mass precipitation accounts for 30–60 % of the overall ExP frequency in spring and autumn and for 30–50 % in winter. Except summer, the only Mediterranean areas

Table 3.3 Descriptive statistics of the frequency (%) of frontal extreme precipitation types in Europe for December 1950–February 2008

ExPT	Season	Average (±SE)		Confidence intervals −95 %	+95 %	Min	Max	SD	CV	ME	Quartiles Lower	Upper
Typ Fc	MAM	15.2	(±0.4)	14.6	15.9	0.0	40.7	8.0	52.7	14.7	9.3	20.2
	JJA	19.3	(±04)	18.5	20.1	0.0	50.0	9.4	48.7	18.6	12.2	25.8
	SON	17.2	(±04)	16.4	18.0	1.1	51.2	9.3	54.2	15.8	10.4	22.7
	DJF	13.4	(±04)	12.7	14.1	0.0	50.8	8.0	59.6	11.9	7.6	17.6
Typ Fw	MAM	11.3	(±0.2)	10.9	11.7	0.0	30.6	4.7	41.6	10.7	8.1	14.0
	JJA	7.6	(±0.2)	7.2	7.9	0.0	26.9	4.6	61.0	6.9	4.4	9.9
	SON	9.3	(±0.2)	8.9	9.7	0.0	32.2	4.4	47.3	8.8	6.3	11.7
	DJF	12.6	(±0.2)	12.1	13.1	0.0	31.5	5.4	42.9	12.0	8.5	15.8
Typ Ff	MAM	31.5	(±0.5)	30.6	32.5	1.6	55.4	11.0	34.8	32.4	23.5	40.4
	JJA	24.5	(±0.4)	23.6	25.3	0.0	51.4	9.8	40.0	25.3	19.1	31.6
	SON	34.5	(±0.5)	33.6	35.4	0.0	59.3	10.8	31.5	36.1	27.6	42.4
	DJF	37.5	(±0.5)	36.5	38.5	4.2	64.2	11.9	31.7	38.5	29.0	46.2
Typ Fo	MAM	20.8	(±0.6)	19.7	21.9	0.0	62.9	12.5	60.2	17.5	12.2	26.6
	JJA	18.9	(±0.6)	17.8	20.1	0.0	71.4	13.3	70.4	16.0	9.0	27.9
	SON	19.3	(±0.6)	18.2	20.4	0.0	58.9	12.9	66.9	16.0	9.8	26.0
	DJF	19.8	(±0.5)	18.7	20.8	0.0	63.0	12.2	61.8	16.3	10.9	25.8
Typ Fs	MAM	3.0	(±0.1)	2.7	3.2	0.0	12.1	2.8	94.0	2.4	0.0	4.9
	JJA	4.0	(±0.2)	3.7	4.4	0.0	18.1	3.9	97.5	2.9	0.9	6.7
	SON	3.1	(±0.1)	2.9	3.4	0.0	16.7	3.1	99.0	2.2	0.7	5.0
	DJF	2.6	(±0.1)	2.4	2.8	0.0	13.6	2.4	93.4	2.1	0.7	4.1
Typ Fd	MAM	0.6	(±0.1)	0.5	0.7	0.0	14.0	1.3	227.1	0.0	0.0	0.9
	JJA	0.8	(±0.1)	0.7	0.9	0.0	14.1	1.5	187.1	0.0	0.0	1.2
	SON	0.3	(±0.0)	0.2	0.3	0.0	5.1	0.7	245.0	0.0	0.0	0.0
	DJF	0.2	(±0.0)	0.1	0.3	0.0	12.5	0.9	389.8	0.0	0.0	0.0

ExPT extreme precipitation types, *MAM* spring, *JJA* summer, *SON* autumn, *DJF* winter, *SE* standard error, *Min* minimum, *Max* maximum, *SD* standard deviation, *CV* variability coefficient, *ME* median

where air-mass precipitation accounts for more than 50 % of extreme precipitation are the eastern islands of Crete and Cyprus in spring and autumn and the Iberian coast (Mediterranean side) in autumn and winter. Generally, winter is the time when air-mass precipitation in southern Europe is at its least frequent because extensive cyclonic activity dominates this region (Rumney 1968; Barry and Perry 1973; Martyn 2000). For example, during that season, the frequency of air-mass precipitation in the Balkan Peninsula drops to below 10 % ExP. According to Llasat et al. (2005) convective precipitation in the western part of the Mediterranean accounts for more than half of the total annual precipitation.

Outside southern Europe, areas with an increased proportion of air-mass precipitation in overall extreme precipitation include the Scandinavian coast and the southeastern coast of the North Sea, which is explained by the more powerful atmosphere dynamics on contact between land and sea and, in some cases, also by the breeze effect (North Sea), as well as by the orographic barrier of the Scandinavian Mountains. The frequency of air-mass precipitation in these areas ranges from more than 20 % to nearly 60 % ExP depending on the season and location (Fig. 3.12). Along the vast tracts of land spanning the British Islands and the eastern boundary of the continent, air-mass precipitation accounts for no more than 20 % ExP in a year. In spring, at many stations (40 % of stations) air-mass precipitation accounts for 10–20 % ExP whereas in other seasons the number of stations where this precipitation type accounts for less than 10 % ExP is even higher (42 % of stations in spring and 46 % stations in winter).

3.3.2 *Frontal Precipitation (Type F)*

A clear majority of European extreme precipitation is linked to weather fronts, as confirmed by studies on a regional scale (Steinacker 1981; Wilby 1998; Frei et al. 2000; Hagen et al. 2000; Kljun et al. 2001; Pradier et al. 2004; Mätlik and Post 2008). In each season, the average proportion of frontal (type F) precipitation in the number of days with extreme precipitation was several times higher than that of air-mass events (Table 3.2). Frontal precipitation accounts for the largest proportion of extreme precipitation in winter, at 86 % ExP, when cyclonic activity increases over the continent (Martyn 2000). In spring and autumn they account, respectively, for 82 % ExP and 84 % ExP, with the summer number the lowest at 75 % ExP.

In winter, the spatial variability of frontal precipitation in Europe is the lowest. Specifically, the proportion of type F precipitation in the number of days with extreme precipitation varies over the range of 63 % (Table 3.2). Frontal precipitation accounted for between 90 and 100 % ExP at 47 % of stations, the majority of which were located: in Western and Central Europe from France to Poland; at the southern tip of the Scandinavian Peninsula; and in Eastern Europe between the Black Sea and the Bay of Riga. At some French stations, all extreme precipitation events recorded during the study period were associated with weather fronts. Extreme frontal precipitation in Western and Central Europe arrives from the Atlantic Ocean; extreme frontal events in Eastern Europe can also involve low-pressure systems originating in the Mediterranean and the Black Sea, as reported by Mätlik and Post (2008) in their study of synoptic influences on extreme precipitation in Estonia. At 30 % of stations in the remaining parts of Eastern Europe, frontal precipitation accounted for somewhat fewer extreme events, 80–90 % ExP (Fig. 3.13). Overall, it is concluded that a clear majority of extreme precipitation in winter is associated with weather fronts. These numbers only drop to 60–80 % ExP at a few stations located in the south of the continent, along the southeastern coast of the North Sea, and the Scandinavian oceanic coast (Fig. 3.13). Also, stations located in mountainous areas had a lesser frequency of frontal precipitation. The lowest proportion of frontal precipitation in the overall extreme precipitation of this season is accounted by events in the Iberian Mediterranean coast (30–60 % ExP). Elsewhere in this peninsula, wintertime Atlantic cyclones arriving from higher moderate latitudes (Alonso et al. 1994; Romeo et al. 1998) bring frontal precipitation to 80 % ExP.

In transitional seasons, the proportion of frontal precipitation ranges from 80 to 100 % ExP at a majority of stations from the Atlantic islands to Eastern Europe (70 % of stations in autumn, 69 % in spring, and 49 % in summer) (Fig. 3.13). A very high contribution of type F (more than 90 % ExP) is recorded at stations near the southwestern Baltic coast in spring; in France and the western part of Eastern Europe in autumn; and, admittedly at a much smaller number of stations compared to the other seasons, in Estonia, Lithuania, and Latvia, near the Jutland Peninsula, in Iceland and in the British Islands in summer.

Southern Europe stands out with the lowest frequency of frontal extreme precipitation in the whole year. In spring and autumn, this type accounts for just 60–80 % ExP in the Iberian Peninsula and in an area between Italy and the Caspian Sea south of the parallel 48°N. In summer, when the climatic conditions are more conducive to the development of free convection, this zone stretches further north to southern Poland and southern Germany. In that season, the frequency of frontal precipitation in the Mediterranean decreases southwards from just less than 50 % ExP to below 10 % ExP and is the lowest in Europe all year round. This low frequency of type F precipitation in southern Mediterranean is explained by an influence of the northern edge of the subtropical high-pressure zone. Partly corroborating these findings is a study by Millán et al. (2005), who found a low proportion of frontal precipitation around Valencia on the Iberian Peninsula.

Relatively little frontal precipitation, from 50 to 80 % ExP, was also found in the northern and western Scandinavian coast in summer and on southeastern North Sea coast and the Polish Baltic coast in autumn (Fig. 3.13). The lower frequency of type F precipitation along the North Sea coast is explained by a relatively high frequency of air-mass precipitation that develops there in response to increased dynamics of the atmosphere at the contact of land and sea surfaces. Low-pressure systems with their weather fronts that travel through Europe from their areas of cyclogenesis over the Atlantic Ocean and in the Mediterranean gradually evolve and finally disappear. It is therefore to be expected that various types of fronts will occur over the continent at different frequencies and, as a consequence, the frequency of precipitation types over the continent should display some spatial pattern.

3.3.3 Cold Front Precipitation (Type Fc)

In Western Europe dominated by western winds, the largest group of extreme precipitation associated with the passage of cold fronts occurs in summer (19 % ExP on average), when polar-maritime air arrives over the continent from the Atlantic

Ocean, which is cooler at that time of the year. At this time precipitation associated with the passage of a cold front accounts for the largest proportion of extreme precipitation (30–50 % ExP) along a Western European belt stretching from the northwestern tip of the Iberian Peninsula through France, to the Alps, and on to southeastern Europe south of the East European Lowland between the Black Sea and the Caspian Sea (Fig. 3.14).

The mountainous part of Western Europe has a greater number of days with type Fc extreme precipitation then other parts of the continent. This is explained by the existence of an orographic barrier in the form of the Massif Central and the Alps that slow down the speed of cold fronts arriving in the area (Schneider 1996) and intensify the cold front's convection mechanism. Most extreme precipitation events in the Alps are associated with a low-pressure trough in the troposphere and an accompanying surface cold front (Massacand et al. 1998; Pradier et al. 2004). Extreme cold front precipitation in the southeastern part of the continent peaks in summer, which is explained by a pressure distribution pattern over Europe at this time of the year. A large-scale low with its centre located to the southeast of the Caspian Sea causes continental air to flow into the area. Meanwhile, either a powerful Azores high develops a ridge that reaches deep into the continent or several smaller highs developing over continental Europe cause air from the northeast to flow behind the cold fronts, and as this air mass reaches the Caucasus Mountains it is pushed up and generates increased precipitation in the mountain foreland. The contact between these two air masses, especially when the northern or northwestern flows are powerful, contributes to the consolidation of cold fronts linked with cyclones travelling over northern Europe. Sample synoptic charts illustrating the inflow of cold air into the Alps and into an area around the Black and Caspian Seas are shown on Figs 3.15 and 3.16.

In areas just described, cold front extreme precipitation often occurs also in autumn and spring (Fig. 3.14). From autumn to spring, a relatively high frequency (30–40 % ExP) of precipitation associated with cold fronts moving from the northwest and west is also recorded on the Iberian Peninsula, except its Mediterranean coast. An example of a synoptic chart illustrating such situation is provided on Fig. 3.17. In summer, frontal precipitation accounts for between 10 and 30 % ExP in Central Europe, whereas in the other seasons these figures vary between 10 and 20 % ExP. In the same season, most of the stations in the east of the continent record 10–20 % type Fc ExP, whereas in the other seasons, especially in the northern and central sections of this area, cold front precipitation tends to be rare at less than 10 % ExP.

In the Scandinavian Peninsula and in Iceland the number of days with type Fc precipitation displays no obvious seasonality; along the peninsula's western coast the frequency ranges between 10 and 20 % ExP although at the other stations it falls short of 10 % ExP. In winter, the contribution of type Fc precipitation to the number of days with extreme precipitation is the lowest at approximately 13 % (Table 3.3). At that time of the year, type Fc frequency decreases in Western Europe from west and southwest towards the central part of the continent, as it does in Eastern Europe from the south to the north. Relatively low frequencies of wintertime type Fc are also recorded in areas located between Estonia and the Ukraine (Fig. 3.14).

3.3.4 Warm Front Precipitation (Type Fw)

Extreme precipitation associated with warm fronts is less frequent in Europe than cold front precipitation. Its average contribution to the overall number of days with extreme precipitation peaks in winter (13 % ExP on average; Table 3.3), when the continental landmass is at its coldest, and in spring, when the temperature of the landmass remains colder than that of the surrounding oceanic and sea waters. In winter, the highest frequencies of warm front precipitation of no more than 20–30 % ExP are recorded at a relatively small number of stations (approximately 10 % of stations) in central parts of Southern Europe, in Eastern Europe, especially its southern part, and at isolated stations of Central Europe and the Scandinavian Peninsula (Fig. 3.18).

In spring, a zone with a similar frequency of type Fw extreme precipitation includes stations located along a belt stretching from the eastern Polish border through the middle of Eastern Europe towards the northeast (only 4 % of stations). Elsewhere, in both winter and spring, warm front precipitation accounts for no more than 20 % ExP, and less than 10 % ExP at the majority of stations in Western Europe between Frisian Islands and the Bay of Finland plus, in spring only, southwestern Europe (Fig. 3.18). The frequency of extreme precipitation accounted for by type Fw does not exceed the 10 % ExP mentioned previously in winter at 35 % of stations and in spring at 44 % of stations. In spring, low proportions of type Fw precipitation are also recorded at stations in the southern part of Eastern Europe, especially between the Black Sea and the Caspian Sea (Fig. 3.18).

In summer and autumn, the most numerous is a group of stations where type Fw precipitation remains within 10 % ExP (71 % of stations and 62 % of stations, respectively); these are located in Western Europe, southern Europe, and at the southern tip of the Norwegian Scandinavian Peninsula. In the Mediterranean and in a southern part of the Iberian Peninsula extreme precipitation of this type did not occur at all during the study period in summer. In Eastern Europe the frequency of ExP developing on warm fronts in summer and autumn remained within 10–20 % ExP.

3.3.5 *Precipitation Linked to the Passage of Several Fronts (Type Ff)*

Precipitation associated with the passage of several different weather fronts stands out from the frontal precipitation types by their highest proportion of the days with extreme precipitation. On average, this type of precipitation ranges between 24 % of all extreme events in summer and 37 % in winter, when the speed of movement of fronts over Europe is at its highest (Table 3.3). The high average seasonal frequencies of occurrence of type Ff in comparison with other precipitation types and their relatively low standard deviation values suggest moderate variability in the frequency of this type of precipitation in Europe. Indeed, fundamental characteristics of the spatial distribution of type Ff occurrence display no significant variability during the year.

Seasonality is found within smaller regions, especially in the continental core of Western Europe, in Central Europe, and in southern Europe (Mediterranean and areas between the Black Sea and the Caspian Sea). Extreme daily totals resulting from the passage of several different fronts are the most frequent in Western Europe, specifically in the Atlantic islands, along the western continental coast between the Bay of Biscay and the Bay of Finland, as well as the western Scandinavian coast, especially in its southern section. In these areas the proportion of type Ff precipitation peaks in winter at 50–60 % ExP, after which it falls in autumn to approximately 40–50 % ExP and reaches its lowest point in summer, when it rarely exceeds 40 % ExP (Fig. 3.19).

The autumn and winter maximums of the frequency of precipitation associated with the passage of various weather fronts along the western continental coast (Fig. 3.20) are explained by the speed of the fronts, which is higher than in the warm half of the year because of higher pressure gradients and intensive cyclogenesis, which, in turn, stems from stronger thermal gradients in that half of the year in the Northern Hemisphere.

In summer, precipitation associated with the passage of several different weather fronts accounts for 40–50 % ExP at isolated few stations located in the central section of the western European coast (North Sea) and at some stations in the British Islands (Fig. 3.19). The most numerous group of stations that record type Ff precipitation at 20–30 % ExP in that season is scattered on the continent north of the parallel 45°N. In the far north of the continent, this frequency drops again to 10–20 % ExP. In Mediterranean islands and peninsulas, as well as the southern part of Eastern Europe, type Ff precipitation is the most rare throughout the year. In summer, this type of precipitation does not occur in the far south of the continent.

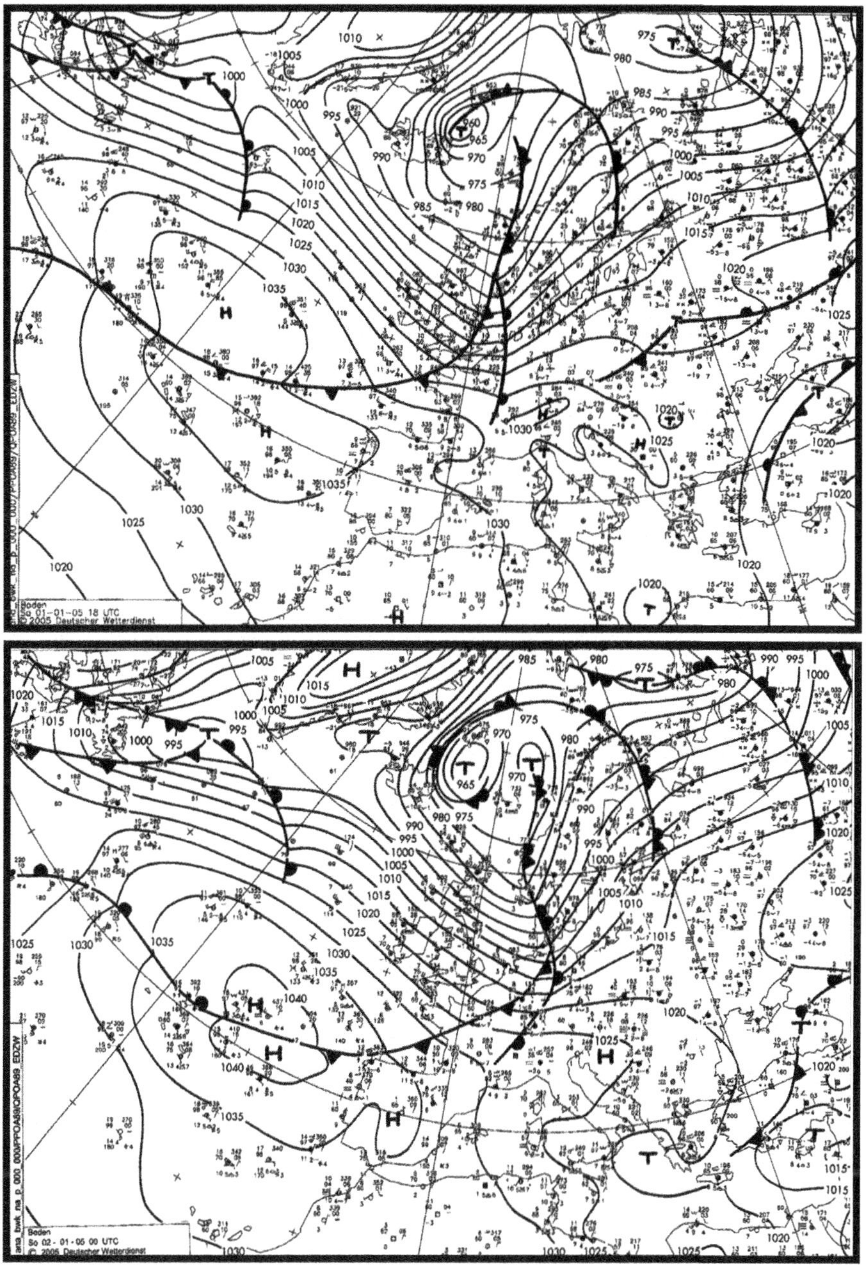

Fig. 3.20 System of meteorological (various) fronts moving across Western Europe. *Upper* 01 January 2005, 18 UTC. *Lower* 02 January 2005, 00 UTC (*Source:* http://www.wetter3.de)

3.3.6 Occluded Front Precipitation (Type Fo)

The frequency of extreme precipitation associated with an occluded front (type Fo) follows a zonal pattern with little seasonal change. Depending on the season, type Fo precipitation accounts for approximately between 19 and 21 % ExP (Table 3.3). This proportion tends to be the highest in the north of the continent (the Baltic Sea and to the east of it) and on the Atlantic islands, where it exceeds 30 % ExP in every season and often more than 40 % ExP seasonally (Fig. 3.21). This high frequency of type Fo precipitation in northern Europe is linked with the life cycle of cyclones over the northern Atlantic Ocean and their travel over Europe. As cyclones transit from over the smooth sea surface to the rough continent, they achieve a mature stage in their development, which manifests itself in the formation of an occlusion. In the central section of the southern end of the Scandinavian Peninsula, this effect is compounded by an orographic component producing an area where occluded fronts are responsible for more than half of all extreme precipitation days in a year. As weather fronts are slowed down in their movement by orography, the process of occlusion development accelerates. Figure 3.22 presents a synoptic chart illustrating an occluded front over a central section of the southern end of the Scandinavian Peninsula.

In a central part of the continent, mainly covering Poland, type Fo accounts for 10–20 % ExP in a year. In summer and in spring, this zone extends further west to the continental coast. Similar proportions are also recorded at relatively coherent groups of stations in the southern part of Eastern Europe; the northern limit of their occurrence varies from season to season. Scattered across southern Europe a number of stations also had the same percentage of occluded front precipitation (10–20 %) (Fig. 3.21).

Southern Europe has the lowest overall frequency of type Fo precipitation (less than 10 % ExP). In summer, Mediterranean stations south of the parallel 40°N did not record any precipitation extrema generated on occluded fronts. In winter and autumn, the lowest frequency of type Fo precipitation is recorded in southwestern Europe and Western Europe, whereas in spring it is found on the Iberian Peninsula and in areas between the Black Sea and the Caspian Sea and the northern edge of southern Europe (Fig. 3.21).

The frequency of extreme precipitation generated on occluded fronts increases from the south to the north, with the exception of the western Scandinavian coast where it remains within 20 % ExP year round.

3.3.7 Precipitation Linked to a Stationary Front (Type Fs) and Discontinuity Line (Type Fd)

Extreme precipitation associated with a stationary front and a discontinuity line is among the least frequent types in Europe. On average, type Fs precipitation constitutes between about 3 % ExP in spring and autumn and 4 % ExP in summer; this

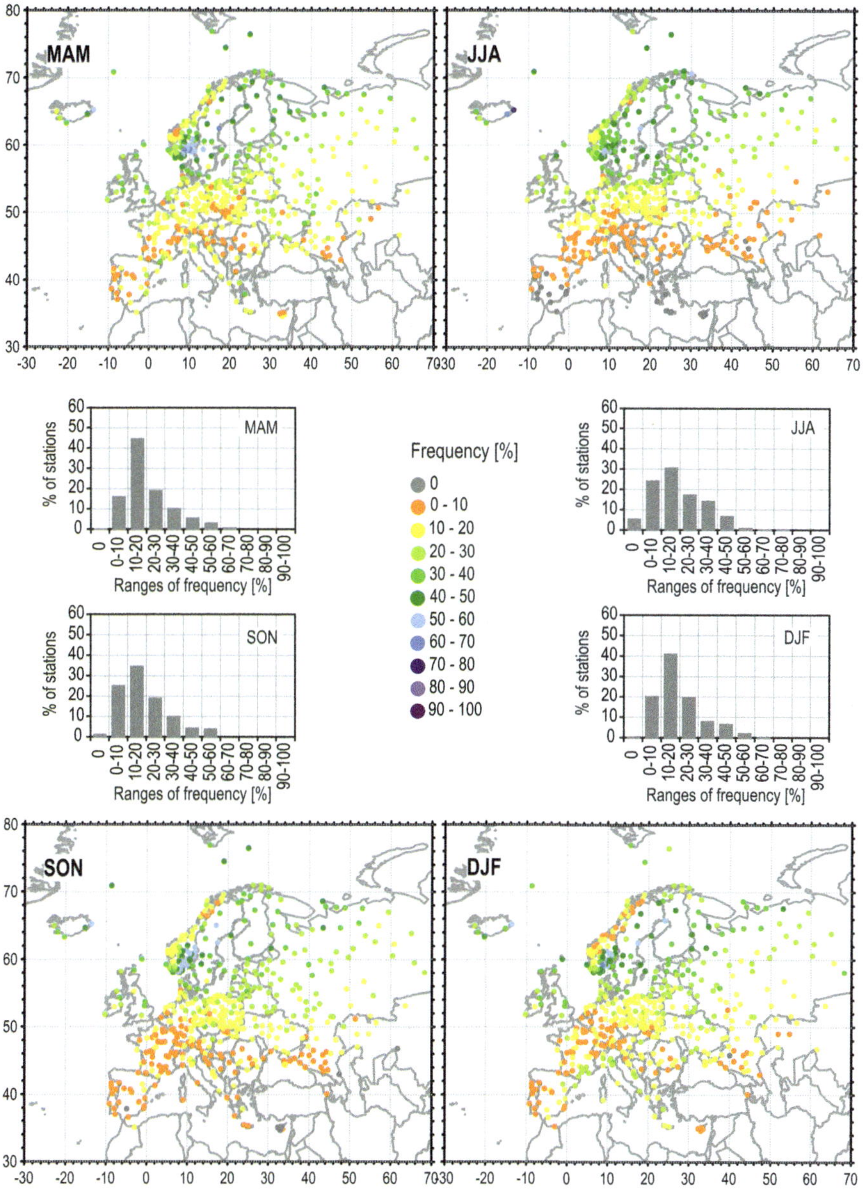

Fig. 3.21 Frequency of occluded front extreme precipitation (type Fo): spatial distribution and histograms for December 1950–February 2008. Right closed intervals

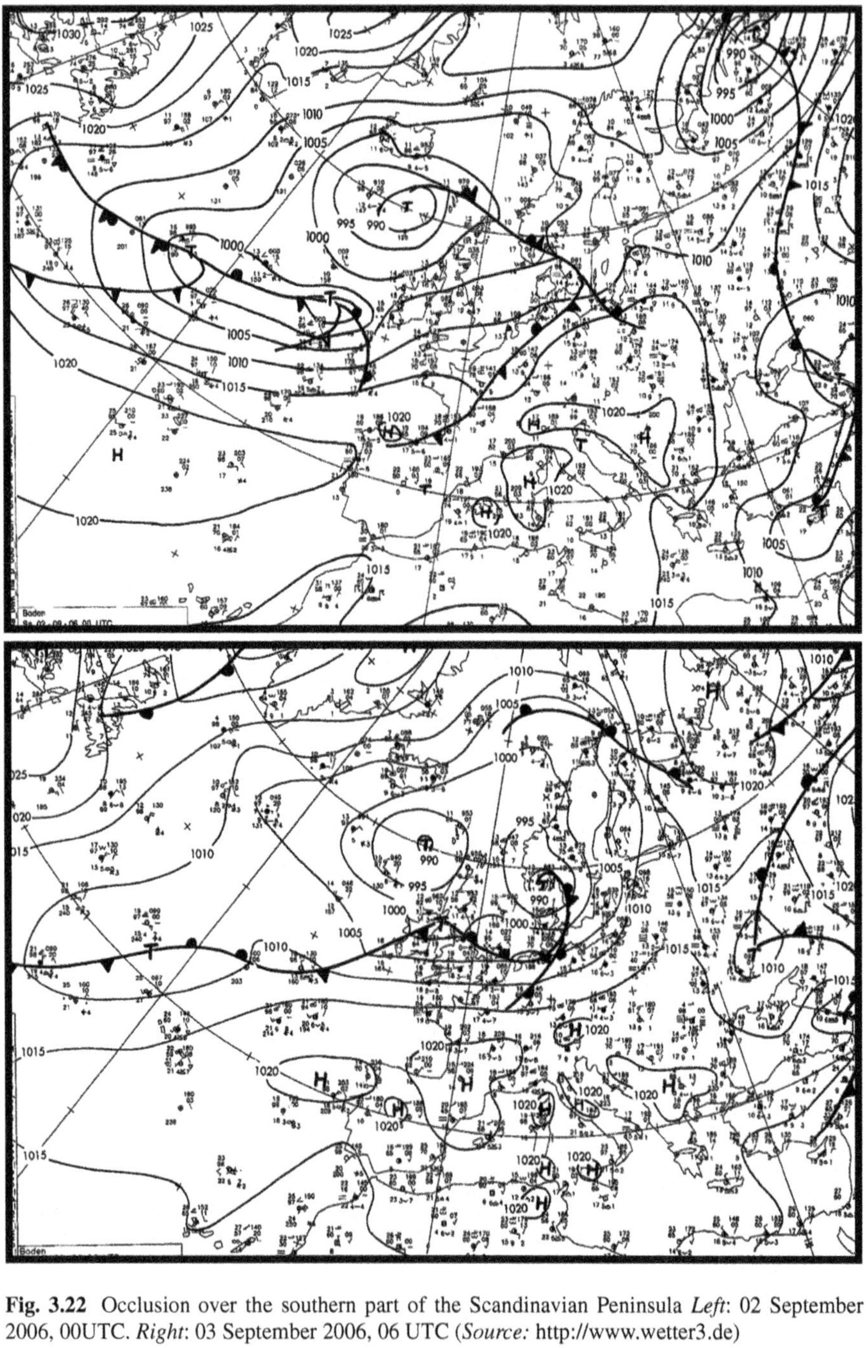

Fig. 3.22 Occlusion over the southern part of the Scandinavian Peninsula *Left*: 02 September 2006, 00UTC. *Right*: 03 September 2006, 06 UTC (*Source:* http://www.wetter3.de)

frequency peaks at between approximately 12 % ExP in spring and approximately 18 % ExP in summer (Table 3.3). There are very few stations, especially in a central part of Western Europe and in Central Europe, where the proportion of type Fs varies between 10 and 20 % ExP. In summer, the number of stations recording type Fs precipitation with the frequency of 10–20 % ExP is the largest at 8 % of stations, whereas in the remaining seasons this group remains within 4 % of stations. Numerous stations in the south and north of the continent record no precipitation of type Fs (28 % of stations in spring, 21 % of stations in summer, 23 % of stations in autumn and winter) (Fig. 3.23).

Extreme precipitation associated with a discontinuity line is particularly rare in Europe. Its average contribution to the number of days with extreme precipitation does not exceed 1 % in any season (Table 3.3). Most of the stations did not record this type during the study period at all (73 % of stations in spring, 63 % of stations in summer, 81 % of stations in autumn and 83 % of stations in winter). The only area where type Fd precipitation is slightly more frequent in spring and summer is a wide band along the western European coast and in the southern part of Eastern Europe (Fig. 3.24).

3.4 Conclusions

Two major origin-based types of precipitation were identified, that is, air-mass (type A) and frontal (type F). They were distinguished considering basic processes leading to the development of precipitation (free convection, forced convection, and less vertical overrunning movements, such as observed on the warm front) and their identification potential on synoptic maps. The types were further subdivided taking into account dynamic processes depending on the type of the weather front, and this produced precipitation associated with the passage of different fronts (type Ff), precipitation associated with the passage of a cold front (type Fc), precipitation associated with the passage of a warm front (type Fw), precipitation associated with an occluded front (type Fo), precipitation associated with a stationary front (type Fs), and precipitation associated with a discontinuity line (type Fd).

One of the fundamental differences between air-mass precipitation and frontal precipitation is the size of the precipitation zone. An analysis of the spatial distribution of the major origin-based types of extreme precipitation (type A and type F) demonstrated that during the whole year air-mass precipitation was simultaneously (i.e., on the same precipitation day) recorded by a much smaller number of stations than was frontal precipitation. Two or less stations simultaneously recording air-mass precipitation accounted for about 38–40 % of days with extreme precipitation whereas the same figure for frontal precipitation was just 15 %.

The frequency of origin-based precipitation types in Europe follows clear spatial and seasonal regularities. These regularities depend on the varying pace of the cyclone life cycle during the year and on ground relief. Ground relief also impacts the pace of cyclogenesis, thus influencing the spatial variability of origin-based types of extreme precipitation.

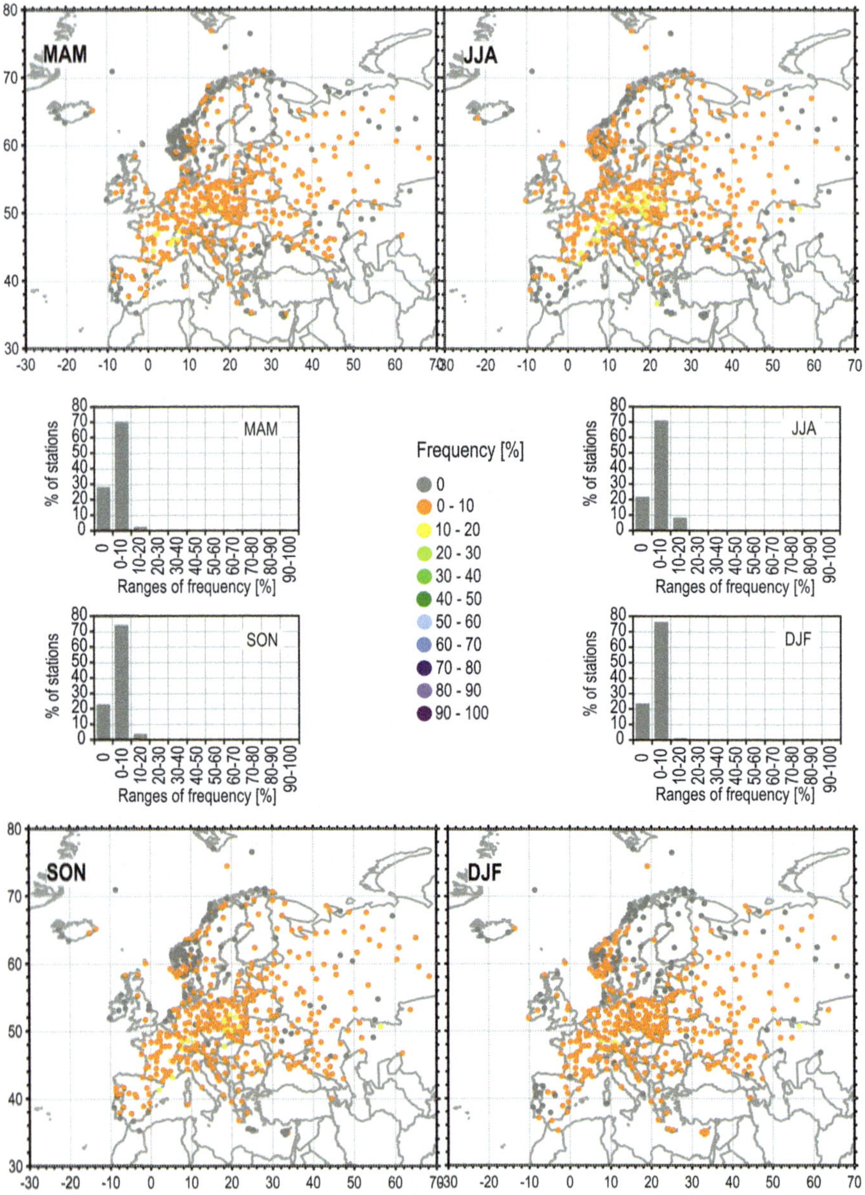

Fig. 3.23 Frequency of stationary front extreme precipitation (type Fs): spatial distribution and histograms for December 1950–February 2008. Right closed intervals

The average proportion of air-mass precipitation (type A) in the overall number of days with extreme precipitation ranged from 14 % in winter to 25 % in summer. In all seasons, type A extreme precipitation is the most frequent in the south of the continent. In summer, when conditions for the development of convection in this area are favourable, it accounts for more than 50 % of extreme precipitation. This

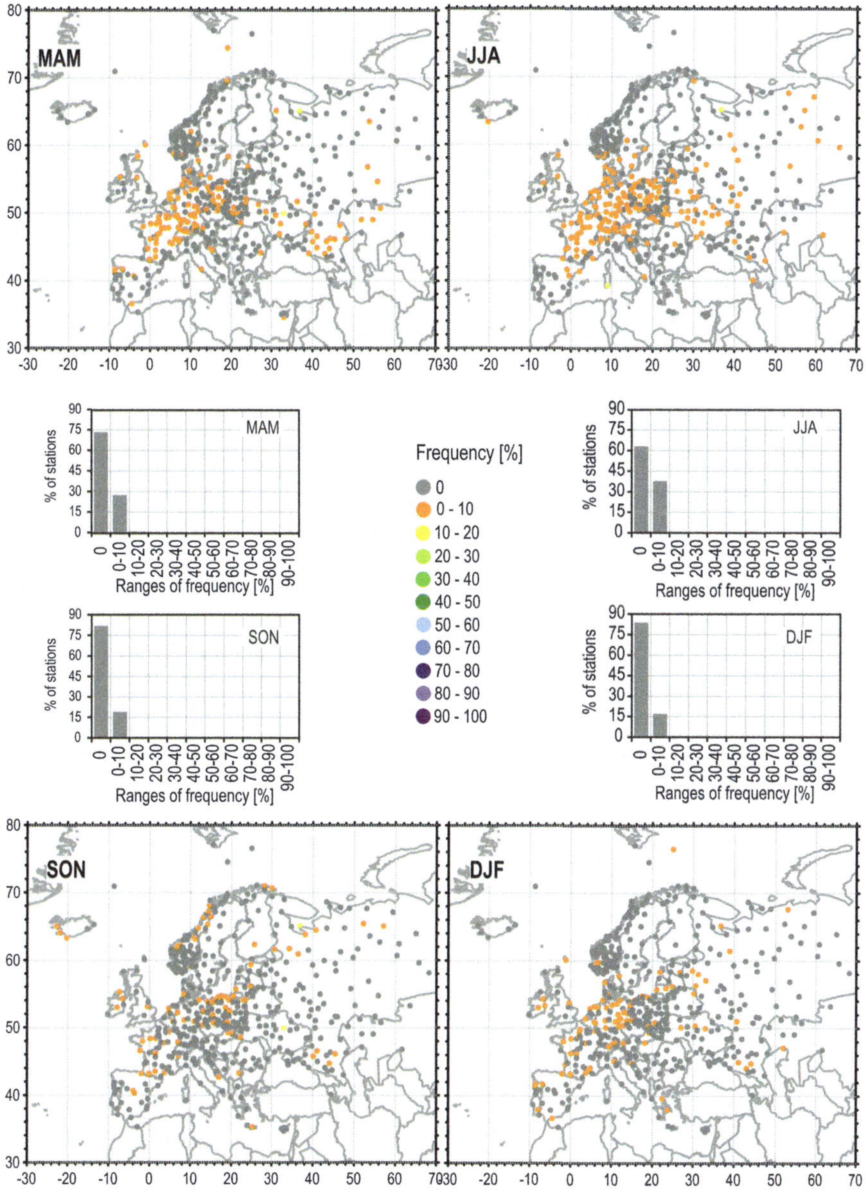

Fig. 3.24 Frequency of extreme precipitation associated with discontinuity line (type Fd): spatial distribution and histograms for December 1950–February 2008. Right closed intervals

frequency declines northwards. Elsewhere in the continent it only reaches 10–20 % of extreme precipitation in summer, while in other seasons its share typically drops to no more than 10 % ExP. Type A precipitation is somewhat more frequent in upland and mountainous areas and along the southeastern North Sea coast (~30–50 % ExP), which is explained by orographic barrier effect or by sea–land transition effect that boosts atmosphere dynamics and leads to the convective movement which produces heavy precipitation.

In Europe most of the days with extreme precipitation are associated with the occurrence of weather fronts. In every season, the average proportion of frontal precipitation in the overall number of days with extreme precipitation was several times higher than that of air-mass precipitation. Type F precipitation accounts for the largest portion of extreme precipitation in winter (on average, 86 % ExP), as cyclonic activity increases on the continent, and for the lowest in summer (on average, 75 % ExP), when conditions favourable for free convection are far more frequent. At the scale of the whole of Europe the lowest frequency of frontal precipitation is found in southern Europe, especially in summer, when this type accounts for about 10 % ExP. Both the highest frequencies of air-mass precipitation and the lowest frequencies of frontal precipitation that characterise this area in summer are an effect of the influence of the northern edge of the subtropical anticyclone zone. In winter, the frequency of frontal precipitation in southern Europe increases with the area's intensive seasonal cyclonic activity. Frontal precipitation accounts for between 50 and 80 % of extreme precipitation along the southeastern North Sea coast as well as along the northern and the western Scandinavian coast in summer and on the Polish Baltic coast in autumn. The lower frequency of type F extreme precipitation and the higher frequency of type A along the southeastern North Sea coast in summer are partly linked to a strong breeze effect in that area.

Extreme precipitation associated with the passage of a cold front (type Fc) is the most frequent in summer, when Europe receives cooler air from over the Atlantic Ocean. In all seasons, the highest frequencies of this origin type of extreme precipitation occurs in Western Europe (~30–50 % ExP in summer and 30–40 % in other seasons), along a belt extending from the northwestern tip of the Iberian Peninsula to France and the Alps, as well as in southeastern Europe. The higher number of days with type Fc extreme precipitation in Western Europe's mountainous areas is linked to the orographic barriers of the Massif Central and the Alps that slow down and intensify the oncoming cold fronts. The summer maximums of type Fc extreme precipitation in the southeastern part of the continent are linked with a vast low-pressure system centred to the southeast of the Caspian Sea, which causes advection of continental air over the heated ground surface and with a strong anticyclone activity, as either the Azores high extending deep into the continent with its wedge or smaller highs forming over the continental Europe cause an inflow of air from the northwest. In Central Europe, the frequency of type Fc precipitation reaches 20–30 % ExP only in summer, in other seasons recording lower figures.

Extreme precipitation associated with a warm front (type Fw) occurs in Europe less frequently than that of cold front precipitation. Its contribution to the overall number of days with extreme precipitation peaks in winter (on average, 13 % ExP), when the ground is the coldest, and in spring, when the temperature of the landmass, cooled down during winter, remains lower than that of the surrounding seas. In general, the frequency of type Fw precipitation increases towards the continental part of Europe, and the rate of this change is the highest in winter and the lowest in summer. In winter, the highest frequencies of type Fw precipitation exceed 20 %

ExP in the northern part of southern Europe, in Eastern Europe, at isolated stations of Central Europe and the Scandinavian Peninsula, and, in spring, also in Eastern Europe.

Extreme precipitation associated with the passage of various weather fronts (type Ff) is the most frequent of all frontal precipitation types. On average it constitutes between about 24 % ExP in summer and about 37 % ExP in winter, when the speed of front travel across Europe is the highest. The main features of the spatial distribution of type Ff remain largely unchanged during the year. Type Ff extreme precipitation most frequently occurs in Western Europe, including the Atlantic islands; the western continental coast between the Bay of Biscay and Bay of Finland; and along the western Scandinavian coast, especially its southern section. In this area the proportion of type Ff precipitation in winter ranges from 50 to 60 % ExP, whereas in summer it rarely exceeds 40 % ExP. In winter, the zone of high frequency of type Ff precipitation reaches deeper into the continent up to and including Central Europe. The autumn/winter maxima of extreme precipitation associated with the passage of various weather fronts along the western continental coast are explained by the speed of atmospheric fronts, which is higher than in the warm half of the year. This effect is a result of higher pressure gradients and of intensive cyclogenesis, which, in turn, is caused by stronger thermal gradients in this half of the year in the Northern Hemisphere. In summer, Mediterranean stations recorded no extreme precipitation associated with the passage of various fronts during the study period.

The proportion of ExP associated with an occluded front (type Fo) in the overall number of days with extreme precipitation follows a regular zonal pattern across Europe. During the year, main features of this pattern vary little, as the proportion of type F precipitation remains relatively stable at approximately 20 % ExP. It is the most frequent in northern Europe (more than 40 % ExP), except the western Scandinavian coast, where its proportion in the overall number of days with extreme precipitation during a year does not exceed 20 % ExP. The central part of the southern tip of the Scandinavian Peninsula is especially prone to the occurrence of type Fo precipitation, which accounts for 50 % ExP. This local peak of type Fo occurrence has an orographic background. Upon reaching the Scandinavian Mountains weather fronts slow down, which accelerates the occlusion development process. The frequency of type Fo precipitation gradually decreases from the north to the south. In summer, many southern European stations did not record any type Fo precipitation during the study period.

Extreme precipitation associated with a stationary front (type Fs) and a discontinuity line (type Fd) is among the least frequent in Europe. Type Fs accounts on average for between approximately 3 % ExP (in spring and in autumn) and 4 % ExP (in summer) annually, whereas type Fd does not exceed 1 % ExP even in a single season. Type Fs precipitation is the most frequent in Western Europe and in Central Europe (~10–20 % ExP at individual stations), but numerous stations in southern and northern Europe record no such events. Summer is the peak season of type Fd precipitation, which is rare in winter. In summer, type Fd precipitation is concentrated mainly in Western Europe and in Central Europe.

References

Aguado E, Burt JE (1999) Understanding weather and climate. Prentice-Hall, Upper Saddle River

Alonso S, Portela A, Ramis C (1994) First considerations on the structure and development of the Iberian thermal low. Ann Geophys 12:457–468

Barry RG, Carleton AM (2001) Synoptic and dynamic climatology. Routledge, London

Barry RG, Chorley RJ (1998) Atmosphere, weather & climate. Routledge, London

Barry RG, Perry AH (1973) Synoptic climatology. Methods and applications. Methuen, London

Berry G, Jacob CH, Reeder M (2011) Recent global trends in atmospheric fronts. Geophys Res Lett 38, L21812

Blüthgen J (1966) Allegemeine klimageographie. In: Obst E (ed) Lehrbuch der allegemeine Geographie 2. de Gruyter, Berlin

Chomicz K (1971) Struktura opadów atmosferycznych w Polsce (The structure of precipitation in Poland). Prace Państwowego Instytutu Hydrologiczno-Meteorologicznego 101:25–67

Chromow SP (1977) Meteorologia i klimatologia (Meteorology and Climatology). PWN, Warszawa

Djurić D (1994) Weather analysis. Prentice Hall, Englewood Cliffs

Doswell CA (1993) Flash-flood producing convective storms: current understanding and research. In: Report on the proceedings of the US-Spain workshop on natural hazards, Barcelona, 8–11 June 1993

Federico S, Avolio E, Bellecci C, Lavagnini A, Colacino M, Walko RL (2008a) Numerical analysis of an intense rainstorm occurred in southern Italy. Nat Hazards Earth Syst Sci 8:19–35

Federico S, Avolio E, Pasqualoni L, Bellecci C (2008b) Atmospheric patterns for heavy rain events in Calabria. Nat Hazards Earth Syst Sci 8:1173–1186

Font Tullot I (1983) Climatología de España y Portugal. Instituto Nacional de Meteorología, Apartado

Frei CH, Davies HC, Gurtz J, Schär CH (2000) Climate dynamics and extreme precipitation and flood events in Central Europe. Integr Assess 1:281–299

Hagen M, Schiesser H-H, Dorninger M (2000) Monitoring of mesoscale precipitation systems in the Alps and the northern Alpine foreland by radar and rain gauges. Meteorol Atmos Phys 72:87–100

Hand WH, Fox NI, Collier CG (2004) A study of twentieth-century extreme rainfall events in the United Kingdom with implications for forecasting. Meteorol Appl 11:15–31

Hobbs PV (1981) Mesoscale structure in mid-latitude frontal system. In: Proceedings of the IAMAP symposium nowcasting: mesoscale observations and short-range prediction. European Space Agency Publications SP–165:29–36

Houze RA Jr, Hobbs PV (1982) Organization and structure of precipitating cloud systems. Adv Geophys 24:225–315

Kane RJ, Chelius CR, Fritsch JM (1987) Precipitation characteristics of mesoscale convective weather systems. J Clim Appl Meteorol 26:1345–1357

Karagiannidis AF, Karacostas T, Maheras P, Makrogiannis T (2012) Climatological aspects of extreme precipitation in Europe, related to mid-latitude cyclonic systems. Theor Appl Climatol 107:165–174

Kljun N, Sprenger M, Schär C (2001) Frontal modification and lee cyclogenesis in the Alps: a case study using the ALPEX reanalysis data set. Meteorol Atmos Phys 78:89–106

Kożuchowski K (1998) Atmosfer, klimat, ekoklimat (The atmosphere, climate and eco-climate). Wydawnictwo Naukowe PWN, Warszawa

Little MA, Rodda HJE, McSharry PE (2008) Bayesian objective classification of extreme UK daily rainfall for flood risk applications. Hydrol Earth Syst Sci Discuss 5:3033–3060

Llasat MC, Ramis C, Barrantes J (1996) The meteorology of high intensity rainfall events over the West Mediterranean region. Remote Sens Rev 14:51–90

Llasat MC, Rigo T, Ceperuelo M, Barrera A (2005) Estimation of convective precipitation: the meteorological radar versus an automatic rain gauge network. Adv Geosci 2:103–109

Martyn D (2000) Klimaty kuli ziemskiej. Wydawnictwo Naukowe PWN, Warszawa

Mason BJ (1955) The physics of natural precipitation processes. Arch Meteorol Geophys Bioklimatol 8:159–179

Mason BJ (1957) The physics of clouds. Clarendon, Oxford, p s. 481

Massacand AC, Wernli H, Davies HC (1998) Heavy precipitation on the Alpine southside: an upper-level precursor. Geophys Res Lett 25:1435–1438

Mätlik O, Post P (2008) Synoptic weather types that have caused heavy precipitation in Estonia in the period 1961–2005. Est J Eng 14(3):195–208

Millán MM, Estrela MJ, Miró J (2005) Rainfall components: variability and spatial distribution in a Mediterranean area (Valencia region). J Climate 18:2682–2705

Niedźwiedź T (2003) Słownik meteorologiczny (Meteorological dictionary). Polskie Towarzystwo Geofizyczne, IMGW, Warszawa

Okołowicz W (1969) Klimatologia ogólna. PWN, Warszawa

Pradier S, Chong M, Roux F (2004) Characteristics of some frontal stratiform precipitation events south of the alpine chain during MAP. Meteorol Atmos Phys 87:197–218

Raddatz RL, Hanesiak JM (2008) Significant summer rainfall in the Canadian Prairie Provinces: modes and mechanisms 2000–2004. Int J Climatol 28:1607–1613

Ramis C (1995) Las observaciones de la atmósferalibre en Mallorca: una breve historia y algunos resultados. Revista de Ciència 17:41–58

Riosalido R (1990) Characterization of mesoscale convective systems by satellite pictures during PREVIMET MEDITERRANEO – 89', Segundo Simposio Nacional de Predicción, Instituto Nacional de Meteorología, Apartado 285, 28071, Madrid

Romeo R, Guijarro JA, Ramis C, Alonso S (1998) A 30-year (1964–1993) daily rainfall data base for the Spanish Mediterranean regions: first exploratory study. Int J Climatol 18:541–560

Rumney GR (1968) Climatology and the world's climates. Macmillan, New York

Schneider SH (1996) Encyclopedia of climate and weather, 2-vols. Oxford University Press, New York, p 875

Steinacker R (1981) Analysis of the temperature and wind field in the Alpine region. Geophys Astrophys Fluid Dyn 17:51–62

Strangeways I (2007) Precipitation. Theory, measurements and distribution. Cambridge University Press, New York

Trewartha GT (1968) An introduction to climate. McGraw-Hill, New York

Wheeler DA (1988) The Barcelona storm: 1–5th October, 1987. J Meteorol 13:78–85

Wheeler DA (1991) Majorca's severe storms of September 1989: a reminder of Mediterranean uncertainty. Weather 46:21–26

Wheeler DA (1995) Heavy rain in Catalonia, October 1994. Weather 50:362–367

Wilby RL (1998) Modelling low-frequency rainfall events using airflow indices, weather patterns and frontal frequencies. J Hydrol 212-213:380–392

Chapter 4
Regionalisation of Air-Mass and Frontal Precipitation Occurrence in Europe

Abstract Spatial and seasonal variability in precipitation-prone factors in Europe leads to a variable frequency of each origin-based extreme precipitation type. This chapter synthesises the results of the occurrence of each of these. Cluster analysis of *k*-means method and carefully selected grouping variables were used to identify six groups of stations (regional groups) in each season characterised by different patterns of occurrence of origin-based extreme precipitation types. A clear spatial order of their occurrence suggests an existence of regularities that govern the occurrence of origin-based precipitation types in Europe. Some of the groups involve scattered stations, but such distribution, far from random, is a result of specific local influences on the processes leading to the development of extreme precipitation and on their volume. In summer, air-mass precipitation dominates three regional groups in Sothern Europe. In the southernmost of them the air-mass type accounts for nearly 70 % of extreme precipitation. Precipitation associated with the passage of different fronts represents the largest proportion of extreme precipitation in winter, when cyclone travel reaches its highest speeds.

Keywords Extreme precipitation • Precipitation regions • Cluster analysis precipitation types • Europe

As demonstrated in Chap. 3, Air-Mass and Frontal Precipitation Types in Europe, the frequency of the occurrence of origin-based extreme precipitation types (ExPTs) is characterised by spatial regularities and seasonal changes. This chapter distils these regularities by grouping together stations, and areas they represent, with similar occurrence structures. The term "occurrence structure" refers to the pattern of the frequency of all the variables used in the grouping process.

E. Łupikasza, *The Climatology of Air-Mass and Frontal Extreme Precipitation*, Springer Atmospheric Sciences, DOI 10.1007/978-3-319-31478-5_4

4.1 Method for Regionalisation of the Types of Extreme Precipitation Origin

Grouping the stations according to the structure of the types of extreme precipitation origin in Europe was done by cluster analysis of the *k*-mean method. Cluster analysis is a tool for exploratory data analysis aimed at sorting a set of objects in a way that the degree of association between two objects is maximal if they belong to the same group and minimal otherwise. Before a cluster analysis is started, the variables describing the objects must be selected carefully, because they form the basis for the clustering process (Stanisz 2006, 2007). A cluster analysis may be performed using several methods: agglomeration, *k*-means clustering, and grouping of objects and attributes (Stanisz 2007). The nonhierarchical *k*-means method, which was developed by MacQueen in 1967, is the most frequently used taxonomic grouping method, which involves moving objects between the number of clusters identified by the user to minimise intragroup variation and maximise intergroup variation. The method is an iterative procedure, which means that with each subsequent iteration some objects are moved to other clusters. As a rule, the method is used with no a priori hypotheses, at exploratory phases of research. An inconvenience of the *k*-means method is the need to take a subjective decision on the number of clusters to be formed. The method has been widely applied in climatology studies (Gadgil and Joshi 1983; Bartholy 1992; Dorling et al. 1992; Bednorz et al. 2003).

In the grouping process, the frequencies of the extreme precipitation types distinguished were used as the grouping variables describing the objects (weather stations). Because both the occurrence and amount of precipitation in Europe are characterised by a high diversity, use was made of additional variables: number of wet days with precipitation ≥1 mm (NWD) and daily mean from maximum precipitation (TOTExP), calculated using all the extreme precipitation events selected in each season for a given station throughout the study period. Based on the number of wet days with precipitation ≥1 mm, the threshold values required to distinguish extreme precipitation were determined (see Sect. 2.4). The frequency of days with such precipitation determines, at the same time, the number of extreme precipitation events identified. To choose the grouping variables (origin-based extreme precipitation types, daily mean extreme precipitation, number of wet days with precipitation ≥1 mm) that best characterise the objects (weather stations), cluster analysis was carried out several times, using various configurations (variants) of the variables:

- *Variant 1*: all types of extreme precipitation origin (type A, type Ff, type Fc, type Fw, type Fo, type Fs and type Fd), and additionally, the cumulative frequency of all frontal types (F = Ff + Fc + Fw + Fo + Fs + Fd).
- *Variant 2*: selected types of extreme precipitation origin (type A, type Ff, type Fc, type Fw, type Fo), excluding infrequent types (type Fs, type Fd).
- *Variant 3*: basic types of extreme precipitation origin (type A, type F – cumulative frequency of frontal types).
- *Variant 4*: selected frontal types (type M, type Ff, type Fc, type Fw, type Fo), excluding infrequent types (type Fs, type Fd) plus daily mean extreme precipitation (TOTExP), and number of wet days with precipitation ≥1 mm (NWD).

All the foregoing sets of variables were subjected to a grouping procedure, using, successively, each of the methods for identifying initial centres of clusters, that is, the method for maximising the initial distances between clusters, fixed interval method, and *n*-first observation method. Every time, the grouping was started by dividing the objects into three clusters and was continued, each time, by increasing the number of clusters by 1. The procedure was discontinued when the objects were divided into eight clusters. The calculations were conducted separately for each season of the year.

The choice of variables best describing the differences in ExPT occurrence in Europe was made using the results of the analysis of variance (*F* test, intracluster variability, intercluster variability). Analysis of variance permits identifying the variables being the main criterion for distinguishing the different clusters. Furthermore, based, above all, on the knowledge (as presented in the previous chapter) of the regularities in the spatial distribution of the individual origin-based extreme precipitation types in Europe and its seasonal variability, as well as the knowledge of the spatial variability of the other variables used in the cluster analysis, I selected the best (according to my subjective judgement) results of the grouping in terms of the number of clusters and preliminary cluster centre identification method. Thus, the best results of the grouping were selected based on an analysis of the spatial distribution of the clusters distinguished and comparing the values of the grouping variables averaged for each cluster. In effect, variant 4 of the set of variables was selected (type A, type Ff, type Fc, type Fw, type Fo, TOTExP, NWD), the initial cluster centres were determined using the distance sorting method, and the objects were selected using fixed intervals.

As a result, in each season, six clusters describing the key characteristics of the spatial diversity of the ExPT structure in Europe were distinguished. The distribution of the stations that form the individual clusters does not meet the spatial continuity criterion, which should characterise regional units. For this reason, referring to them as "regions" might raise justified objections. In fact, the clusters are typological units differing in terms of the ExPT occurrence structure. However, in the present study, the term "type" has already been used twice in other contexts (types of extreme precipitation, types of synoptic situations/circulation types). Introducing another typology with respect to the results of the grouping would most probably lead to terminological confusion. Despite the spatial noncontinuity of the clusters, the groups distinguished clearly describe regional characteristics of the aforementioned structure of ExPT occurrence on a continental scale, as a result of which they are referred to as "*regional groups*" (RGs). The spatial distribution of the regional groups identified, which differ in terms of the ExPT occurrence structure, is depicted in Fig. 4.1. In each season, the distribution of these groups is clearly organised, except for the area of the Alps and Southwestern Europe, especially in winter, where the ExPT occurrence structure is diversified (Fig. 4.1). Table 4.1 presents the values of the *F* statistic, indicating how well the respective variable discriminates between clusters. The *F* statistic is an element of the variance analysis: It was calculated on the basis of intergroup and intragroup sum of squares (Table 4.2). The grouping variables characterised by a higher inter- than intragroup sum of squares are more important in the grouping process than the variables for which the magnitudes of the values are reverse. The resultant analysis of variance is summarised by the *F* statis-

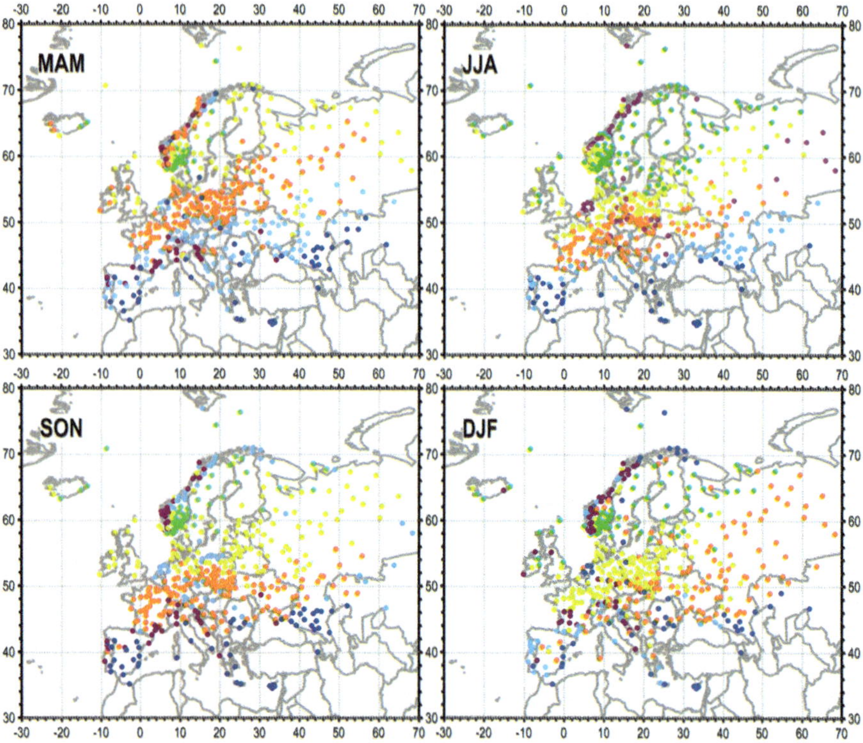

Fig. 4.1 Distribution of the regional groups differing in the structure of the origin-based extreme precipitation occurrence in the December 1950–February 2008 period. The stations marked with the same colour belong to the same regional group of origin-based extreme precipitation types occurrence. The extensive description of each of the regional groups distinguished in particular seasons is included in the following chapter. The only aim of this graph is to demonstrate the spatial distribution of the regional groups regardless of their individual features that is why no legend is included

Table 4.1 F statistic for the objects grouping variables

	F statistic			
Variable	MAM	JJA	SON	DJF
Type A	224.0	480.6	198.3	211.1
Type Ff	205.5	159.9	133.4	227.8
Type Fo	358.5	489.8	390.4	213.9
Type Fw	9.7	6.8	2.9	37.2
Type Fc	91.4	136.2	79.5	40.0
NWD	22.7	176.1	72.3	47.7
TOTExP	149.4	20.2	158.7	200.4

F statistic gives information as to which variable was the crucial criterion of clustering, *type A*, … *type Fc* types of extreme precipitation (abbreviations explained in Table 3.1), *NWD* number of wet days (with precipitation ≥1 mm), *TOTExP* average daily extreme precipitation total, *MAM* spring, *JJA* summer, *SON* autumn, *DJF* winter

Table 4.2 Selected results of variance analysis: inner group and intergroup square sums

	MAM		JJA		SON		DJF	
Variable	Inter SS	Inner SS	Inter SS	Inner SS	Inter SS	Inner SS	Inter SS	Inner SS
Type A	48,715.6	22,052.1	113,656.8	23,980.0	53,461.1	27,333.2	47,110.9	22,626.9
Type Ff	41,381.0	20,414.6	30,097.1	19,087.5	34,240.5	26,023.2	50,179.9	22,338.3
Type Fo	6469.7	17,668.0	75,334.8	15,594.8	67,861.2	17,625.7	51,714.8	24,518.3
Type Fw	988.6	10,326.7	684.5	10,165.8	277.3	9684.1	4023.9	10,979.2
Type Fc	15,657.3	17,374.5	26,033.5	19,375.2	19,610.8	25,018.8	9254.1	23,467.2
NWD	5024.2	22,463.0	30,273.2	17,436.3	21,677.7	30,411.1	12,950.4	27,539.7
TOTExP	33,635.3	22,834.1	7059.0	35,396.4	78,270.3	49,996.0	67,886.3	34,357.6

Inter SS intergroup square sums, *Inner SS* inner group square sums, *type A*, … *type Fc* types of extreme precipitation (abbreviations are explained in Table 3.1), *NWD* number of wet days (with precipitation ≥1 mm), *TOTExP* average daily extreme precipitation total, *MAM* spring, *JJA* summer, *SON* autumn, *DJF* winter tic, which is used for comparing variance estimations. High *F* statistic values indicate a significant discriminative variable in the grouping process (Stanisz 2007; electronic statistic textbook; http://www.statsoft.pl/textbook).

In spring and autumn, the key discriminative variable was the frequency of occluded front precipitation (type Fo), and in summer, these were the frequency of air-mass precipitation (type A) and the frequency of occluded front precipitation (type Fo), whereas in winter, four variables were equally discriminative, namely, the frequency of air-mass precipitation (type A), precipitation associated with the passage of several fronts (type Ff), occluded front precipitation (type Fo), and average extreme precipitation total (TOTExP).

In each season, the variable least differentiating the clusters was the frequency of warm front precipitation (type Fw). A full characteristic of the regional groups distinguished in each season is presented in the next chapter of this study (see Sect. 4.2).

4.2 Spatial Distribution and Description of Regional Groups of Origin-Based Extreme Precipitation Types Occurrence

Before going on to characterise the regional groups of origin-based extreme precipitation types occurrence, it must be stressed, once again, that precipitation is among the most changeable elements of the climate in both temporal and spatial terms. Being a noncontinuous phenomenon, it can cover vast areas or occur locally. The "local nature" characterises, in particular, extreme precipitation. Clear spatial patterns of precipitation totals and frequencies are seen for its seasonal and monthly values and are much less pronounced for extreme daily precipitation. Thus, the clearly organised spatial distribution of regional groups in Europe obtained as a result of this study proves the existence of factors governing the occurrence of the types of extreme precipitation origin on the continent. Many of the extreme precipitation events studied were local in nature (Sect. 3.2); hence, the spatial distribution obtained is an effect of accumulating a number of events, many of which were isolated in nature. The weather stations forming some of the regional groups are dispersed across the continent, with the distribution not being accidental, but attributable to the substantial impact of local conditions on the processes that lead to extreme precipitation and determine its level. The conclusion made by Okołowicz (1969), that the occurrence of fronts and their movement in space is not significantly correlated with the season, does not apply to fronts generating heavy precipitation.

The location of each of the regional groups distinguished in the successive seasons is shown in Figs. 4.2, 4.4, 4.6, and 4.8. The ExPT occurrence structure in each of the RGs is shown by the graphs in these figures, which depict the averaged values of the grouping variables, that is, selected types of extreme precipitation origin (types A, Ff, Fo, Fw, Fc), as well as the number of wet days with precipitation $\geq$1 mm (NWD), and daily mean extreme precipitation total (TOTExP). In addition, Figs. 4.3, 4.5, 4.7, and 4.9 include box plots containing descriptive statistics of the

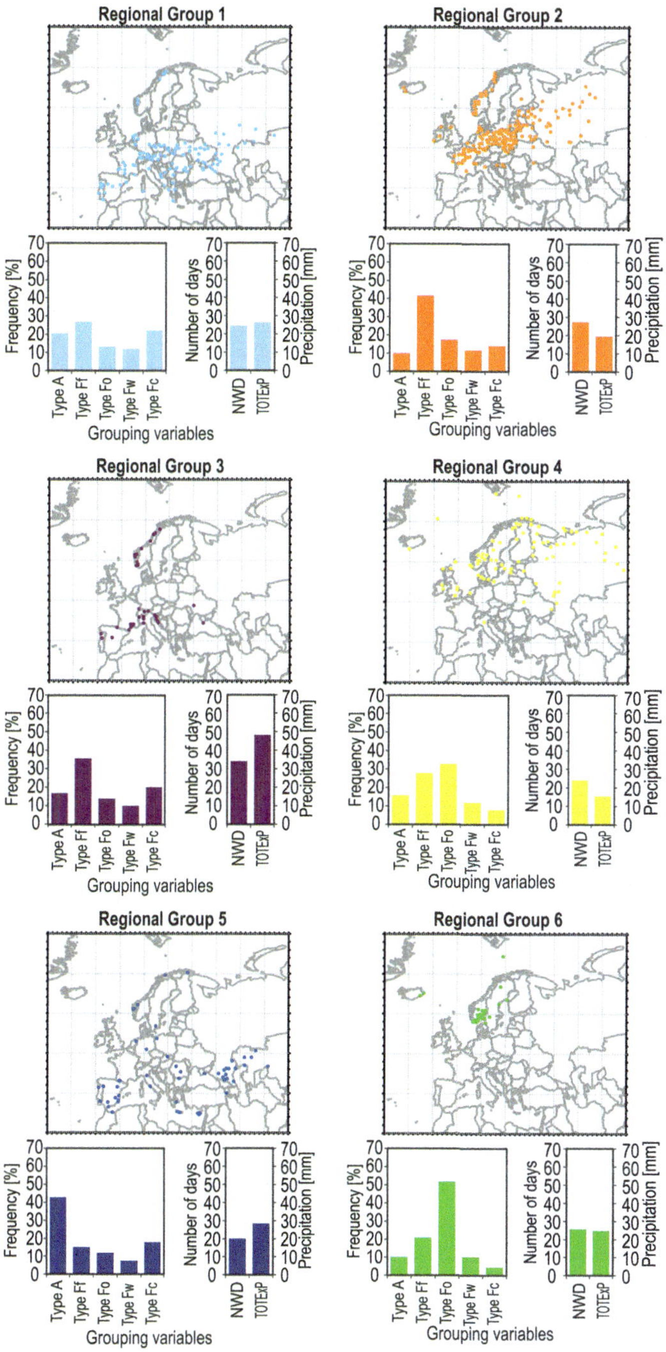

Fig. 4.2 Distribution of the regional groups of origin-based extreme precipitation types occurrence (*maps*) and the averaged grouping variables for the regional groups (*graphs*), spring. *Type A*, ... *Type Fc* origin-based extreme precipitation types (abbreviations explained in Table 3.1), *NWD* number of wet days (with precipitation ≥1 mm), *TOTExP* average daily extreme precipitation total

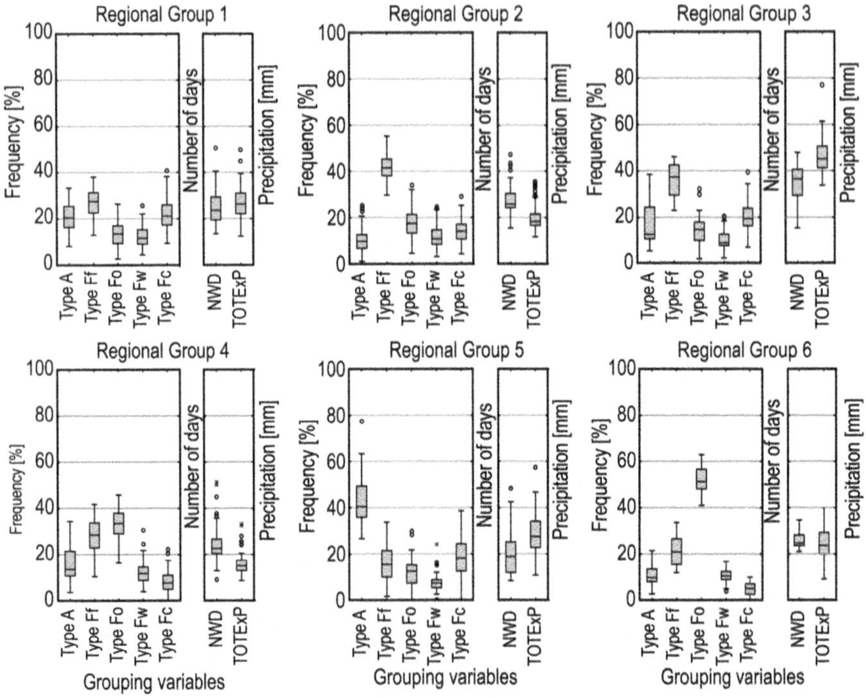

Fig. 4.3 Descriptive statistics of grouping variables for the regional groups of origin-based extreme precipitation types occurrence, spring. *Type A, … Type Fc* origin-based extreme precipitation types (abbreviations explained in Table 3.1), *NWD* number of wet days (with precipitation ≥1 mm), *TOTExP* average daily extreme precipitation total

grouping variables for each of the regional groups distinguished. The plots provide information about the justifiability of the regional grouping and about the similarity of the structure of ExPTs occurrence at the stations within the regional group concerned. Tables 4.3–4.6. show descriptive statistics for the types of extreme precipitation origin that were not taken into account as grouping variables in the cluster analysis, and several other indicators adding to obtaining a fuller picture of precipitation within the regional groups.

4.2.1 Spring

In spring, the variables that most strongly discriminate the regional groups include the frequencies of type Fo, type A, and then type Ff precipitation. TOTExP was also highly relevant in the grouping process. Three of the six regional groups distinguished (RG1, RG2, RG3) are characterised by the highest frequency of precipitation associated with the passage of several fronts (type Ff), which ranges among

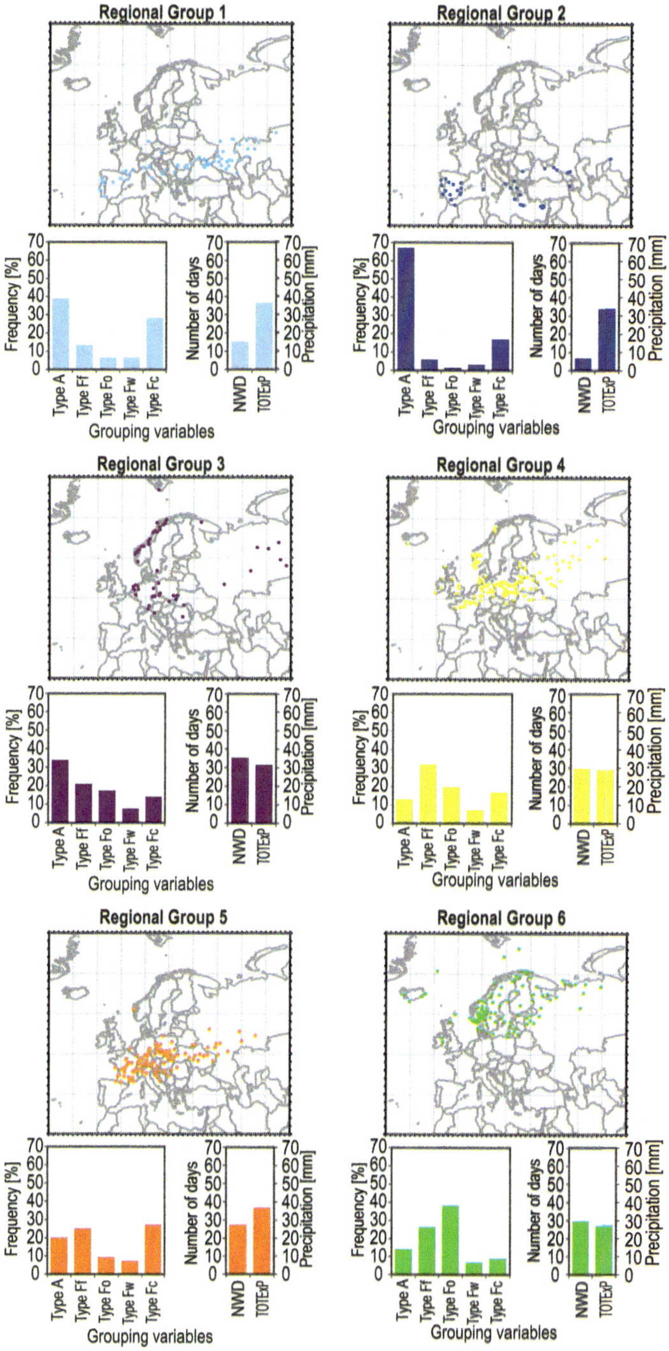

Fig. 4.4 Distribution of the regional groups of origin-based extreme precipitation types occurrence (*maps*) and the averaged grouping variables for the regional groups (*graphs*), summer. *Type A*, ... *Type Fc* origin-based extreme precipitation types (abbreviations explained in Table 3.1), *NWD* number of wet days (with precipitation ≥1 mm), *TOTExP* average daily extreme precipitation total

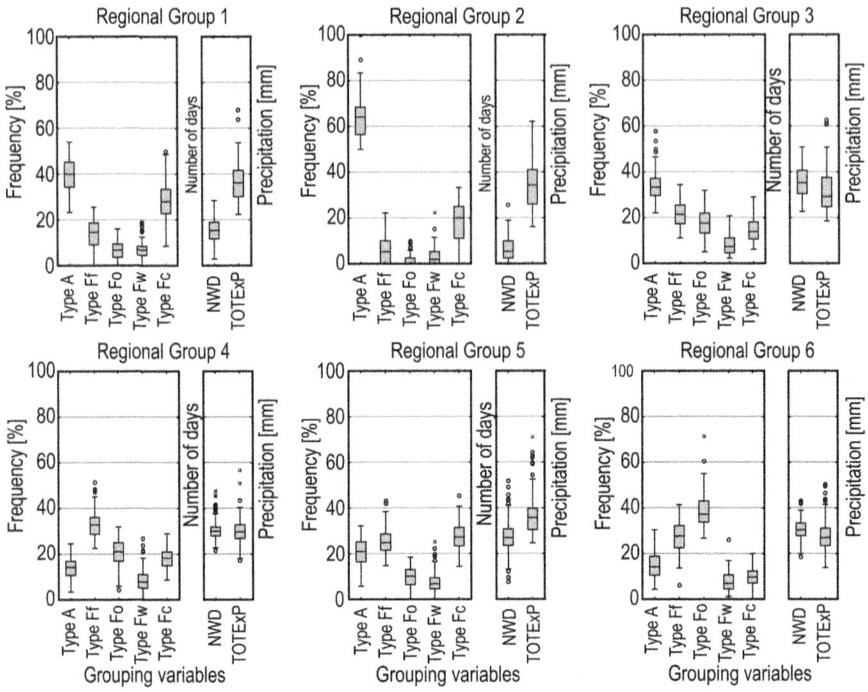

Fig. 4.5 Descriptive statistics of grouping variables for the regional groups of origin-based extreme precipitation types occurrence, summer. *Type A, ... Type Fc* origin-based extreme precipitation types (abbreviations explained in Table 3.1), *NWD* number of wet days (with precipitation ≥1 mm), *TOTExP* average daily extreme precipitation total

these groups between approximately 27 % in RG2 and approximately 42 % in RG3. In the other two regional groups (RG4 and RG6), extreme precipitation is mainly associated with an occluded front, representing slightly more than 30 % of extreme precipitation in RG4 and over 50 % in RG6. On the other hand, RG 5 demonstrates the highest share of air-mass precipitation (more than 40 %) in the overall number of days with extreme precipitation (Fig. 4.2).

Regional Group 1 (RG1) includes 117 stations, located mostly in the south of Europe. The northern border of this compact regional group in the western part of the continent reaches approximately 45°N, and farther east, it gradually extends northwards, covering the southern part of Poland and reaching 55°N in the areas located to the north of the Caspian Sea. This group also includes stations located on the southeast coasts of the North Sea and isolated stations on the coasts of the Scandinavian Peninsula (Fig. 4.2). RG1 demonstrates the least pronounced individual features, which is manifested by the lowest diversity in the frequency of the individual origin-based precipitation types of all the regional groups distinguished in that season. The frequency of the prevalent precipitation, that is, that associated with the passage of several fronts (type Ff), averages 27 %, whereas the least fre-

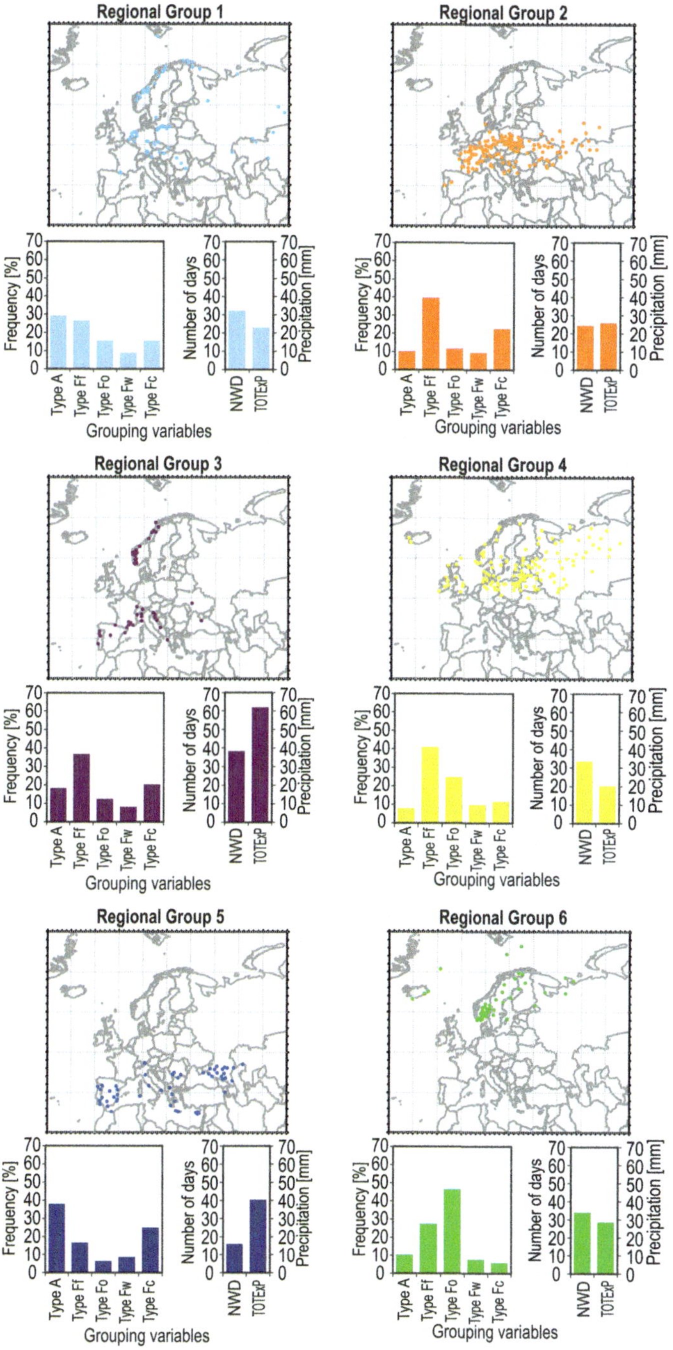

Fig. 4.6 Distribution of the regional groups of origin-based extreme precipitation types occurrence (*maps*) and the averaged grouping variables for the regional groups (*graphs*), autumn. *Type A*, … *Type Fc* origin-based extreme precipitation types (abbreviations explained in Table 3.1), *NWD* number of wet days (with precipitation ≥1 mm), *TOTExP* average daily extreme precipitation total

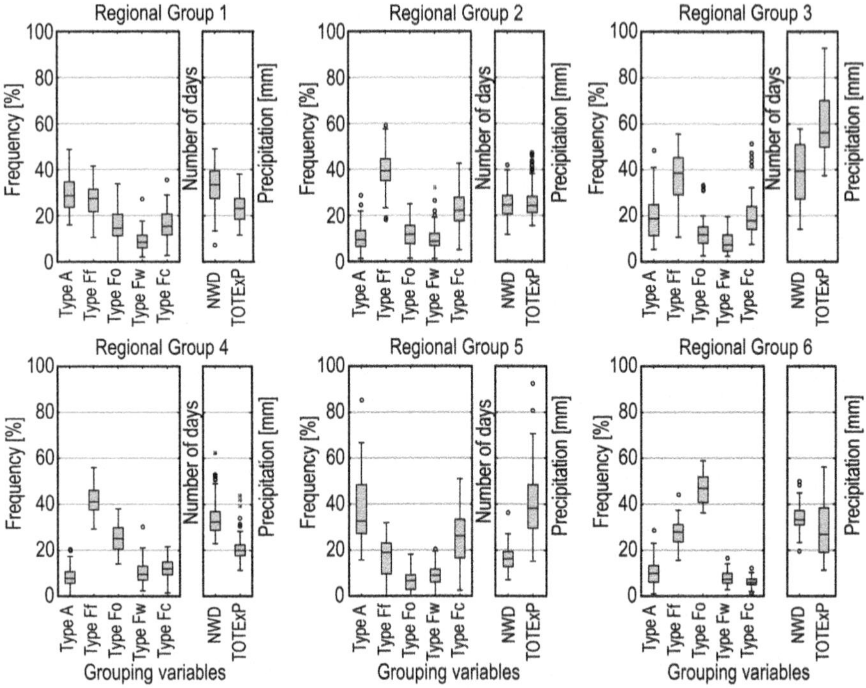

Fig. 4.7 Descriptive statistics of grouping variables for the regional groups of origin-based extreme precipitation types occurrence, autumn. *Type A, … Type Fc* origin-based extreme precipitation types (abbreviations explained in Table 3.1), *NWD* number of wet days (with precipitation ≥1 mm), *TOTExP* average daily extreme precipitation total

quent precipitation, that is, that associated with a warm front (type Fw) represents 10 % of ExP on average, which translates into a frequency difference of 17 % (Fig. 4.2). This group shows relatively frequent air-mass precipitation (type A), which represents approximately 21 % of ExP. The daily mean extreme precipitation recorded by RG1 stations reaches 26.6 mm, which represents 17 % of the season's total. In general, precipitation ≥1 mm is recorded on 25 days during the season (28 % of the days) (Table 4.3). The number of outliers, which are nontypical and infrequent in the group, is low, and identified only for two stations for the frequency of warm front precipitation (type Fw), and two stations for TOTExP, as well as for one station for the frequency of cold front precipitation (type Fc), and for NWD (Fig. 4.3). Thus, assigning the individual stations to the cluster is justified, because only individual variables at isolated stations are highly distanced from the cluster centre. Given the small range of changes in the frequency of the individual ExPTs averaged for RG1, and the location of the weather stations that constitute it, the group may be considered transitory between the other regional groups, which show more pronounced individual features (Fig. 4.2).

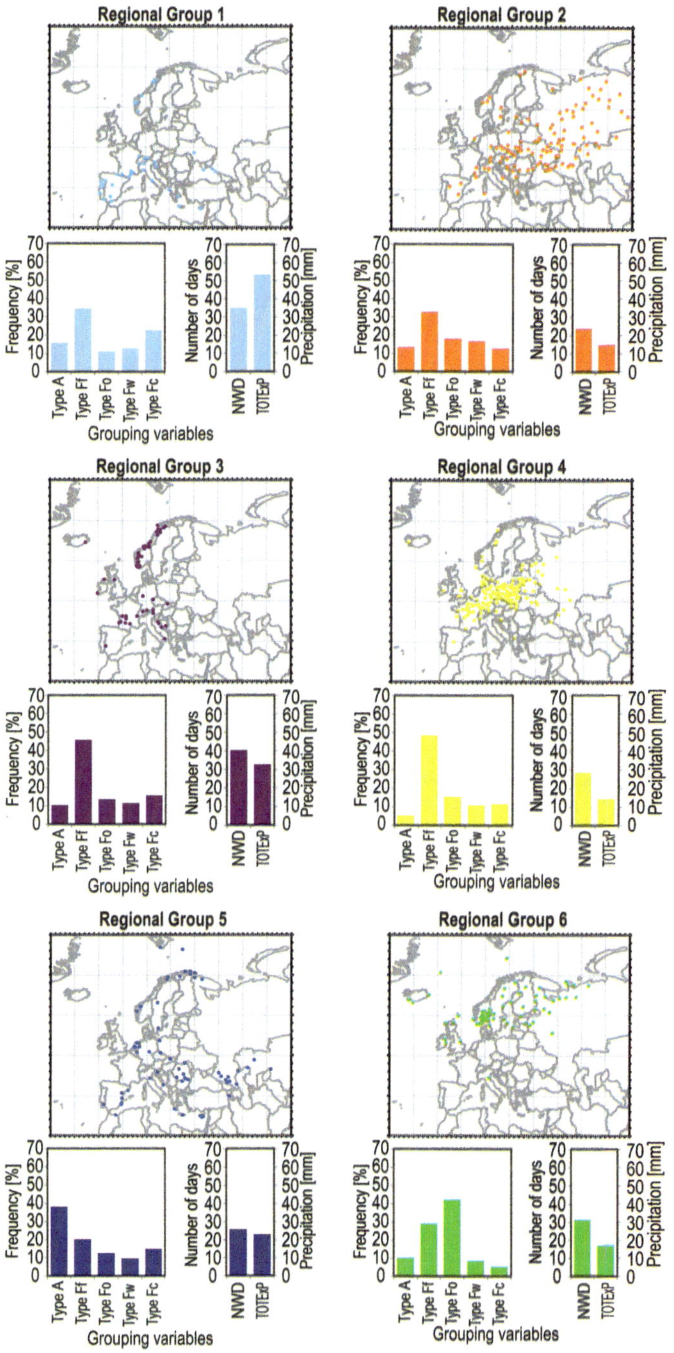

Fig. 4.8 Distribution of the regional groups of origin-based extreme precipitation types occurrence (*maps*) and the averaged grouping variables for the regional groups (*graphs*), winter. *Type A*, ... *Type Fc* origin-based extreme precipitation types (abbreviations explained in Table 3.1), *NWD* number of wet days (with precipitation ≥1 mm), *TOTExP* average daily extreme precipitation total

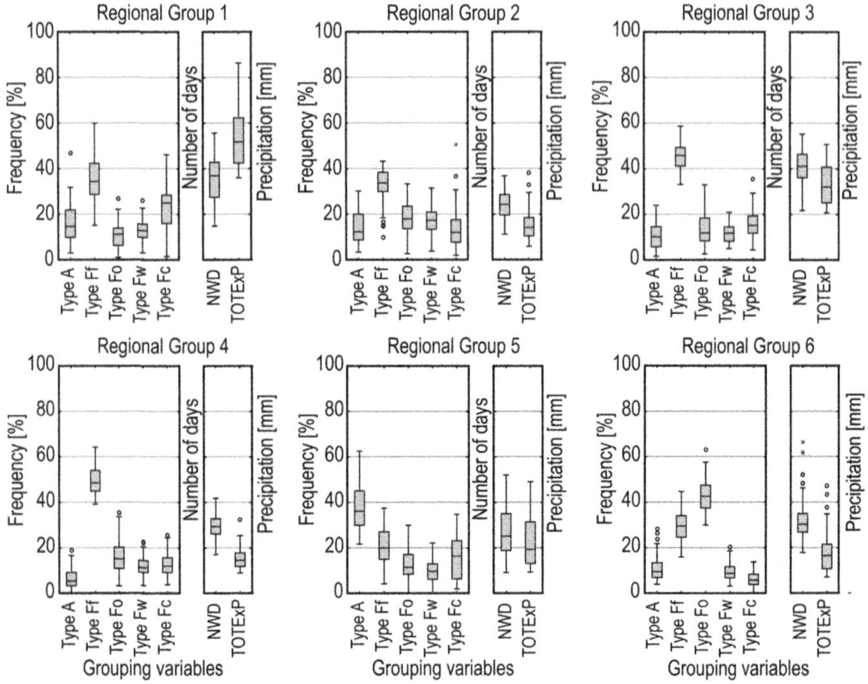

Fig. 4.9 Descriptive statistics of grouping variables for the regional groups of origin-based extreme precipitation types occurrence, winter. *Type A, … Type Fc* origin-based extreme precipitation types (abbreviations explained in Table 3.1), *NWD* number of wet days (with precipitation ≥1 mm), *TOTExP* average daily extreme precipitation total

Regional Group 2 (RG2) consists of 184 weather stations, the great majority of which are located to the north of RG1, within the European lowlands (including nearly all the territory of Poland) and areas located to the northeast of Poland, stretching across the centre of the East European Plain. RG2 stations also lie on the western coasts of the Scandinavian Peninsula, in the westernmost area of Iceland, and on the British Isles (Fig. 4.2). RG2 is characterised by a much higher frequency of extreme precipitation associated with the passage of several fronts (type Ff) compared to the frequency of the other extreme precipitation types. It accounts for approximately 42 % of ExP, whereas the other types represent from approximately 10 % of ExP (type A) to 18 % of ExP (type Fo). In total, frontal precipitation accounts for as much as 90 % of all extreme events recorded by RG2 stations, which means the largest share among all the regional groups. The high predominance of precipitation associated with the passage of several fronts at some RG2 stations results from their proximity to areas with cyclogenesis (the Atlantic Ocean) and the lowland nature of most of the areas covered by RG2, which allows relatively free flow of low-pressure systems and their slower development, not accelerated by land surface roughness leading to occlusion.

Table 4.3 Descriptive statistics of selected grouping variables for regional groups, spring

Index	Average (±SE)		Confidence intervals		SD	ME	Quartiles		Min	Max
			−95 %	+95 %			Lower	Upper		
Regional group 1										
Type Fs	3.6	(±0.3)	3.0	4.1	2.8	3.4	1.4	5.2	0.0	11.8
Type Fd	0.7	(±0.1)	0.4	1.0	1.6	0.0	0.0	1.0	0.0	14.0
Type F	79.1	(±0.6)	77.9	80.3	6.4	79.6	74.5	83.7	56.0	91.8
NWD%	27.0	(±0.7)	25.6	28.3	7.6	25.7	21.2	32.0	14.7	55.1
TOTSsP	157.0	(±6.0)	145.2	168.8	64.7	140.8	113.5	181.5	60.6	447.1
Regional group 2										
Type Fs	3.9	(±0.2)	3.4	4.3	2.9	3.8	1.2	5.6	0.0	12.1
Type Fd	0.6	(±0.1)	0.5	0.8	1.1	0.0	0.0	1.1	0.0	5.1
Type F	90	(±0.3)	89.1	90.5	4.7	90.3	87.4	93.1	67.9	98.9
NWD%	30	(±0.4)	29.0	30.5	5.3	28.0	26.4	32.9	16.8	51.2
TOTSsP	145.1	(±3.7)	137.8	152.4	50.4	126.2	114.2	159.5	55.0	322.6
Regional group 3										
Type Fs	2.6	(±0.5)	1.6	3.5	2.9	1.7	0.0	4.4	0.0	9.8
Type Fd	0.2	(±0.1)	0.1	0.3	0.4	0.0	0.0	0.0	0.0	1.3
Type F	83.0	(±1.5)	79.9	86.2	9.4	87.3	75.7	89.2	61.5	94.4
NWD%	37.3	(±1.6)	34.0	40.5	9.7	39.6	32.0	44.1	16.8	52.1
TOTSsP	379.9	(±21.9)	335.4	424.4	133.4	366.5	303.1	476.0	138.8	605.4
Regional group 4										
Type Fs	1.6	(±0.2)	1.2	2.1	1.9	1.3	0.0	2.4	0.0	8.5
Type Fd	0.4	(±0.2)	0.0	0.7	1.5	0.0	0.0	0.0	0.0	12.7
Type F	83.3	(±0.9)	81.6	85.0	8.0	86.1	77.7	89.0	65.6	94.6
NWD%	26.4	(±0.8)	24.7	28.1	7.7	24.6	22.2	29.0	10.1	55.8
TOTSsP	104.8	(±5.0)	94.7	114.8	46.3	96.5	79.4	124.6	33.0	328.4
Regional group 5										
Type Fs	2.5	(±0.3)	1.8	3.2	2.6	2.2	0.0	3.9	0.0	10.8
Type Fd	1.0	(±0.2)	0.5	1.4	1.5	0.0	0.0	1.9	0.0	7.1
Type F	56.5	(±1.3)	53.8	59.2	9.9	59.0	50.0	64.1	22.5	73.3
NWD%	22.2	(±1.4)	19.4	25.0	10.4	20.5	12.9	27.3	9.1	52.7
TOTSsP	139.6	(±12.5)	114.6	164.6	92.5	120.4	75.9	167.6	39.1	496.5
Regional group 6										
Type Fs	0.9	(±0.2)	0.4	1.3	1.3	0.0	0.0	1.5	0.0	5.6
Type Fd	0.1	(±0.1)	0.0	0.3	0.4	0.0	0.0	0.0	0.0	1.5
Type F	89.5	(±0.8)	87.9	91.2	4.9	90.3	86.6	92.0	78.6	97.4
NWD%	28.2	(±0.6)	26.9	29.5	3.8	26.8	25.6	30.6	22.9	37.6
TOTSsP	173.4	(±9.7)	153.6	193.2	58.5	157.7	133.1	204.4	63.8	321.7

Type Fs frequency of extreme precipitation linked to stationary front, *type Fd* frequency of extreme precipitation linked to discontinuity line, *type F* frequency of frontal extreme precipitation (the total of all frontal events), *NWD%* frequency of days with precipitation ≥1 mm expressed as a percent of days in season, *TOTSsP* seasonal precipitation total (mm), *SE* standard error, *SD* standard deviation, *ME* median, *Min* minimum value, *Max* maximum value

Table 4.4 Descriptive statistics of selected grouping variables for regional groups, summer

Index	Average (±SE)		Confidence intervals −95 %	+95 %	SD	ME	Quartiles Lower	Upper	Min	Max
Regional group 1										
Type Fs	3.9	(±0.5)	2.9	4.8	3.8	3.2	0.0	6.0	0.0	15.6
Type Fd	0.5	(±0.1)	0.2	0.7	1.2	0.0	0.0	0.0	0.0	4.7
Type F	60.5	(±0.9)	58.7	62.3	7.2	60.0	54.5	65.6	45.9	76.7
NWD%	17.1	(±0.7)	15.6	18.5	5.7	16.7	12.7	20.6	3.1	30.9
TOTSsP	127.4	(±6.3)	114.8	140.0	50.3	123.6	87.3	155.0	13.1	248.1
Regional group 2										
Type Fs	1.9	(±0.6)	0.7	3.0	3.5	0.0	0.0	3.1	0.0	14.3
Type Fd	0.9	(±0.4)	0.1	1.7	2.3	0.0	0.0	0.0	0.0	10.0
Type F	32.8	(±2.4)	27.9	37.6	14.6	35.8	31.6	44.0	0.0	50.0
NWD%	7.9	(±1.1)	5.6	10.2	6.8	5.8	2.7	10.8	0.3	27.8
TOTSsP	59.9	(±10.5)	38.6	81.2	63.8	42.8	17.6	66.3	2.2	256.4
Regional group 3										
Type Fs	3.1	(±0.5)	2.1	4.1	3.6	1.6	0.0	5.6	0.0	16.0
Type Fd	0.6	(±0.1)	0.4	0.9	1.0	0.0	0.0	1.2	0.0	3.8
Type F	66.2	(±1.1)	64.1	68.4	7.9	67.0	63.5	70.4	42.4	88.0
NWD%	38.9	(±1.0)	36.9	41.0	7.6	38.3	33.1	44.2	24.7	55.4
TOTExP	283.7	(±17.4)	248.8	318.7	129.3	225.3	200.1	332.2	113.3	649.7
Regional group 4										
Type Fs	4.6	(±0.3)	4.0	5.2	3.5	3.5	1.6	7.4	0.0	14.6
Type Fd	0.9	(±0.1)	0.7	1.1	1.2	0.0	0.0	1.3	0.0	5.9
Type F	85.7	(±0.4)	85.0	86.5	4.4	85.8	83.0	89.2	75.4	96.6
NWD%	33.6	(±0.5)	32.7	34.6	5.3	32.7	30.5	34.6	23.3	51.8
TOTSsP	222.1	(±6.3)	209.6	234.5	71.2	210.9	190.0	230.2	124.8	659.1
Regional group 5										
Type Fs	7.1	(±0.4)	6.3	7.8	4.2	6.8	3.9	9.9	0.0	18.1
Type Fd	1.4	(±0.2)	1.1	1.8	1.8	1.0	0.0	2.1	0.0	9.1
Type F	79.2	(±0.6)	78.0	80.4	6.3	79.0	74.7	83.4	57.1	94.3
NWD%	30.3	(±0.7)	28.8	31.7	8.0	29.3	25.6	33.4	8.3	56.4
TOTSsP	243.7	(±10)	224.0	263.5	107.5	216.1	177.9	260.5	66.0	780.4
Regional group 6										
Type Fs	1.5	(±0.1)	1.2	1.8	1.5	1.1	0.0	2.1	0.0	6.7
Type Fd	0.4	(±0.1)	0.1	0.6	1.4	0.0	0.0	0.0	0.0	14.1
Type F	85.0	(±0.5)	83.9	86.1	5.7	85.7	81.1	89.5	69.6	95.7
NWD%	33.1	(±0.5)	32.2	34.0	5.0	33.0	30.3	36.3	20.3	46.7
TOTSsP	210.3	(±6.0)	198.5	222.2	63.6	204.8	172.1	246.3	76.7	440.1

Type Fs frequency of extreme precipitation linked to stationary front, *type Fd* frequency of extreme precipitation linked to discontinuity line, *type F* frequency of frontal extreme precipitation (the total of all frontal events), *NWD%* frequency of days with precipitation ≥1 mm expressed as a percent of days in season, *TOTSsP* seasonal precipitation total [mm], *SE* standard error, *SD* standard deviation, *ME* median, *Min* minimum value, *Max* maximum value

Table 4.5 Descriptive statistics of selected grouping variables for regional groups, autumn

Index	Average (±SE)		Confidence intervals		SD	ME	Quartiles		Min	Max
			−95 %	+95 %			Lower	Upper		
Regional group 1										
Type Fs	2.4	(±0.4)	1.6	3.2	3.3	1.0	0.0	3.8	0.0	16.7
Type Fd	0.4	(±0.1)	0.2	0.6	0.8	0.0	0.0	0.0	0.0	3.4
Type F	70.5	(±0.9)	68.8	72.2	7.1	71.3	65.3	76.4	51.2	84.0
NWD%	35.3	(±1.2)	33.0	37.7	9.7	36.3	29.9	42.8	7.8	53.3
TOTSsP	201.5	(±10.5)	180.6	222.4	86.3	195.5	126.4	256.1	31.1	391.6
Regional group 2										
Type Fs	5.2	(±0.3)	4.7	5.7	3.1	5.1	3.0	7.1	0.0	15.6
Type Fd	0.3	(±0.1)	0.2	0.4	0.6	0.0	0.0	0.0	0.0	3.0
Type F	89.7	(±0.4)	88.8	90.5	5.2	90.5	86.6	93.5	71.2	98.6
NWD%	26.7	(±0.5)	25.7	27.7	5.9	26.7	22.5	31.1	12.8	45.4
TOTSsP	160.1	(±4.7)	150.9	169.3	56.6	143.5	116.9	192.5	49.7	359.4
Regional group 3										
Type Fs	2.5	(±0.4)	1.6	3.4	3.2	1.2	0.0	4.7	0.0	14.0
Type Fd	0.2	(±0.1)	0.0	0.3	0.5	0.0	0.0	0.0	0.0	2.4
Type F	81.4	(±1.3)	78.9	83.9	9.1	81.3	75.2	88.6	51.8	94.6
NWD%	41.9	(±2.0)	37.8	46.0	14.8	42.8	29.6	55.4	15.4	62.9
TOTSsP	553.9	(±33.4)	486.9	621.0	243.2	528.0	418.5	692.1	186.5	1165.8
Regional group 4										
Type Fs	2.3	(±0.2)	2.0	2.7	2.2	1.9	0.8	3.4	0.0	14.3
Type Fd	0.3	(±0.1)	0.2	0.4	0.7	0.0	0.0	0.0	0.0	5.1
Type F	91.6	(±0.4)	90.8	92.3	4.2	92.2	89.4	94.4	79.5	100
NWD%	36.8	(±0.7)	35.3	38.2	8.4	35.2	31.3	40.0	24.9	67.9
TOTSsP	187.2	(±7.5)	172.3	202.0	86.6	158.8	141.0	196.5	90.8	627.3
Regional group 5										
Type Fs	3.5	(±0.4)	2.7	4.3	2.9	2.9	1.3	5.9	0.0	9.5
Type Fd	0.5	(±0.1)	0.2	0.7	1.0	0.0	0.0	0.0	0.0	4.5
Type F	61.8	(±2.0)	57.7	65.8	14.8	67.4	51.6	72.7	14.8	84.3
NWD%	17.7	(±0.8)	16.2	19.3	5.8	17.7	13.8	21.1	7.6	39.4
TOTSsP	146.3	(±10.1)	126.1	166.4	73.9	135.4	95.6	166.9	38.7	401.4
Regional group 6										
Type Fs	0.7	(±0.1)	0.4	0.9	0.8	0.0	0.0	1.1	0.0	3.0
Type Fd	0.0	(±0.0)	0.0	0.1	0.1	0.0	0.0	0.0	0.0	1.1
Type F	89.4	(±0.7)	87.9	90.9	5.6	90.1	86.8	93.9	71.1	99.0
NWD%	37.2	(±0.8)	35.6	38.9	6.2	36.1	33.8	40.4	21.3	54.2
TOTSsP	260.9	(±18.1)	224.7	297.2	136.6	233.7	146.1	361.3	77.3	661.4

Type Fs frequency of extreme precipitation linked to stationary front, *type Fd* frequency of extreme precipitation linked to discontinuity line, *type F* frequency of frontal extreme precipitation (the total of all frontal events), *NWD%* frequency of days with precipitation ≥1 mm expressed as a percent of days in season, *TOTSsP* seasonal precipitation total [mm], *SE* standard error, *SD* standard deviation, *ME* median, *Min* minimum value, *Max* maximum value

Table 4.6 Descriptive statistics of selected grouping variables for regional groups, winter

Index	Average (±SE)		Confidence intervals −95 %	+95 %	SD	ME	Quartiles Lower	Upper	Min	Max
Regional group 1										
Type Fs	1.8	(±0.3)	1.2	2.4	2.0	1.0	0.0	2.7	0.0	6.2
Type Fd	0.2	(±0.1)	0.0	0.4	0.5	0.0	0.0	0.0	0.0	2.5
Type F	84.0	(±1.2)	81.5	86.5	8.6	85.4	78.1	90.2	53.0	97.0
NWD%	39.2	(±1.8)	35.5	42.9	12.6	41.0	30.5	47.6	16.5	61.8
TOTSsP	466.2	(±35.2)	395.4	537.0	241.1	423.6	272.3	591.0	146.5	1057.3
Regional group 2										
Type Fs	3.6	(±0.2)	3.1	4.1	2.7	3.3	1.7	5.3	0.0	13.6
Type Fd	0.3	(±0.1)	0.0	0.6	1.6	0.0	0.0	0.0	0.0	12.5
Type F	86.1	(±0.6)	85.0	87.3	6.4	87.7	80.0	91.3	69.8	96.7
NWD%	27.1	(±0.6)	25.9	28.2	6.2	27.0	21.8	31.8	12.4	40.9
TOTSsP	100.2	(±3.2)	94.0	106.5	34.5	96.9	82.2	117.0	36.1	235.9
Regional group 3										
Type Fs	1.4	(±0.2)	0.9	1.8	1.7	0.7	0.0	1.8	0.0	6.7
Type Fd	0.1	(±0.0)	0.1	0.2	0.3	0.0	0.0	0.0	0.0	0.9
Type F	89.3	(±0.8)	87.7	90.9	6.1	89.8	85.4	94.2	76.0	98.3
NWD%	45.2	(±1.2)	43.0	47.3	8.3	45.6	40.1	51.5	24.1	61.4
TOTSsP	361.8	(±17.1)	327.4	396.1	130.6	328.5	248.9	465.9	168.4	666.6
Regional group 4										
Type Fs	3.4	(±0.2)	3.0	3.7	2.4	3.2	1.5	4.7	0.0	11.0
Type Fd	0.2	(±0.0)	0.2	0.3	0.5	0.0	0.0	0.0	0.0	1.8
Type F	93.8	(±0.3)	93.1	94.4	4.1	94.7	90.9	96.7	81.0	100
NWD%	32.7	(±0.4)	31.9	33.6	5.3	32.7	29.0	36.5	18.9	46.5
TOTSsP	131.1	(±3.4)	124.5	137.7	41.1	122.4	97.7	159.6	54.9	288.4
Regional group 5										
Type Fs	2.4	(±0.3)	1.8	3.0	2.3	1.9	0.0	3.8	0.0	9.8
Type Fd	0.3	(±0.1)	0.1	0.4	0.6	0.0	0.0	0.0	0.0	2.1
Type F	61.6	(±1.4)	58.7	64.4	11.2	63.8	54.9	70.0	37.5	78.3
NWD%	29.0	(±1.5)	26.1	31.9	11.4	28.0	21.0	38.9	10.1	57.8
TOTSsP	155.4	(±12.6)	130.2	180.6	98.4	124.2	78.6	201.9	25.5	556.2
Regional group 6										
Type Fs	1.2	(±0.2)	0.8	1.5	1.5	0.9	0.0	2.2	0.0	6.8
Type Fd	0.1	(±0.0)	0.0	0.2	0.3	0.9	0.0	0.0	0.0	1.7
Type F	89.2	(±0.6)	88.0	90.5	5.5	90.5	86.7	93.2	71.7	96.1
NWD%	35.5	(±1.0)	33.5	37.6	9.1	33.7	29.9	38.9	19.9	73.8
TOTSsP	161.9	(±10.2)	141.8	181.9	89.1	132.7	99.2	198.6	56.6	461.0

Type Fs frequency of extreme precipitation linked to stationary front, *type Fd* frequency of extreme precipitation linked to discontinuity line, *type F* frequency of frontal extreme precipitation (the total of all frontal events), *NWD%* frequency of days with precipitation ≥1 mm expressed as percent of days in season, *TOTSsP* seasonal precipitation total ([mm), *SE* standard error, *SD* standard deviation, *ME* median, *Min* minimum value, *Max* maximum value

Table 4.7 Seasonal distribution of regional groups (GR) of origin-based types of extreme precipitation

GR	Spring	Summer	Autumn	Winter
GR1	Southern Europe between the Atlantic Ocean and the Caspian Sea and southern part of Central Europe	Northern part of southern Europe	Western Scandinavian coast, coastal stations in Western Europe and Central Europe (Baltic coast) and in mountains	Northern Europe and western coast of the Scandinavian Mountains
GR2	Western, Central, and Eastern Europe and western Scandinavian coast	Southern Europe	Northern part of southern Europe and southern part of Western and Central Europe	Northern part of southern Europe, Eastern Europe
GR3	Western Scandinavian coast and northern part of southwestern Europe	Western Scandinavian coast, mountain stations and Central European coastal stations, isolated stations in Eastern Europe	Western Scandinavian coast, northern part of southwestern Europe	Scandinavian coast, northern part of southwestern Europe
GR4	Northern Europe	Northern part of Western, Central and Eastern Europe and south-western part of the Scandinavian Peninsula	Western Europe, northern part Central Europe, and Eastern	Western and Central Europe
GR5	Eastern and Western part of southern Europe	Southern part of Western and Central Europe and the northern part of southern Europe	Southern Europe	Scattered stations in southern, central, and northern Europe
GR6	Southern part of the Scandinavian Peninsula	Northern Europe	Northern Europe (lee side of the Scandinavian Mountains, Atlantic islands)	Northern Europe on the lee side of the Scandinavian Mountains, Iceland, and Great Britain

GR regional groups

RG2 stations are located deep inside the continent, and at the same time, at quite a distance from the centres of the low-pressure systems crossing Northern Europe. In those systems, occlusion occurs in the northern, possibly central, parts of the weather fronts, and does not reach their southern ends. Some of the fronts generating heavy precipitation recorded by the southernmost stations also come from the Mediterranean Sea Region. On average, in RG2 precipitation ≥1 mm appears on average on 27 days in the season (30 % of the days). The daily average amount of extreme precipitation in the group is approximately 20 mm (13 % of the seasonal total) and ranges between 11.7 and 35.6 mm (Table 4.3). Across the RG2 stations, the cluster shows a characteristic feature, namely a clear predominance of the frequency of type Ff precipitation over the other types. At all the stations, type Ff shows no outliers. For each of the other grouping variables, outliers exist (Fig. 4.3). The greatest number of outliers is generated by TOTExP, eight stations located mainly in the Scandinavian Peninsula, and NWD, five stations, three of which lie on the British Isles, and the others in the northern part of the southern coasts of the Scandinavian Peninsula and Germany (a mountain site). During the study period, a large majority of the RG2 stations saw rare stationary front precipitation (82 % of the stations), which represents, on average, 4 % of the days with extreme precipitation in the group. Precipitation generated along the discontinuity line was recorded by 32 % of the stations. On average, it represented only 0.6 % of extreme precipitation (Table 4.3).

Regional Group 3 (RG3) consists of 37 stations located mainly in the western coast of the Scandinavian Peninsula, as well as in the western and central part of Southern Europe at mountain (e.g., alpine) and coastal stations (e.g., area of Istria, areas to the north of the Ligurian Sea, northeastern coast of the Adriatic Sea) (Fig. 4.2).

Similarly to RG2, the distribution of the ExPT frequency averaged for RG3 reveals the highest share of precipitation associated with the passage of several fronts, whereby such precipitation represents slightly less ExP than in RG2, 36 %. RG3 sees more air-mass precipitation (17 % of ExP) than in RG2 and more cold front precipitation (20 % of ExP on average). At some RG3 stations located in Spain, in the area of Istria and at isolated stations in the southern part of Eastern Europe, the highest share of extreme precipitation is of the air-mass type (type A). On the other hand, the daily mean precipitation totals (48.8 mm) are markedly higher than in the other groups, representing approximately 13 % of the seasonal total, as is the number of wet days with precipitation ≥1 mm, which amounts to 34 days (37 % of days in the season) (Table 4.3). Thus, RG3 is the cluster richest in precipitation, in terms of both its occurrence and its amount. Generally, in spring, the amount of precipitation here is 380 mm, whereas in the other RGs the seasonal totals are several times lower, ranging between 95 mm in RG4 and 154 mm in RG6. Outliers were identified for four grouping variables (type Fo, type Fw, type Fc, and TOTExP) and are recorded at most by two stations (Fig. 4.3). The synoptic causes of the occurrence of extreme precipitation associated with the passage of several fronts are similar to RG1. However, RG3 stations are marked by a stronger—than in RG1 and RG2—impact of local conditions on precipitation amounts, notably the landform (elevation above sea level) and location in the zone of sea–land contact.

As a result of these local factors, the group receives the highest daily mean extreme precipitation totals and the most frequent precipitation ≥1 mm compared to the other RGs.

Regional Group 4 (RG4) consists of 84 weather stations, most of which lie in the northern part of Europe, on the British Isles and in Eastern Europe on the Oksko-Donskaya Plain, and further south reaching the Sea of Azov (Fig. 4.2). In RG4, the highest share in the number of days with extreme precipitation characterises occluded front precipitation (approximately 33 % of ExP). Precipitation associated with the passage of several fronts is also frequent (28 % of ExP). The frequency of the other ExPTs ranges between 16 % of ExP (type A) and approximately 8 % of ExP (type Fc). A detailed analysis of the share of the individual origin-based precipitation types in the overall number of days with extreme precipitation at RG4 stations indicates that type Fo is the most frequent at approximately 60 % of the stations, and Ff at 35 % of the stations. At the few remaining stations, the largest proportion of extreme precipitation is that of the air-mass type. A large majority of the stations characterised by a predominance of type Fo are located in the Scandinavian Peninsula, in the northern part of Eastern Europe, and on Atlantic islands. Across the RG4 stations, the share of types A, Ff, and Fo is pronouncedly higher than that of types Fw and Fc (Fig. 4.2).

Northern Europe is crossed by east bound or northeast bound low-pressure systems (system centre). When such a system approaches the coast of the continent and enters the land, occlusion starts in its centre, resulting from classic cyclogenesis, which may be accelerated by increased roughness of the surface.

RG4 is specific for the lowest (compared to the other groups) daily mean precipitation totals (averaging 16 mm), which represent, however, 15 % of the seasonal total, which does not make them stand out from the other RGs (Table 4.3). This distinction means that the lower daily mean precipitation totals than in the other classes are attributable to the generally low precipitation in spring recorded by the RG4 stations (only 95 mm on average). The low precipitation totals at RG4 stations are an effect of their location in the rain shadow of the Scandinavian Mountains (the Scandinavian Peninsula) or being distanced from the source of humidity (Eastern Europe). At all the RG4 stations, the share of type A, type Ff, and type Fo precipitation in the overall number of days with extreme precipitation across shows no outliers. As concerns each of the other two types (type Fw and type Fc), only two values per each type are highly distanced from the cluster centre (Fig. 4.3). At six stations located mainly on islands and on the Atlantic coast (Iceland, Great Britain), the daily mean extreme precipitation totals, and at seven stations located both in Western Europe and in Eastern Europe (the Oksko-Donskaya Plain) the number of wet days with precipitation ≥1 mm, deviates statistically from the distribution of the other values in the group.

Regional Group 5 (RG5) consists of 55 stations, most of which are located in the southern part of Europe and at single stations in Central Europe and in the Scandinavian Peninsula. In the south of Europe, two major RG5 areas are distinguished: the Iberian Peninsula and the areas surrounding the Caspian Sea from the west, northwest, and north, that is, covering the foreland of the Caucasus, including the Stavropol Upland and the Caspian Lowland (Fig. 4.2). The RG5 stations in Central Europe and Northern Europe are either located in coastal areas or are highly elevated above sea level. A feature that distinguishes RG5 from the other regional groups is the clear predominance of air-mass precipitation, which accounts for an average of 43 % of extreme precipitation. This characterises nearly all (except two) weather stations in RG5. In the other groups, type A precipitation accounts for at most 21 % of ExP (RG3). Some of the air-mass precipitation at mountain stations is likely to be shaped by orography, and thus it is orographic precipitation.

The share of the other ExPTs in the overall number of days with extreme precipitation is lower by at least 25 % and ranges between around 8 % of ExP for type Fw and around 18 % of ExP for type Fc. The daily mean extreme precipitation at RG5 stations is relatively high (29 mm on average) against the background of the other regional groups, while the number of wet days with precipitation ≥1 mm is among the lowest (20 days, 22 % of days in the season) (Table 4.3). The daily mean extreme precipitation represents as much as 21 % of the total amount of spring precipitation and represents the highest share in the season. Outliers were identified for most of the ExPTs except for types Ff and Fc. The highest number of outliers is demonstrated by type Fw (Fig. 4.3).

The least numerous group in spring, that is, *Regional Group 6 (RG6)*, includes only 36 weather stations, located in the Scandinavian Peninsula, mainly in its southern part, on the eastern side of the Scandinavian Mountains. This class also includes stations in the east of Iceland and the Bear Island (Fig. 4.2). This class is distinguishable for a high share of precipitation associated with occlusion in the total number of days with extreme precipitation, reaching 52 % of ExP and ranging across the stations between 41 % and 63 % of ExP. In the group, 21 % of extreme precipitation is type Ff, approximately 10 % is types A and Fw, and the lowest share, 5 % of ExP, is represented by type Fc. Neither the frequency of precipitation ≥1 mm nor the daily mean amount of extreme precipitation stands out from the other groups, reaching 26 days (28 % of days in the season) and 25 mm (Table 4.3). In RG6 only one grouping variable value (type Fw frequency at the Halden station in Norway) was recognised as an outlier (Fig. 4.3). The high frequency of precipitation associated with occlusion at stations located in the central part of the southern end of the Scandinavian Peninsula is an effect of merging of incoming fronts over the area, most of which arrive from the south or southwest, stopping or slowing down when faced with the orographic barrier of the Scandinavian Mountains. Similar mechanisms are responsible for the high frequency of type Fo precipitation in Iceland. Most of the low-pressure systems reach the Bear Island in a mature state, which is characterised by occlusion.

The regularities in the spatial distribution of the regional groups of origin-based extreme precipitation types occurrence and the determinants of its distribution in spring also apply, although in different degrees, to the other seasons.

4.2.2 Summer

In the summer season, the distribution of the regional groups of origin-based types of extreme precipitation occurrence demonstrates the strongest spatial order, with the stations in the distinguished groups forming more compact areas across the continent than in the other seasons. In general, in summer three (RG1, RG2, and RG3) of the six regional groups show the highest share of air-mass precipitation in the overall number of days with extreme precipitation. Although the cumulative share of all frontal precipitation types highly exceeds the frequency of air-mass precipitation, summer is undoubtedly the season when extreme air-mass precipitation is the most frequent.

Regional Group 1 (RG1) comprises 64 stations, located within a relatively narrow belt, which extends from the Iberian Peninsula eastwards as far as the Caspian Lowland (Fig. 4.4). At the RG1 stations, the greatest share in the total number of days with extreme precipitation is that of the air-mass type (39 % of ExP). A large proportion is represented by precipitation associated with the passage of a cold front, type Fc (29 % of ExP).

The frequency of type Fc precipitation is the highest in the group compared to the other regional groups. In general, across the RG1 stations, type A and type Fc precipitation is the most frequent, with air-mass precipitation predominating at 84 % of the stations and type Fc at 6 % of the stations. On average, in RG1 the share of the other origin-based precipitation types reaches 14 % for type Ff and 7 % for Fo and Fc each. RG1 is also distinguished by high daily extreme precipitation totals, reaching 37 mm on average. This regional group is characterised by a low number of days with precipitation $\geq$1 mm, which is 16 days on average (17 % of days in the season) (Table 4.4). The daily mean extreme precipitation represents as much as 29 % of the precipitation occurring in the part of Europe in the summer. At RG1 stations, only three grouping variables showed outliers, yet they are only occasional. Only the frequency of precipitation associated with the passage of a warm front in summer at seven weather stations diverges from the other values in the group (Fig. 4.5). The structure of the ExPT occurrence in RG1 is similar to that characterising RG2. In RG2, the individual characteristics of the group are more pronounced, with wider-ranging changes in the frequency of the occurrence of origin-based precipitation types, notably an exceptionally high frequency of type A (Fig. 4.4). The causes of such high-frequency air-mass precipitation in RG1 and RG2 are similar and are explained in the section dedicated to RG2. Type Fc extreme precipitation, which is frequent in RG1, is linked to the location of Europe within a zone of westerly winds, which causes air masses to inflow to the continent from over the Atlantic, which is cooler in the season than the heated surface of the land. The thermal contrasts between the inflowing and stationary air start to be clearly observable only at some distance from the coast, where the impact of the ocean on the climate is not permanent. The orographic barriers present in the area drive convection processes along cold fronts and cause them to come to a halt or slow down.

Regional Group 2 (RG2) includes 37 stations in Southern Europe (Fig. 4.4). Notwithstanding the small number of stations, the group is characterised by a high frequency of air-mass precipitation, which accounts for an average of 67 % of ExP. At some stations located on Mediterranean islands, all extreme precipitation is caused by free convection, that is, is type A. Similarly to RG1, RG2 demonstrates a slightly higher, compared to the other types, frequency of cold front precipitation (type Fc), yet its share in the number of days with extreme precipitation is lower than in RG1 and amounts to 18 % of ExP. RG2 sees the other frontal precipitation types relatively rarely, ranging between 2 % of ExP (type Fo) and 6 % of ExP (type Ff). The cumulative share of all frontal types in RG2 represents 33 % of ExP on average and is the lowest compared both to the other regional groups and to seasons.

The daily mean extreme precipitation at RG2 stations is slightly lower (34.4 mm on average) than on stations located farther north (RG1), but much lower than the number of wet days with precipitation ≥1 mm (7 days, which translates into 8 % of days in the season). The daily mean extreme precipitation is as high as 57 % of the seasonal precipitation total (Table 4.4), which means that at RG2 stations, precipitation is rare in the summer, yet when it does occur, its daily totals are very high. The high frequency of air-mass precipitation in the south of Europe is attributable to the seasonal shift in the position of the subtropical high-pressure zone. In the summer, when the zone moves north, its effect extends to include also the Mediterranean Basin. The high-pressure systems, which predominate at this time of the year, generate radiative, usually rain-free, weather. The occasional rainfall is an effect of free convection. In mountain areas, convection is forced (enhanced) by orography. The farther north, the effect of the subtropical high-pressure zone declines, hence a drop in the frequency of air-mass precipitation, which is reflected by a lower frequency of type A precipitation in RG2. In RG1 and RG2, the farther north, the higher the frequency of extreme precipitation associated with the passage of a cold front. In RG2, the number of outliers is low. They appear at isolated stations for the frequency of type A and type Fw precipitation and for the number of wet days with precipitation ≥1 mm. Only the frequency of occluded front precipitation deviates from the statistical distribution of those values in the group at a higher number of stations (five stations) (Fig. 4.5).

Regional Group 3 (RG3) consists of 55 weather stations located on the west coasts of the Scandinavian Peninsula, on the southeast coasts of the North Sea, in Central Europe, and in Eastern Europe (Fig. 4.4). Most of the RG3 stations in Central Europe are located at high altitudes. RG3 is the third and last group in the summer season where air-mass precipitation (type A) is the predominant type of extreme precipitation, whereby its average share, 34 % of ExP, is the lowest compared to the frequency of such precipitation in RG1 and RG2. The group-average share of the individual origin-based types in the overall number of days with extreme precipitation reflects the actual situation at a vast majority of the stations, with type A occurring more frequently than any other type at 95 % of the stations in the group. In RG3, also cold front precipitation is less frequent (15 % of ExP) than in the other two groups. By contrast, the share of precipitation associated with the passage of different fronts (21 % of ExP) and with the passage of an occluded front (18 % of ExP) increases. Another feature distinguishing RG3 from RG1 and RG2 is the much higher frequency of precipitation ≥1 mm (36 days or 39 % of days in the season). The daily mean extreme precipitation in the above groups is similar, 32 mm, and represents merely 11 % of the seasonal total (Table 4.4). The number of wet days with precipitation ≥1 mm is the highest in the season compared to the other regional groups. A large proportion of air-mass precipitation recorded by mountain stations or by stations located in the foreland of the mountains on the windward side is orographic in nature. At inland stations, a proportion of precipitation is most probably associated with free convection, which develops within the low-pressure systems that lie over that part of the continent in the summer. In RG3, the values of the grouping variables deviate from their statistical distribution in the group only at four stations for type A and at two stations for TOTExP (Fig. 4.5).

Regional Group 4 (RG4) is formed by 128 weather stations located in a belt stretching from the British Isles across northern France, the Benelux countries, and Poland as far as the northern and central part of Eastern Europe. The stations clustered in RG4 are also located in the Scandinavian Peninsula, especially in its southwest end, and on Iceland (Fig. 4.4). In RG4, extreme precipitation is most often associated with the passage of several fronts (33 % of ExP), and its frequency in each of the other regional groups is lower. A large proportion of extreme precipitation is also associated with an occluded front (21 % of ExP on average) and with a cold front (18 % of ExP). Type Ff extreme precipitation predominates at 93 % of the RG4 stations. At individual stations in the group, the frequency maximums were demonstrated by type Fc (2 % of the stations), whereas at 5 % of the stations located in the western part of Eastern Europe, the largest proportion of extreme precipitation was accounted by type Fo. The daily average extreme precipitation total in RG4 is the lowest of all the regional groups distinguished (30.2 mm on average), ranging at the individual stations between 16.9 mm and 56.6 mm. Precipitation in excess of 1 mm is recorded on approximately 30 days in the season. Daily extreme precipitation represents on average 14 % of the seasonal precipitation total (Table 4.4).

Most of the stations observed no grouping variable outliers. The share of type Ff and type Fw precipitation and TOTExP at four weather stations deviates from the distribution of the other values in the cluster. By contrast, a large number of outliers characterises the number of days with precipitation ≥1 mm (Fig. 4.5).

The large share of extreme precipitation associated with the passage of several fronts in Western Europe, Central Europe, and Northern Europe is attributable to the location of the cluster stations on islands or coasts or near the Atlantic Ocean, and the relatively early stage of the low-pressure systems inflowing over these parts of the continent.

Regional Group 5 (RG5) is a cluster of 116 weather stations located to the south of the afore-described RG4. It includes stations in Central Europe, in Mediterranean countries, and in the south of Eastern Europe (Fig. 4.4). Compared to RG4, which lies to the north, RG5 is characterised by a lower share of precipitation associated with the passage of several fronts in the overall number of days with extreme precipitation (26 % of ExP on average) and a distinctly higher share of cold front precipitation, which is the most frequent type spotted here (28 % of ExP on average). A large percentage of extreme precipitation is also represented by air-mass precipitation (21 % of ExP on average). Type Fc precipitation represents the largest proportion of extreme precipitation at 54 % of the stations in the group, type Ff at 34 %, and type A at 11 %. Most of the stations characterised by the predominance of cold front extreme precipitation (type Fc) are located in mountain or highland areas of Western Europe and Central Europe, as well as at isolated stations in Eastern Europe. In RG5, extreme precipitation is rarely associated with an occluded front (10 % of ExP on average). Generally, the share of frontal precipitation in the overall number of days with extreme precipitation exceeds 79 %. In RG5, the daily mean extreme precipitation, which amounts to 37 mm, represents 15 % of the seasonal total (Table 4.4). At most stations, three grouping variables (type A, type Ff, type Fc precipitation frequency) show no outliers, yet attention must be drawn to the high number, compared to the other regional groups, of outliers as regard the frequency of type Fw precipitation, number of wet days with precipitation ≥1 mm, and TOTExP (Fig. 4.5). The mechanisms responsible for increased frequency of cold front precipitation in RG5 are similar to those described in RG1.

Regional group 6 (RG6), which is the last group distinguished in the season, consists of 113 weather stations, covering northern parts of the continent and individual stations on the British Isles (Fig. 4.4). In RG6, the largest proportion of extreme precipitation is associated with occluded fronts (39 % of ExP on average), with a high share attributable to type Fc (27 % of ExP on average). RG6 is also distinguishable for the smallest share of precipitation associated with the passage of a cold front (8 % of ExP). The daily mean extreme precipitation, which ranks among the lowest compared to the other groups (28 mm), represents 13 % of the mean seasonal precipitation (Table 4.4). At RG6 stations, precipitation ≥1 mm is registered on 30 days during the season on average. Outliers are sporadic, with the highest number observed for the daily mean extreme precipitation (Fig. 4.5). The RG6 stations are located along the track of low-pressure systems, in which occlusion starts, accelerated by the surface roughness, which increases as the low enters the land. Therefore, the frequency of occlusion and associated extreme precipitation is higher in this regional group than elsewhere in Europe.

The spatial distribution of the regional groups, and thus also the ExPT patterns characterising them in the summer, suggests clear regularities, such as a gradual decline in the share of air-mass precipitation from northwards and a marked decline in the share of type Fo precipitation, type Ff precipitation, and the daily mean extreme precipitation total southwards. The extreme precipitation totals averaged for RG6 also indicate a slight decline in the north–south direction.

4.2.3 Autumn

In autumn, the spatial distribution of the regional groups of extreme precipitation occurrence in Europe still shows a clear order, whereby of all the six regional groups distinguished in the season, two groups (RG1 and RG3) may be referred to as nonzonal, three groups (RG2, RG3, RG4) are characterised by the largest share of precipitation associated with the passage of several fronts, and two groups are distinguishable for a high share of air-mass precipitation (RG1, RG5).

Regional Group 1 (RG1) includes 68 weather stations, most of which are located in Central Europe, including northern Poland, and the coast of the Scandinavian Peninsula. The group also includes isolated stations in southeast Europe and Central Europe. Thus, RG1 comprises stations located in areas where water and land surfaces come into direct contact (coasts), in mountain and highland areas (Central Europe), or inland (Eastern Europe) (Fig. 4.6). The largest proportion of extreme precipitation in RG1 is represented by air-mass precipitation (30 % of ExP). Precipitation associated with the passage of several fronts is also frequent (27 % of ExP). Indeed, air-mass precipitation represents the largest share of extreme precipitation at 50 % of the stations in the group, and precipitation associated with the passage of several fronts at 45 %. At each station in the group, the frequency of both type A and type Ff precipitation is high compared to the other precipitation types. Some of the extreme air-mass precipitation at mountain and highland stations is fostered, in addition to atmospheric circulation, by orography, thus becoming orographic precipitation. The share of the other origin-based precipitation types (type Fo, type Fw, type Fc) ranges between 9 % and 16 % of ExP. The daily mean extreme precipitation, which amounts to 23 mm, represents 12 % of the season's total (Table 4.5). In RG1, precipitation ≥1 mm is recorded on average on 32 days during the season. Outliers are infrequent. Their greatest number is seen by precipitation associated with the passage of a warm front, which is the least frequent in this group (Fig. 4.7). In RG1, the difference between the most and least frequent origin-based types, which amounts to 20 %, is the least pronounced compared to the remaining regional groups in autumn.

Regional Group 2 (RG2) is formed by 133 stations lying within the belt between 45° and 55° north latitude, stretching from the western coasts of the continent to Eastern Europe (Fig. 4.6). RG2 is distinguishable for a distinct predominance of extreme precipitation associated with the passage of several fronts, which represents as much as 40 % of all extreme precipitation recorded in autumn. In RG2 cold front precipitation (type Fc) is also relatively frequent. Its share in the overall number of days with extreme precipitation is on average 23 % of ExP, while the share of the three other types (type A, type Fo, type Fw) ranges between 10 % of ExP (type Fw) and 12 % of ExP (type Fo). In the regional group in question, frontal precipitation constitutes as much as 91 % of ExP, proving, at the same time, that the conditions conducive to free convection occur relatively rarely (Table 4.5). Precipitation ≥1 mm occurs, on average, on 25 days in the season, which translates into 27 % of autumn days. The daily mean extreme precipitation in RG2, reaching 26 mm, represents 16 % of the total amount of precipitation in the season. In RG2, outliers are more numerous than in RG1 and are observed for daily mean extreme precipitation (TOTExP) (Fig. 4.7).

Regional Group 3 (RG3), which is another nonzonal group in the season, consists of 53 stations located mainly on the west coast of the Scandinavian Peninsula and in the western and southern parts of Southern Europe (Fig. 4.6). In RG3, similarly to most of the RG2 and RG4 stations, extreme precipitation is recorded most often on days with the passage of several fronts: it (type Ff) represents 37 % of ExP on average. Another 40 % of ExP is represented by types A (19 % of ExP) and Fc (21 % of ExP). The frequency of extreme occluded front precipitation is 13 % of ExP on average. The share of warm front precipitation is, similarly to other regional groups and seasons, the smallest (8 % of ExP). The most characteristic feature of RG3 is a very high daily mean extreme precipitation of 62 mm. The number of days with precipitation ≥1 mm is also high (39 days, i.e., 40 % of days in the season).

The daily mean extreme precipitation represents only approximately 11 % of the season's total (Table 4.5). The RG3 distinguished in autumn is the most "humid" regional group of all the RGs distinguished in the individual seasons; this is determined by the highest values of both the number of days with precipitation and the daily mean precipitation total resulting from the heavy precipitation in the season. The unusually heavy precipitation at RG3 stations is attributed to their location in highly elevated or coastal areas.

Regional Group 4 (RG4) is composed of 133 weather stations located above 50°N, most of which lie in the northern and central parts of Eastern Europe and in the Scandinavian Peninsula. RG4 also incorporates certain stations in Poland, Germany, Denmark, on the British Isles, and Iceland (Fig. 4.6). The most pronounced characteristic of RG2 is the high frequency of precipitation associated with the passage of several fronts compared to the other types of precipitation. Type Ff represents as much as 41 % of ExP. RG4 borders from the south with the territory of RG2, which is also distinguishable for the predominance of type Ff precipitation. However, RG4 differs from RG2 in that it shows greater frequency of type Fo precipitation (RG4, 25 % of ExP; RG3, 12 % of ExP) and smaller frequency of type Fc (RG4, 12 % of ExP; RG3, 23 % of ExP). The grouping variables averaged for RG4 reflect the actual contribution of the individual types to the overall number of days with extreme precipitation at 95 % of the stations where the maximum frequency is that of type Ff. At 5 % of the stations, the highest share is that of type Fo. In RG4, warm front precipitation (type Fw) and air-mass precipitation (type A) represent only 10 % of ExP and 8 % of ExP, respectively. Moreover, a comparison of the grouping variables, as averaged for the classes, reveals a pronouncedly higher average number of days with precipitation ≥1 mm (34 days or 37 % of days in the season; Table 4.5) in RG4, compared to RG2, and decidedly lower daily mean extreme precipitation (20.5 mm) (Fig. 4.6). The number of outliers ranges between 5 and 6 for the number of days with precipitation ≥ 1 mm (NWD) and the daily mean extreme precipitation (TOTExP) (Fig. 4.7).

Regional Group 5 (RG5), which consists of 54 weather stations, extends over southern Europe (Fig. 4.6). This group distinguishes itself from the other groups identified in autumn by clear predominance of air-mass precipitation, which constitutes 38 % of ExP, and relatively high daily mean extreme precipitation totals, reaching 40.7 mm, and representing as much as 28 % of the seasonal mean precipitation total (Table 4.5). Air-mass precipitation is the most frequently occurring type at 65 % of the stations in the group. The large contribution of daily mean extreme precipitation in the seasonal total results both from exceptionally high daily extreme precipitation totals and exceptionally low seasonal precipitation totals. In this group, also the frequency of precipitation ≥1 mm is low, and it is recorded on approximately 16 days in the season (18 % of days in the season). In RG5, precipitation in autumn is rare compared to other regional groups, but when it does occur, it is heavy (as are daily totals in RG3). The second most frequent precipitation type in RG5 is type Fc, which represents 25 % of ExP. This origin-based type of precipitation predominates at 31 % of the RG5 stations. Most of these stations are located on the western coasts of Southern Europe. In RG5, the frequency of precipitation associated with the passage of several fronts is much lower (17 % of ExP) than in Central Europe or Northern Europe (e.g., RG2 and RG3). In RG5, warm front precipitation is more frequent (9 % of ExP) than occluded front precipitation (7 % of ExP). Against the background of the other regional groups, RG5 is also distinguishable for the lowest frequency of frontal precipitation (type F), which represents, on average, only 62 % of ExP (Table 4.5). RG5 shows few outliers, most of which are observed for type Fo (Fig. 4.7).

Regional Group 6 (RG6) consists of 54 weather stations located mostly in the Scandinavian Peninsula. This group also includes stations on North Atlantic islands and isolated stations in the northern part of Western Europe (Fig. 4.6). In autumn, similarly to the seasons already discussed, Northern Europe is dominated by extreme precipitation associated with occlusion. Type Fo represents 47 % of ExP occurring at RG6 in autumn. It is only at 1 station that the maximum frequency is that of type Ff. The frequency of precipitation associated with occlusion is particularly high (usually more than 50 % of ExP) at stations in the central part of the southern end of the Scandinavian Peninsula. The second most frequent type, Ff, accounts for 28 % of ExP on average. That part of Europe rarely sees warm front extreme precipitation (type Fw) and cold front precipitation (type Fc), which represent 8 % and 6 % of extreme events, respectively. Altogether, the frontal types constitute 89 % of extreme precipitation. Type A is more frequent (11 % of ExP) than types Fw and Fc. The daily mean extreme precipitation at RG6 stations amounts to 28.9 mm, and varies widely between 11.2 mm and 56.2 mm, similarly to the average number of days with precipitation ≥1 mm (34 days), ranging among the stations from 20 to 50 days in the season. Despite such a huge variability of TOTExP and NWD at RG6 stations, outliers for these indices are recorded at most by 5 weather stations (Fig. 4.7).

4.2.4 Winter

Winter is characterized by the greatest dynamics of the atmosphere, caused, among other things, by the strongest thermal gradients of all the seasons. The spatial diversity of air temperatures, which determines the gradients, also contributes to movement faster than that in the other seasons of low-pressure systems and the weather fronts that accompany them. Winter extreme precipitation in Europe is associated with the passage of several fronts, which account for the largest proportion of extreme precipitation in four of the six regional groups distinguished in the group (RG1, RG2, RG3, RG4). It must also be noted that in winter the distribution of the RGs is characterised by lower spatial coherence than in summer and autumn, especially in the countries of Southern Europe (Fig. 4.8).

Regional Group 1 (RG1) is formed by 47 weather stations located in Southern Europe in a belt stretching from the Iberian Peninsula to the northeast coasts of the Black Sea and in the western coasts of the Scandinavian Peninsula (Fig. 4.8). As is shown by the averaged values of the grouping variables, in RG1 the largest proportion of extreme precipitation is represented by precipitation associated with the passage of several fronts (35 % of ExP). Type Ff represents the largest share of extreme precipitation at 68 % of the stations in the group. At the remaining 32 % of the stations, type Ff appears frequently, yet the largest share of extreme precipitation is that of cold front precipitation (type Fc). These stations are mainly located in the Iberian Peninsula. In RG1, type Fc precipitation accounts, on average, for 23 % of ExP, which is the largest share of the type in the overall number of days with extreme precipitation of all the groups distinguished in winter. RG1, compared to the three other groups characterised by predominance of type Ff precipitation (Gr2, RG3, RG4), is distinguished for slightly higher frequency of air-mass precipitation (type A), which represents 16 % of ExP. Type Fw accounts for an average of 13 % of ExP and type Fo for 11 %. The most characteristic feature of RG1 is the highest of all the regional groups daily mean extreme precipitation, reaching 53.8 mm, which changes from 36.0 to 86.4 mm, and represents, on average, 12 % of the exceptionally high seasonal precipitation total (Table 4.6). Precipitation ≥1 mm occurs on 35 days in the season. The cumulative share of frontal precipitation in the number of days with extreme precipitation in RG1 is 84 % of ExP on average. Outliers are few, and identified for three grouping variables only (Fig. 4.9).

Regional Group 2 (RG2) is the second largest RG, with 119 weather stations. At the same time, it is the most widely spread group in the continent, covering vast areas of Eastern Europe, the southern part of Central Europe, and the south of the continent, reaching slightly below 40°N. Isolated RG2 stations were identified in Northern Europe (Fig. 4.8). RG2 is another winter group in which the largest percentage of extreme precipitation is associated with the passage of several fronts (33 % of ExP). RG2 is also notable for the lowest range of variability in the averaged values of grouping variables, which is 20 %. This is attributable, in part, to the large spread of the RG2 stations across the continent. The frequency of the origin-based types at individual stations indicates that extreme precipitation is most frequently associated with the passage of several fronts at 86 % of the stations in the group, and for the remaining 14 % of the stations, with Fo (6 % of the stations, mainly in the east ends of the continent), type Fc (5 % of the stations in southeast Europe, in areas lying to the north and northeast of the Caspian Sea), and occasionally with types Fw and A.

The average share of the other origin-based precipitation types in the overall number of days with extreme precipitation in RG2 varies between 13 % of ExP for type Fc and 19 % of ExP for type Fo. In general, frontal precipitation represents 86 % of ExP occurring in this RG.

Notably, RG2 also shows some of the lowest (in addition to RG4) daily mean extreme precipitation (15.4 mm), which, however, compared to the other regional groups, represents a large part of the mean seasonal precipitation (15 % of the winter precipitation total) (Table 4.6). In RG2, the low daily extreme precipitation and low seasonal totals are accompanied by the smallest number of days with precipitation ≥1 mm (on average, 24 days, which translates into 27 % of days in the season). In spite of the large number of the weather stations in the group and their dispersal across the continent, the number of grouping variable outliers is low (at most 2). Only the share of type Fc precipitation in the total number of days with extreme precipitation at four weather stations deviates significantly from the distribution of the other values in the group (Fig. 4.9).

Regional Group 3 (RG3), which consists of only 58 weather stations and is non-zonal in nature, comprises Western Europe, reaching not farther than 20°E. Most of the stations forming the regional group lie in the western coasts of the Scandinavian Peninsula and along the border between Central and Southern Europe. Only a few stations in RG3 are located on the British Isles and in Iceland (Fig. 4.8). In RG3, extreme precipitation associated with the passage of several fronts is much more frequent (46 % of ExP) than any of the other origin-based types, which represent from 11 % of ExP (type A) to 16 % of ExP (type Fc). In this group, the frequency of all the frontal types accounts for 89 % of extreme precipitation events in winter. Furthermore, RG3 is distinguishable for the largest number of days with precipitation ≥1 mm (41 days, i.e., 45 % of days in the season). In RG3, the daily mean extreme precipitation is also high (33 mm), representing, however, a relatively low proportion of the seasonal total (9 % of the winter total) (Table 4.6). The high seasonal precipitation total, which averages 362 mm, and the substantial, compared to the continent as a whole, daily mean extreme precipitation total, which is actually low relative to the aforementioned seasonal amount of precipitation recorded in that part of Europe, prove that RG2 is among the most "precipitation reach" regional groups, meaning a high frequency of moderate daily precipitation compared to the other seasons. The inland stations in the group are located in elevated areas (highlands and mountains). Isolated outliers are recorded for most of the grouping variables (Fig. 4.9).

Regional Group 4 (RG4), which is the largest group in the season, comprising 150 weather stations, occupies the central part of the continent from France to Poland, and stretches as far as 45°E. Isolated stations belonging to the group lie on the British Isles, in Iceland, the Scandinavian Peninsula, and in the south of the continent (Fig. 4.8). Despite being the most numerous group in winter, RG4 is a strongly coherent in spatial term. In RG4, the largest proportion of extreme precipitation is represented by that associated with the passage of several fronts (49 % of ExP). It must be noted that type Ff predominates at all the 150 stations forming the cluster. In RG4, air-mass precipitation is the least frequent (only 6 % of ExP), whereas the shares of the other precipitation types ranges between 12 % of ExP (types Fc and Fw) and 16 % of ExP (type Fo). The daily mean extreme precipitation in RG4 is merely 15.3 mm and is counted among the lowest precipitation in winter. On the other hand, the number of days with precipitation ≥1 mm, which reaches

29 days on average (33 % of days in the season), does not deviate from the frequency of such precipitation in the other RGs (Fig. 4.8). The lowest amount of air-mass precipitation of all the groups distinguished in winter results in the highest share of frontal precipitation, which, altogether, represents as much as 94 % of extreme events in the group (Table 4.6). Nearly all values of the grouping variables, with one exception, fall within the range of non-outlier values (Fig. 4.9).

Regional Group 5 (RG5), which is the least spatially coherent one, consists of 61 weather stations located along a belt running across the centre of Europe from the north to the south and in southern parts of Western Europe and Eastern Europe (Fig. 4.8). Compared to the other RGs, this group is distinguished for the highest frequency of air-mass precipitation, which constitutes 38 % of ExP, and the lowest share of precipitation associated with the passage of several fronts (21 % of ExP). At mountain stations (including Śnieżka and Kasprowy Wierch), and those located in the foreland of the mountains on the side exposed to rain-bearing winds, some extreme air-mass precipitation is orographic. Extreme precipitation associated with the occurrence of the other types of weather fronts ranges from 10 % of ExP (type of Fw) to 15 % of ExP (type Fc). RG5 is characterised by the lowest frequency of frontal precipitation (type F) of all the groups selected in winter; its share in the overall number of days with extreme precipitation amounts to merely 62 %. The amount of daily mean extreme precipitation in RG5, which amounts to 23.3 mm, is moderate compared to the other groups, except that the value represents the largest, compared to the other groups, proportion of seasonal precipitation (15 % of the seasonal total) (Table 4.6). Precipitation ≥1 mm at the RG5 stations is recorded, on average, during 26 days (29 % of days in the season). Isolated outliers (not more than 4) are observed for most of the grouping variables (Fig. 4.9). At the stations belonging to RG5, which (as already noted) is characterised by considerable spatial dispersion, extreme precipitation is largely influenced by specific local conditions. An analogous nonzonal group RG5 was delimited in spring, covering a similar area (Figs. 4.2 and 4.8).

Regional Group 6 (RG6), which is the last regional group, consisting of 78 weather stations, is mainly located in the Scandinavian Peninsula, in the northern part of Western Europe, and on islands of the Atlantic Ocean (Fig. 4.8). Similar groups, although varying in size, were distinguished in all the seasons. In winter, as in the other seasons, the largest proportion of extreme precipitation in RG6 is associated with occlusion. Type Fo precipitation predominates at 85 % of the stations. A significantly large share of type Fo precipitation occurs at stations located in the middle of the southern part of the Scandinavian Peninsula, where occlusion is fostered by landform. In spring, the occurrence of an analogous group to the RG6 identified in winter is limited, except isolated stations, to the central part of the southern end of the Scandinavian Peninsula (Figs. 4.2 and 4.8). In winter, precipitation associated with the passage of various fronts is relatively frequent in RG6, representing 30 % of ExP. At 5 % of the weather stations located in the southern part of RG6, it is type Ff that predominates. In RG6, air-mass precipitation accounts for 11 % of ExP, and warm front precipitation (9 % of ExP) and cold front precipitation (6 % of ExP) is even rarer. All in all, frontal precipitation in RG6 represents 89 % of the extreme events. The daily mean extreme precipitation is 17.8 mm here, representing 11 % of the seasonal total (Table 4.6). Extreme precipitation ≥1 mm occurs, on average, on 32 days, that is, 36 % of days in the season. In RG6, none of the grouping variables shows outliers (Fig. 4.9).

4.3 Conclusions

The identification of strong spatial and seasonal variabilities in the occurrence of the origin-based precipitation types inspired a further study intended to obtain a more synthetic picture. Using cluster analysis of *k*-mean method and a careful selection of the grouping variables, six groups of stations with different structure of the occurrence of origin-based extreme precipitation types were identified (Table 4.7). Although the selected groups of stations fail to meet the spatial continuity criterion required to become true regions, they represent regional features of such a regional structure and hence they have been called regional groups (GR). Considering the enormous spatial and temporal variability of precipitation, which is even more true of extreme precipitation, it must be noted that the spatial regularity in the distribution of the regional groups prove an existence of factors that govern the occurrence of origin-based extreme precipitation types in Europe. Some of these groups include weather stations scattered across the study area, but rather than random their distribution is an effect of specific local influences on the processes leading to the development and controlling the amount of extreme precipitation.

These regional groups synthesise the most important features of the spatial and seasonal variability of origin-based types of extreme precipitation (structure of ExPT occurrence). In the regional groups of stations located in northern Europe (GR6 in all seasons) the largest proportion of extreme precipitation is associated with occluded fronts (~50 % ExP in winter, in autumn, and in spring, and ~40 % ExP in summer). In summer, air-mass precipitation is the dominant origin-based type in three regional groups covering southern Europe (GR1, GR2, GR3). In the southernmost regional group, GR2, type A precipitation accounts for almost 70 % of extreme precipitation.

In autumn, air-mass precipitation is the most frequent origin-based type in two regional groups (GR1, GR5), whereas in spring and in winter it is most frequent only in one of them (GR5). In autumn, one of these groups includes stations in the southern part of the continent, while type A precipitation accounts for a considerably lower proportion of extreme precipitation (just short of 40 % ExP) than in summer. In this regional group, much ExP is also associated with cold fronts (more than 25 % ExP). The other group identified in autumn featuring the highest frequency of type A precipitation, just like similar regional groups identified in spring and in winter, covers coastal mountain stations. An important role in the shaping of precipitation in these groups is played by local factors, mainly ground relief and dynamic process at the contact of land and ocean. In every season, the highest European frequencies of precipitation associated with a cold front are found within regional groups covering stations in the southern parts of Western, Central, and Eastern Europe, but this type accounts for the highest proportion of extreme precipitation (just short of 30 % ExP) only in one regional group identified in summer (GR5). In the other groups the maximum frequencies are accounted for by precipitation associated with the passage of various fronts (type Ff). Type Ff precipitation in general accounts for the largest proportion of extreme precipitation in regional groups covering Western and Central Europe, as well as the northern and central parts of Eastern Europe. This type reaches its highest frequency (~50 % ExP) in the winter group, GR 4, that covers Western and Central Europe. In general, precipitation associated with the passage of various fronts accounts for the largest part of extreme precipitation in winter across Europe, when the rate of cyclone movement in the area is the highest. In this season, type Ff accounts for the largest proportion of extreme precipitation in four regional groups of stations.

References

Bartholy J (1992) Meteorological choices to clustering precipitation data series and a case study for Hungary. Preprints, 12th Conference on Probability and Statistics in the Atmospheric Sciences, Toronto, ON, Canada. Bull Am Meteorol Soc J123–J124

Bednorz E, Kolendowicz L, Bielec-Bąkowska Z, Bokwa A, Żelazny M, Kicińska B, Lewik P, Nowosad M, Ustrnul Z (2003) Regionalizacje, Typologie i wyznaczanie sezonów klimatycznych z zastosowaniem analizy skupień (Regionalization, typology and determination of climatic seasons using cluster analysis). Prze Geof 158(1–2):11–32

Dorling SR, Davies TD, Pierce CE (1992) Cluster analysis: a technique for estimating the synoptic meteorological controls on air and precipitation chemistry: method application. Atmos Environ 26A:2575–2581

Gadgil S, Joshi NV (1983) Climatic clusters of India region. J Climatol 5:487–501

MacQueen JB (1967) Some methods for classification and analysis of multivariate observations. In: Proceedings of 5th Berkeley Symposium on Mathematical Statistics and Probability, University of California Press, Berkeley

Okołowicz W (1969) Klimatologia ogólna. PWN, Warszawa: 395

Stanisz A (2006) Przystępny kurs statystyki z zastosowaniem STATISTICA PL na przykładach z medycyny (Course of Statistics using medical examples). Tom 1. Statystyki podstawowe. StatSoft Polska, Kraków

Stanisz A (2007) Przystępny kurs statystyki z zastosowaniem STATISTICA PL na przykładach z medycyny Course of Statistics using medical examples). Tom 3. Analizy wielowymiarowe. StatSoft Polska, Kraków

Chapter 5
Air-Mass and Frontal Extreme Precipitation Occurrence and the North Atlantic Oscillation (NAO)

Abstract The North Atlantic Oscillation is a significant teleconnection pattern driving weather and climatic conditions in the Northern Hemisphere, including Europe. This chapter discusses relationships between the occurrence of extreme precipitation and daily NAO. The latter's influence on extreme precipitation in Europe was assessed by comparing empirical distribution functions of daily NAO values on days with extreme precipitation with distribution functions on dry days, while also taking into account origin-based precipitation types. The statistical significance of the differences between empirical distribution functions was tested with the Mann–Whitney *U* test (U M–W) and with the Kolmogorov–Smirnov test (K–S). The influence of the NAO phase on the spatial variability of the frequency of extreme precipitation was analysed using conditional probability of the occurrence of extreme precipitation during positive NAO phase (NAO>0, or NAO+) and the negative NAO phase (NAO<0, or NAO−).

Both the strength and the spatial extent of these relationships between NAO and extreme precipitation display seasonality, with a winter peak, and deppend on origin-based precipitation type. In autumn, the influence of NAO is weaker than in winter, but clearly stronger than in summer and spring. In winter, the positive NAO phase accompanies a majority of extreme precipitation in an overwhelming majority of the Northern and Western Europe, as moist air masses move over this part of Europe. A majority of precipitation occurring at that time is associated with either a cold front (Fc) or with an occluded front (Fo), depending on the region. In southern Europe advection from the Atlantic Ocean and the occurrence of extreme precipitation is associated mostly with the negative NAO phase. Extreme precipitation occurring then is associated with the passage of different fronts (Ff) or with an occluded front (Fo).

Keywords Extreme precipitation • NAO • Daily NAO • Positive NAO • Negative NAO • NAO phase

In studying the relationship between extreme precipitation occurrence and atmospheric circulation, consideration must be given to the North Atlantic Oscillation (NAO), which is the most important teleconnection pattern shaping weather and climatic conditions in the Northern Hemisphere, including Europe. The North Atlantic Oscillation and methods for calculating the NAO index have been widely discussed in the literature (Walker and Bliss 1932; van Loon and Rogers 1978;

E. Łupikasza, *The Climatology of Air-Mass and Frontal Extreme Precipitation*, Springer Atmospheric Sciences, DOI 10.1007/978-3-319-31478-5_5

Wallace and Gutzler 1981; Rogers 1984; Hurrell 1995; Jones et al. 1997; Serreze et al. 1997; Hurrell and Deser 2010). Previous research has confirmed that changes in the zonal air flow associated with swings in the NAO index lead to changes in the transport of humidity, and, as a result, to changes in the occurrence of precipitation and its amount across Europe (Hurrell and van Loon 1997: Dickson et al. 2000; Trigo et al. 2004; Zorita et al. 1992; Rodríguez-Puebla et al. 2001; Vicente-Serrano et al. 2009; Casanueva et al. 2014; Lima et al. 2015). The relationships between the occurrence of extreme precipitation and the NAO have been usually assessed by correlating time series of seasonal numbers of days with extreme precipitation with the NAO index (Haylock and Goodess 2004; Casanueva et al. 2014; Lima et al. 2015). However, a question arises whether seasonal values of the NAO index should be correlated with the number of days with extreme precipitation given that ExP occurs merely on a few days during a season.

This chapter investigates the relationship between the NAO teleconnection pattern and extreme precipitation occurrence using daily values of the NAO index, as published on the website of the Climate Prediction Center, National Centers for Environmental Prediction, National Weather Service, NOAA, U.S. Department of Commerce (CPC 2006) (http://www.cpc.ncep.noaa.gov). The relationship between extreme precipitation occurrence and NAO was analysed separately for each weather station and each season.

First, it was checked whether the intensity of the zonal atmospheric circulation described by the NAO index has a significant effect on the occurrence of extreme precipitation in Europe. To this end, the empirical distribution functions of daily NAO index values during days with extreme precipitation were compared with the corresponding distribution functions for days without precipitation. This comparison was also done for the various types of precipitation origin.

The statistical significance of the difference between the distribution functions of daily NAO values during days with and without extreme precipitation was checked using the nonparametric Mann–Whitney U (U M–W) and Kolmogorov–Smirnov (K–S) tests, which do not require any assumptions as to the statistical distribution of the random variable. The samples compared vary in size, which means that the data do not qualify for using parametric methods.

The Mann–Whitney test U (U M–W) is the most powerful nonparametric alternative to the t test used for assessing the statistical significance of differences between means for independent samples (Stanisz 2005). Wilks (2006) calls it a test of location (position) difference, thus making a reference to one of the statistical measures of position, that is, the median. According to Wilks (2006), the power of the U M–W test is close to that of the t test. The U M–W test verifies the null hypothesis that two randomly selected samples are drawn from the same population, or more strictly speaking, from populations with different medians. The null hypothesis is verified by means of a U-statistic, that is, the lower rank-sum calculated for the groups under study. To assign the ranks, the values of both groups taken together are arranged in ascending order, and then are assigned ranks, successive natural numbers. Identical values are assigned tied ranks representing the arithmetic means of the ranks assignable to them (Stanisz 2006). The U statistic is expressed by the following formula:

$$U = n_1 n_2 + \frac{n_1 (n_1 + 1)}{2} - R_1, \tag{5.1}$$

where n_1 and n_2 are sample sizes, and R_1 is the rank-sum of sample 1.

The Kolmogorov–Smirnov test (K–S) allows comparing the distribution of two random variables and is sensitive to differences in both the position and shape of the empirical distribution functions of the samples compared (Smirnov 1939; Kolmogorov 1941; Nikiforov 1994). In this case, the null hypothesis states that the distributions in the groups tested are similar, that is, are drawn from the same population. The empirical distribution function F_n for an n-element sample is defined as

$$F_n(x) = \frac{1}{n} \sum_{i=1}^{n} I_{Xi \le x} \tag{5.2}$$

where Xi is the value of variable x for the ith observation, and $I_{Xi \le x}$ is the characteristic function that takes value 1 here, where $Xi \le x$ and 0 otherwise. The Dn statistic of the K–S test is expressed by the following formula:

$$D_n = \max_x \left| F_n(x) - F(x) \right|, \tag{5.3}$$

where $F(x)$ is the empirical distribution function of the other sample. The value of statistic D_n is equal to the maximum value of the difference between the empirical distribution functions examined (Koronacki and Mielniczuk 2006; Stanisz 2006; Francuz and Mackiewicz 2007). Statistically significant testing results for both these tests (U M–W and K–S) form a basis for rejecting the null hypothesis and indicate the existence of relationships between the occurrence of daily extreme precipitation and the NAO Index.

According to some authors, the power of nonparametric tests is smaller than that of parametric ones (Stanisz 2005; Francuz and Mackiewicz 2007). Thus, the relationship between the occurrence of extreme precipitation and NAO was considered significant when the level of statistical significance (α) of the differences in empirical distributions, as indicated by each test, was $\alpha \le 0.05$. The significance level assumed translates into 5 % probability of error in rejecting the null hypothesis that the samples are drawn from the same population. The median of the NAO Index during days with extreme precipitation at stations characterised by a statistically significant relationship between the variables (extreme precipitation, NAO), as assessed by comparing the aforementioned empirical distribution functions, is shown in Fig. 5.1.

Statistically significant differences between the empirical distribution functions of NAO during days with extreme precipitation and days without extreme precipitation indicate a significant effect of the intensity of zonal circulation on extreme precipitation occurrence, yet they do not identify clearly the effect of the NAO phase (positive, negative) on the occurrence of extreme precipitation events. The impact of the NAO phase on spatial differences in the frequency of extreme precipitation was assessed on the basis of conditional probability of ExP occurrence during the positive (NAO>0 or NAO+) and negative (NAO<0 or NAO−) phase of the NAO (Fig. 5.2). The method for calculating the conditional probability is described in Sect. 6.3, Conditional probability of air-mass and frontal extreme precipitation in circulation types. The values of conditional probability of extreme precipitation occurrence during NAO+ and NAO− are low, resulting from the low number of extreme precipitation events during the individual seasons compared to the seasonal number of days with NAO+ and NAO− in the period under study. However, as shown in Chap. 6, it is a useful method of assessing the relationships between atmospheric circulation and climate elements even for rare extreme events. When calculating the quotient of extreme precipitation conditional probability in opposite NAO phases, the stations were identified where extreme precipitation occurrence depends on NAO. It was assumed that the NAO phase fosters extreme precipitation when the probability quotient exceeds 1.5, which means that the probability of precipitation occurrence during NAO+ (NAO−) is at least 50 % higher than during NAO− (NAO+). The probability of extreme precipitation occurrence during NAO+ and NAO− and the extreme precipitation probability quotient in opposite NAO phases are shown in Figs. 5.2 and 5.3. Similar analyses were carried out for each type of extreme precipitation origin. The results of the analysis of the conditional probability of extreme precipitation types and the probability quotient in opposite NAO phases are shown in Figs. 5.4 and 5.5 for spring, Figs. 5.6 and 5.7 for summer, Figs. 5.8 and 5.9 for autumn, and Figs. 5.10 and 5.11 for winter.

5.1 Relationships Between Extreme Precipitation Occurrence and North Atlantic Oscillation

5.1.1 Empirical Distribution Function of NAO During Days with Extreme Precipitation and During Days Without Precipitation

The effect of the North Atlantic Oscillation on extreme precipitation occurrence in Europe is seasonal in nature. In winter and autumn, the NAO distribution functions during days with extreme precipitation and during days without precipitation differed significantly at more than half the stations in question (Table 5.1). In wintertime, the statistical significance of these differences was <0.001 at 33 % of the stations, which means that the probability of error in null hypothesis rejection did not exceed 1 %. In the other seasons, the intensity of zonal circulation had a much poorer impact on the occurrence of extreme precipitation events in Europe. Both in

Table 5.1 Percentage of station with significant and insignificant relationship between extreme precipitation occurrence and NAO. The relationship between ExP occurrence and NAO assessed by comparison of the empirical distribution of NAO during days with extreme precipitation and during days without precipitation

	Percentage of stations		
Season	α >0.05	0.05> α ≥0.001	α <0.001
MAM	74	13	13
JJA	66	21	13
SON	44	31	25
DJF	41	26	33

$\alpha > 0.05$, $0.05 > \alpha \geq 0.001$, $\alpha < 0.001$ levels of statistical significance

summer and in autumn, the empirical distribution functions of NAO during the days analysed differed at a significance level of $\alpha < 0.001$ only at 13 % of the stations.

The spatial variability in the relationship between NAO and extreme precipitation are illustrated well by the NAO median calculated on the basis of days with extreme precipitation in the individual seasons (Fig. 5.1). In winter, the continent was found to be clearly divided into the northern part, comprising Western, Central, and Northern Europe, where extreme precipitation occurred more often during the positive phase of the NAO, and the southern part, where extreme precipitation occurrence was associated with the negative phase of the NAO. The highest median values and, at the same time, the most significant differences in the empirical distributions considered ($\alpha < 0.001$ for both tests) were found in the southernmost part of the Scandinavian Peninsula and at sites located between the eastern part of the French lowland and the Pomeranian Lakeland in Poland. The highest increase in the winter frequencies of precipitation extremes in that part of the continent is associated with stronger than average surface westerlies across the middle latitudes of the Atlantic onto Europe during the positive phase of the NAO (Hurrell et al. 2003).

Notably, there is a small area in the northernmost part of the Scandinavian Peninsula where the relationship is reverse: precipitation extremes are recorded more often during days with a negative NAO Index. In Southern Europe, the lowest NAO median was observed in the western part of the Iberian Peninsula, which indicates the strongest effect of NAO– on the occurrence of precipitation extremes in that part of the continent (Fig. 5.1). The negative NAO phase leads to intensified westerly circulation over the Mediterranean area and increased frequency of low-pressure systems (Hurrell et al. 2003), and precipitation extremes. During the positive NAO phase, the Mediterranean area sees anomalous northerly flow (Hurrell et al. 2003), resulting in drier conditions. In autumn, the effect of zonal circulation on the occurrence of extreme precipitation in Europe is weaker than in winter, and the nature of the relationship is utterly different. At most of the stations, the median of that circulation index was negative, as was the case in Western Europe, in Central Europe, and at stations located on the eastern side of the Scandinavian Mountains. In autumn, the NAO median was positive only in the western coast of the Scandinavian Peninsula and on Atlantic islands (Fig. 5.1). In Southern Europe, the occurrence of

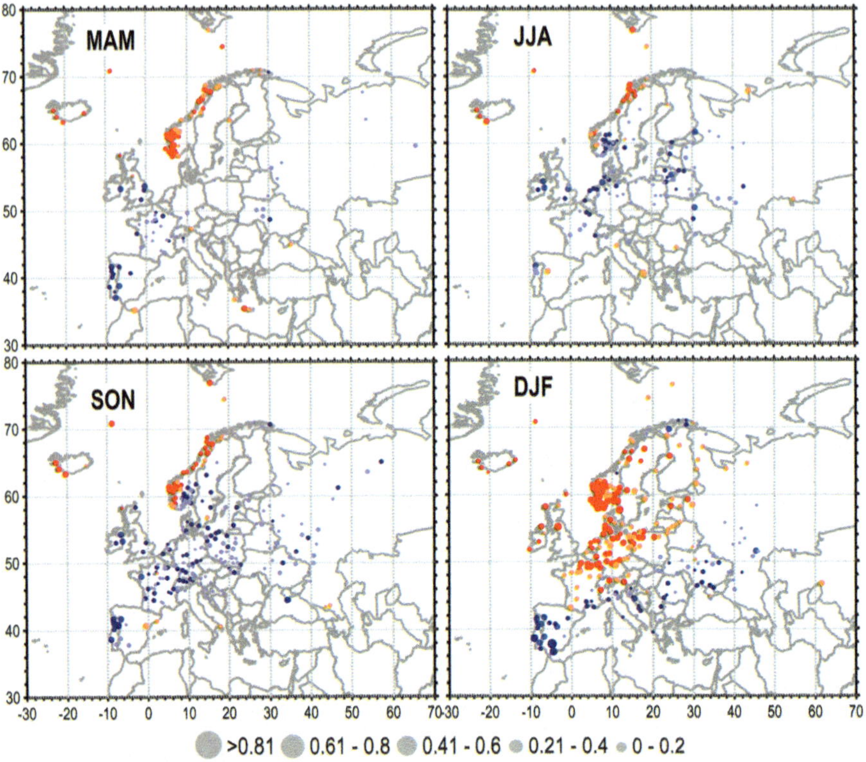

Fig. 5.1 Median of North Atlantic Oscillation (NAO) from days with extreme precipitation at stations with statistically significant relationships between NAO and extreme precipitation occurrence: *red dots* positive NAO median, calculated if majority of extreme precipitation occurred in NAO+ *blue dots* negative NAO median, calculated if majority of extreme precipitation occurred in NAO–. Significance of relationships assessed on the base of differences in empirical distribution functions of NAO at days with extreme precipitation and at days without precipitation, *Light colours* distributions significantly different at $0.05 > \alpha \geq 0.001$, *dark colours* distributions significantly different at <0.001

extreme precipitation was significantly related to increased intensity of zonal circulation only in the western part of the Iberian Peninsula, where precipitation extremes were associated with the negative phase of NAO. Elsewhere in Southern Europe, no significant links were found between the variables in question, except for isolated stations where precipitation extremes were associated with NAO+.

In summer, and even more so in spring, the effect of NAO on extreme precipitation occurrence was much less pronounced than in winter and autumn. In spring, the NAO distribution functions on days with ExP and days without precipitation were significantly different at some stations in the west and south-west of Europe. At western coasts of the Scandinavian Peninsula and on islands in the northern parts of the Atlantic Ocean, most precipitation extremes were recorded during days with NAO+ (positive median). In the western part of the Iberian Peninsula, in France, and on the British Isles most springtime precipitation extremes were recorded on days with NAO– (Fig. 5.1). In summer, the nature of the relationship between extreme precipitation and the North Atlantic Oscillation in Northern Europe was similar to that in spring: most precipitation extremes were recorded during the positive NAO phase. However, the extent of these relationships in the Scandinavian Peninsula was smaller, being limited to the Lofoten area and isolated stations in the southern part of the coast (Fig. 5.1). In summer, the highest possibility of ExP occurrence during NAO– characterises stations located in a belt extending across the central part of the continent from the British Isles to the Gulf of Finland, including the southern part of the Scandinavian Peninsula on the eastern side of the Scandinavian Mountains and single stations in the western part of the East European Plain (Fig. 5.1). Statistically significant relationships between the NAO index and precipitation and the sequence of wet days in summer was also identified by Casanueva et al. (2014). The nature of that relation was opposite to that in the winter season: in Western and Central Europe, the occurrence of precipitation and sequences of wet days was fostered by the negative NAO phase, and by contrast, in the south of Europe, by the positive phase. The probability analysis performed shows that the relationships between NAO and precipitation in summer found by Casanueva et al. (2014) apply also to extreme precipitation (Fig. 5.1).

5.1.2 *Probability of Extreme Precipitation Occurrence in Opposite Phases of North Atlantic Oscillation*

Although in many areas of Europe extreme precipitation occurrence shows a clear relationship with the NAO phase (see Sect. 5.1), in the period under study, most stations recorded extreme precipitation during days with both NAO+ and NAO–. Figure 5.2 illustrates the spatial distribution of the probability of extreme precipitation occurrence during days with NAO+ (right column) and NAO– (left column). Schematic maps showing the quotient of extreme precipitation probability in opposite NAO phases are shown in Fig. 5.3. Warm colours indicate the stations where the probability of precipitation occurrence during days with NAO+ is higher by at least 50 % than during days with NAO–, and cold colours indicate an opposite situation, that is, higher probability during days with NAO– than during days with NAO+. A clear disproportion in the occurrence of ExP in opposite NAO phases allows determining whether and which NAO phase is conducive to such precipitation in Europe.

The distribution of the probability of precipitation occurrence changes depending on the NAO phase and season. In winter, the highest probability of precipitation during NAO– was observed at stations situated in the southern part of Europe and some stations in the western coasts of the Scandinavian Peninsula (more than 2–4 %), whereas during NAO+, stations located in the southwest part of the Scandinavian Peninsula, on the British Isles, and in Iceland (more than 4 %). In the area extending from the Bay of Biscay to the Gulf of Finland, ExP was recorded

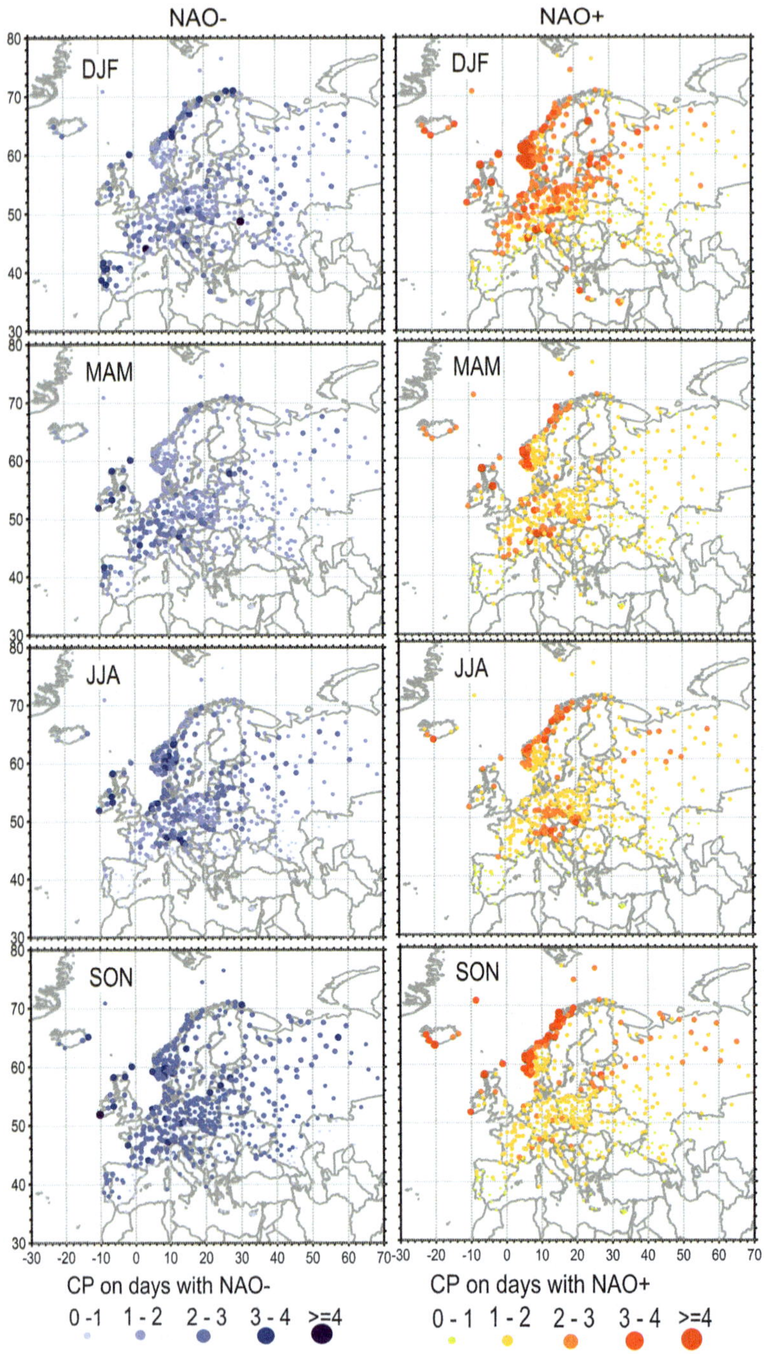

Fig. 5.2 Conditional probability (CP) of extreme precipitation occurrence in positive (NAO+) and negative (NAO−) NAO phases. *Warm colours* probability of extreme precipitation occurrence in NAO+ at least 50 % higher than in NAO−, *cold colours* opposite case

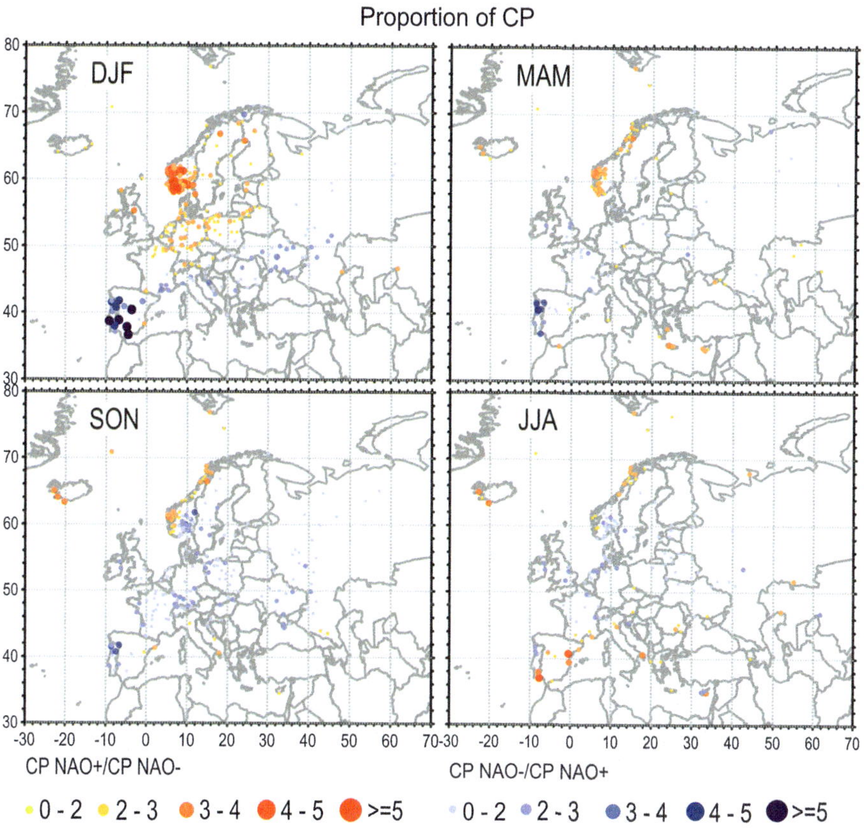

Fig. 5.3 Quotient of conditional probabilities in opposite NAO phases (Quotient CP). *Warm colours* probability of extreme precipitation occurrence in NAO+ at least 50 % higher than in NAO–, *cold colours* opposite case

during 2–3 % of days with NAO+, while elsewhere in Europe its probability did not exceed 1 %. The positive NAO phase fostered extreme precipitation clearly in the southwest part of the Scandinavian Peninsula, where its probability during NAO+ was more than three times, and at some stations, five times, higher than during NAO–. In the west and centre of the Iberian Peninsula, the probability of ExP during NAO– is more than five times higher than during NAO+ (Figs. 5.2, and 5.3). As it was found in previous section (5.1.1), in autumn, the number of stations where ExP occurrence is significantly related to NAO is similar to that in winter; however, the conditional probability indicates that the relationships are poorer. The likelihood of ExP during NAO– exceeds 2–3 % in western and northern Europe. During NAO+ ExP is most likely to occur in the western coast of the Scandinavian Peninsula and on Iceland (>4 %), and at some sites in the northern part of Eastern Europe (Fig. 5.2). Despite the differences in the distribution of probability in NAO+ and NAO–, in fact, the positive NAO phase fosters precipitation occurrence only in the western coasts of the Scandinavian Peninsula and on islands in the northern part of the Atlantic Ocean. In those areas, the occurrence of ExP during NAO+ is more than twice as probable as during NAO–. Even though at the other sites the likelihood of ExP during NAO– is higher than during NAO+, the NAO–/NAO+ ratio at most stations does not exceed 2 (Figs. 5.2 and 5.3).

In spring, a higher probability of precipitation during NAO– is seen by stations in southwest Europe and on the British Isles, while in summer, by stations located in the band stretching from the British Isles to the Gulf of Finland, as well as in the south of the Scandinavian Peninsula, and on the northern side of the Alps. At most of those stations, the probability does not exceed 2 %. In both seasons, the likelihood of ExP in NAO+ is higher at the west coasts of the Scandinavian Peninsula, in the northern foreland of the Alps, and at some sites in Iceland and Great Britain (spring). In summertime, the importance of the Alps and the Carpathians in generating extreme precipitation during the positive NAO phase becomes apparent (Fig. 5.2). In spring and summer, pronounced differences in the probability of extreme precipitation occurrence in opposite NAO phases were found only in the western coasts of the Scandinavian Peninsula and in Iceland (spring and summer), in the western part of the Iberian Peninsula (spring), in the coasts of the North Sea, and at individual stations in the western part of the East European Plain. In those seasons, the positive NAO phase drives extreme precipitation also at some stations in Southern Europe, whereby, in spring, the effect is more pronounced in the eastern, while in summer, in the western and central, parts of Southern Europe (Figs. 5.2 and 5.3).

Generally speaking, the quotient of extreme precipitation conditional probability in opposite NAO phases confirms the results obtained by comparing the distribution functions of the NAO during days with ExP and days without precipitation (Fig. 5.2). NAO drives ExP most strongly in winter. In the other seasons, the importance of the NAO phase is rather poor.

5.2 Relationships Between Occurrence of Origin-based Extreme Precipitation Types and North Atlantic Oscillation

Conditional probability analysis was also used to assess the relationship between the type of extreme precipitation origin and NAO. Obviously, the probability of the occurrence of extreme precipitation types in NAO+ and NAO– is even smaller than for all ExP, but its spatial distribution (Figs. 5.4, 5.6, 5.8 and 5.10) proves the relevance of the type of precipitation origin in relationship with NAO. Similarly to all precipitation, the probability and the quotient of extreme precipitation conditional probability in opposite NAO phases was calculated for each type (Figs. 5.5, 5.7, 5.9 and 5.11).

The distribution of the conditional probability of precipitation types does not change radically from season to season, but the magnitude of probability is subjected to noticeable seasonal variations, especially in the negative NAO phase,

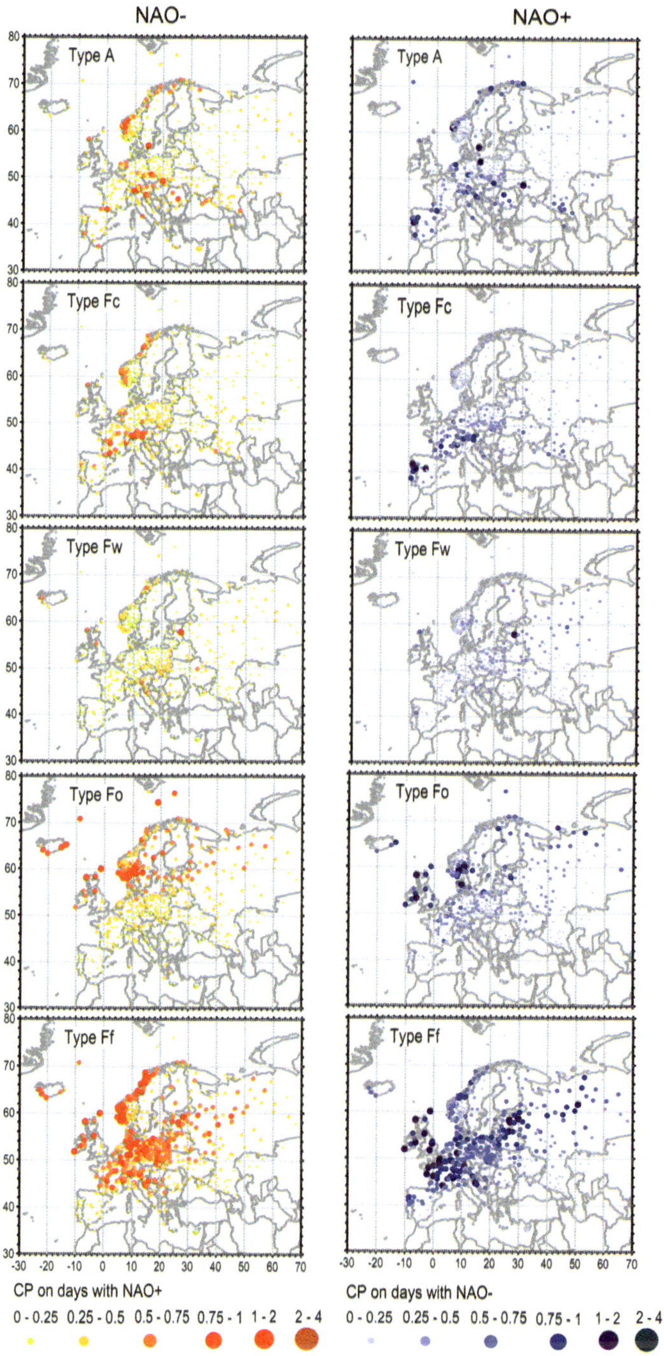

Fig. 5.4 Conditional probability (CP) of origin-based extreme precipitation types in positive (NAO+) and negative (NAO−) NAO phases, SPRING. *Warm colours* probability of extreme precipitation occurrence in NAO+ at least 50 % higher than in NAO−, *cold colours* opposite case

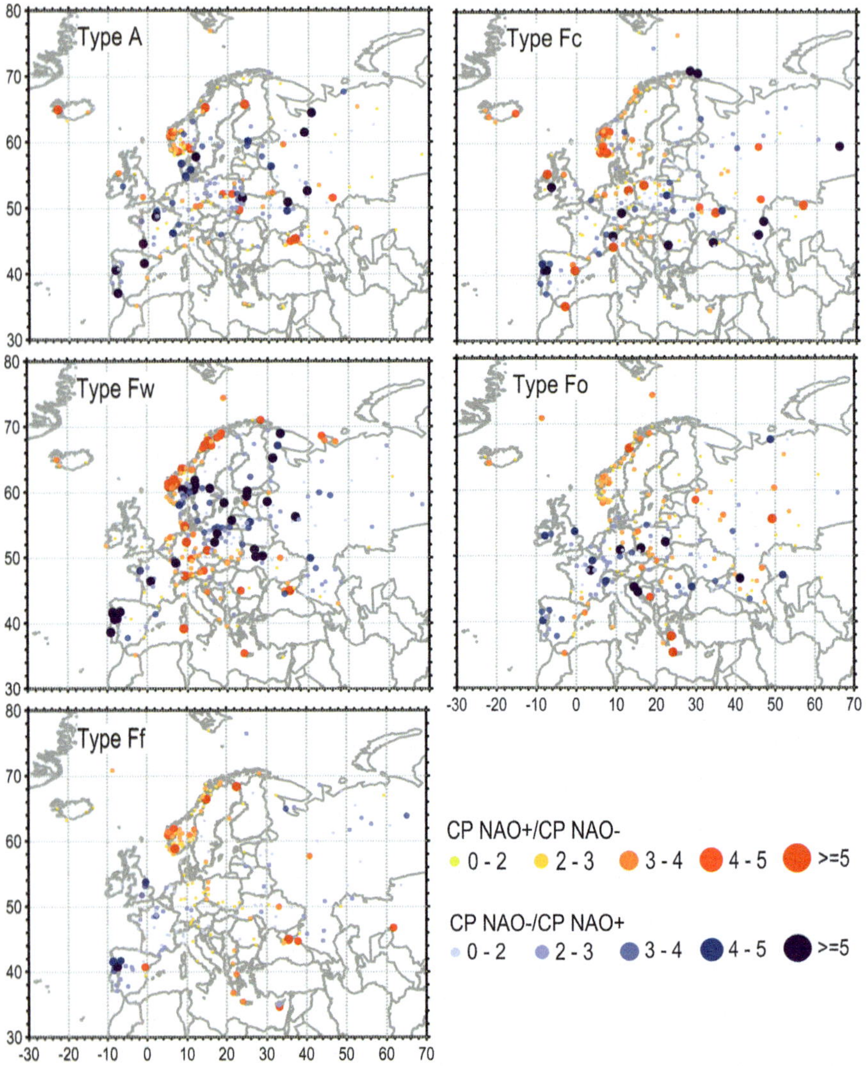

Fig. 5.5 Quotient of conditional probabilities of the occurrence of extreme precipitation types in opposite NAO phases (Quotient CP), SPRING. *Warm colours* probability of extreme precipitation occurrence in NAO+ at least 50 % higher than in NAO−, *cold colours* opposite case

which means that the NAO phase influences the seasonal pattern of extreme precipitation frequency in Europe. Across the seasons, both during the positive and negative phase of the NAO, the highest probability is that for precipitation associated with the passage of several fronts (type Ff). In winter, type Ff precipitation is most likely to occur during NAO+, and in autumn, during NAO−. Intensive westerly circulation (NAO+) increases the probability of such precipitation (type Ff) in the western coastal areas of the Scandinavian Peninsula and on Atlantic islands; this is

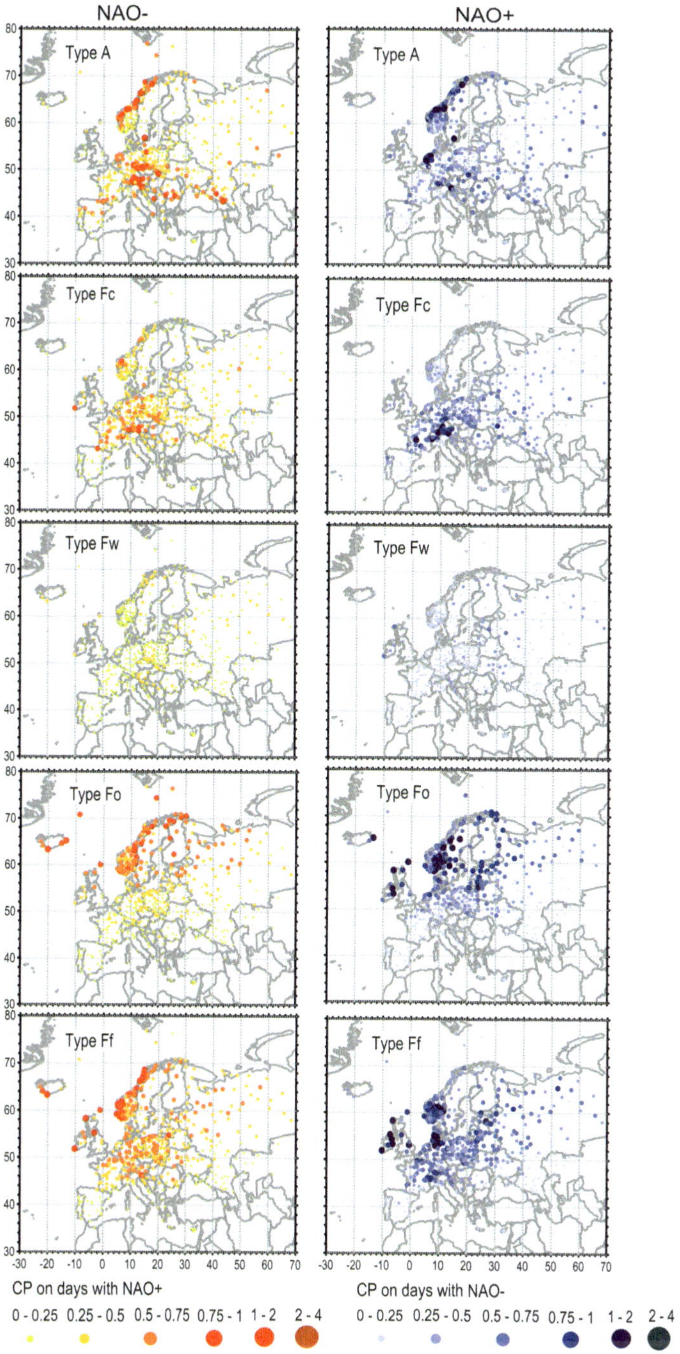

Fig. 5.6 Conditional probability (CP) of origin-based extreme precipitation types in positive (NAO+) and negative (NAO−) NAO phases, SUMMER. *Warm colours* probability of extreme precipitation occurrence in NAO+ at least 50 % higher than in NAO−, *cold colours* opposite case

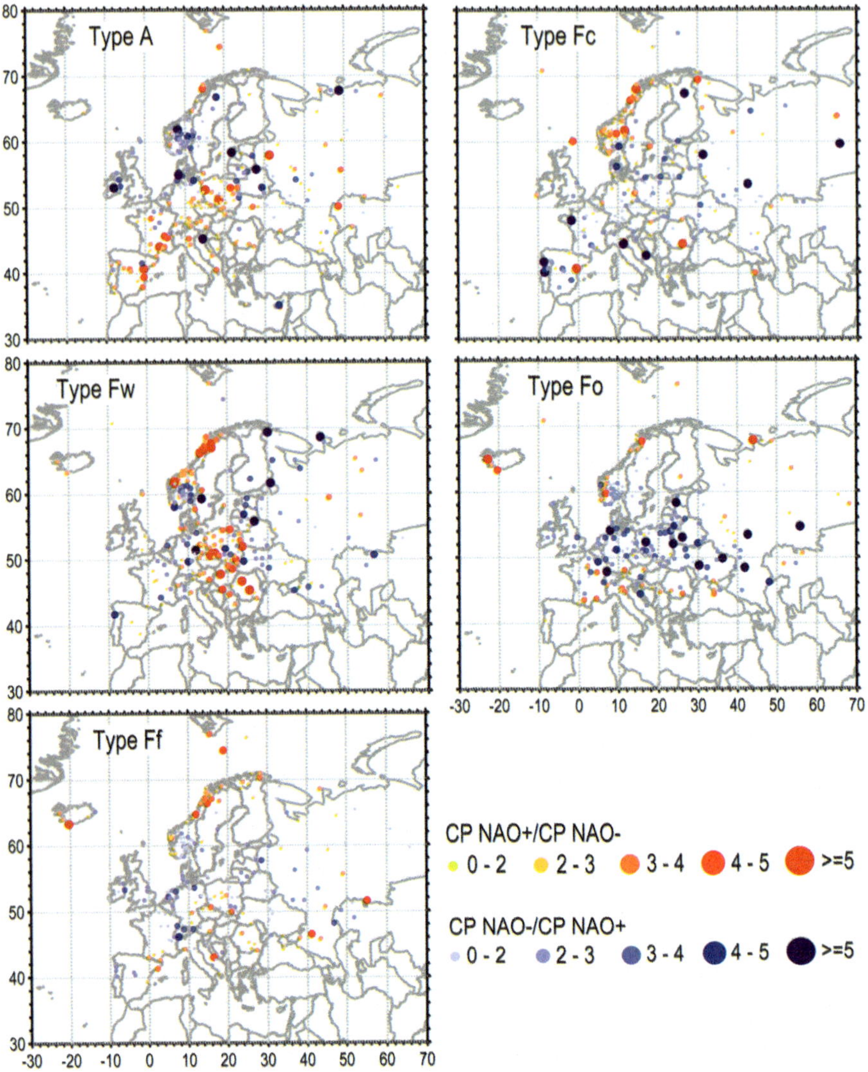

Fig. 5.7 Quotient of conditional probabilities of the occurrence of extreme precipitation types in opposite NAO phases (Quotient CP), SUMMER. *Warm colours* probability of extreme precipitation occurrence in NAO+ at least 50 % higher than in NAO–, *cold colours* opposite case

associated with a northeastward shift in the Atlantic storm activity during positive phase of the NAO with enhanced activity from Newfoundland into northern Europe (Rogers 1990, 1997; Hurrell and van Loon 1997; Serreze et al. 1997; Alexandersson et al. 1998) and intensification and increased frequency of storms in the vicinity of Iceland and the Norwegian Sea (Serreze et al. 1997; Deser et al. 2000). The probability of extreme precipitation of Ff type is also high during negative NAO in Western Europe, the northwest part of Eastern Europe (Figs. 5.4, 5.6, 5.8 and 5.10), and in winter, also in the west coast of the Scandinavian Peninsula and in the western part of the Iberian Peninsula (Fig. 5.10). In Northern Europe, on the eastern side of the Scandinavian Mountains, the highest probability both during NAO+ and NAO− is that of occluded front precipitation (type Fo). However, the spatial distribution of the probability changes from season to season. For NAO+, in spring, the highest values were observed at stations between Iceland and the southernmost part of the Scandinavian Peninsula (Fig. 5.4), in summer, in the central and northern parts of the Scandinavian Peninsula, and on islands in the north of the Atlantic (Fig. 5.6), while in autumn, especially in winter, across Northern Europe (Figs. 5.8 and 5.10). During NAO− the probability of type Fo precipitation in that part of Europe is higher only in summer and in autumn (Figs. 5.6 and 5.8).

The positive NAO phase increases the probability of cold front extreme precipitation (type Fc) in different parts of the western coasts of the Scandinavian Peninsula, depending on the season, as well as on the northern side of the Alps (in spring, summer, and autumn), on the Frisian Islands (in winter) and at some stations in France. During the negative NAO phase, the highest probability of such precipitation was also found on the northern side of the Alps [in spring (Fig. 5.4), summer (Fig. 5.6), and autumn (Fig. 5.8)] and in the western part of the Iberian Peninsula [spring (Fig. 5.4), autumn (Fig. 5.8), and winter (Fig. 5.10)]. In autumn, the entire southwestern Europe is distinguishable for a higher likelihood of type Fc precipitation (Fig. 5.8).

The variability in the probability of warm front precipitation is small in both spatial and seasonal terms as a result of the low frequency of this precipitation type. Increased probability of air-mass precipitation is observed at stations located in the vicinity of the mountains or in mountain areas, and on the southeast coast of the North Sea. In summer, during NAO+ the probability of air-mass precipitation is the highest on the northern and western side of the mountains, which proves that its frequency is fostered by orography. During the negative NAO phase, increased likelihood of air-mass precipitation is observed at sites on the southeast coasts of the North Sea (Frisian Islands) in summer (Fig. 5.6) and autumn (Fig. 5.8), on the west coast of the Scandinavian Peninsula, especially in summer (Fig. 5.6), on the southern coast of the Baltic Sea in autumn (Fig. 5.8), and in the northern end of the Scandinavian Peninsula in autumn (Fig. 5.8) and winter (Fig. 5.10).

The analysis of the probability quotient for origin-based precipitation types demonstrates that in winter NAO+ fosters all types of extreme precipitation in the southern part of the Scandinavian Peninsula. However, in western and central Europe NAO+ drives, above all, cold front precipitation, and at some stations, also warm front precipitation. In the south of the continent, the negative NAO phase fosters mainly occluded front precipitation and precipitation associated with the passage of several fronts (Fig. 5.11). At some sites in the northern part of southern Europe, a higher probability of ExP during NAO− than NAO+ is also observable for warm front extreme precipitation (Fw) (Fig. 5.10).

In autumn, the NAO phase has a significant effect on the occurrence of precipitation associated with an occluded front (type Fo). The negative NAO phase fosters this type of precipitation in western and central Europe. In the south of Norway, on the eastern side of the Scandinavian Mountains, markedly higher probability during NAO– than NAO+ is observable for warm front, cold front, and air-mass precipitation (Fig. 5.9). In the season, NAO+ enhances various types of precipitation in different parts of the west coast of the Scandinavian Peninsula and the occurrence of precipitation associated with the passage of several fronts and an occlusive front on islands in the northern part of the Atlantic Ocean (Fig. 5.9).

In the other seasons, the influence of the NAO on extreme precipitation occurrence declines, yet showing certain regularities. In summer, NAO– contributes to the occurrence of type Fo precipitation in the belt extending from northeastern France to the south of eastern Europe, and NAO+ fosters warm front precipitation in the western coasts of the Scandinavian Peninsula, in Central Europe as far as the northern part of the Balkan Peninsula, and air-mass precipitation in the continental parts of western, central, and southern Europe (Fig. 5.7).

In spring, the relationships in question are even poorer. Some spatial order in the probability quotient in opposite NAO phases is noticeable only for warm front precipitation. The likelihood of this precipitation type during NAO– is clearly higher than during NAO+ at many stations in Northern Europe (on the eastern side of the Scandinavian Mountains), in the north of Central Europe, in the western part of the East European Plain, and at some stations in southwestern Europe. An opposite situation, that is, a greater likelihood of type Fw precipitation during NAO+ than NAO–, was found on the west coast of the Scandinavian Peninsula, along the border between Western and Central Europe (from the Frisian Islands, across Germany and southwards), and in the south of the continent (Fig. 5.5).

The spatial distribution of the probability quotient in opposite NAO phases for air-mass precipitation is complicated, with coherent areas where the NAO phase markedly fosters such precipitation hard to be identified, which is attributable to the higher contribution of local factors to the occurrence of air-mass precipitation than of general atmospheric circulation. However, it can be observed that the probability of type A ExP during NAO– is greater than during NAO+ in the southern part of western and central Europe and in southern Europe in winter (Fig. 5.11), as well as in western Europe and the northern part of southern Europe in autumn (Fig. 5.9). The positive NAO phase drives type A precipitation in Central and southwestern Europe in summer (Fig. 5.7). The poor effect of NAO on the occurrence of occluded front precipitation in the southern part of the Scandinavian Peninsula on the eastern side of the Scandinavian Mountains is also noteworthy, proving the huge significance of landform for the formation of Type Fo precipitation in that part of Europe.

Both the comparison of the empirical distribution functions of NAO during days with and without ExP and the conditional probability of ExP during days with NAO+ and NAO– show a significant effect of the North Atlantic Oscillation on the frequency of extreme precipitation in Europe during winter and autumn. The rela-

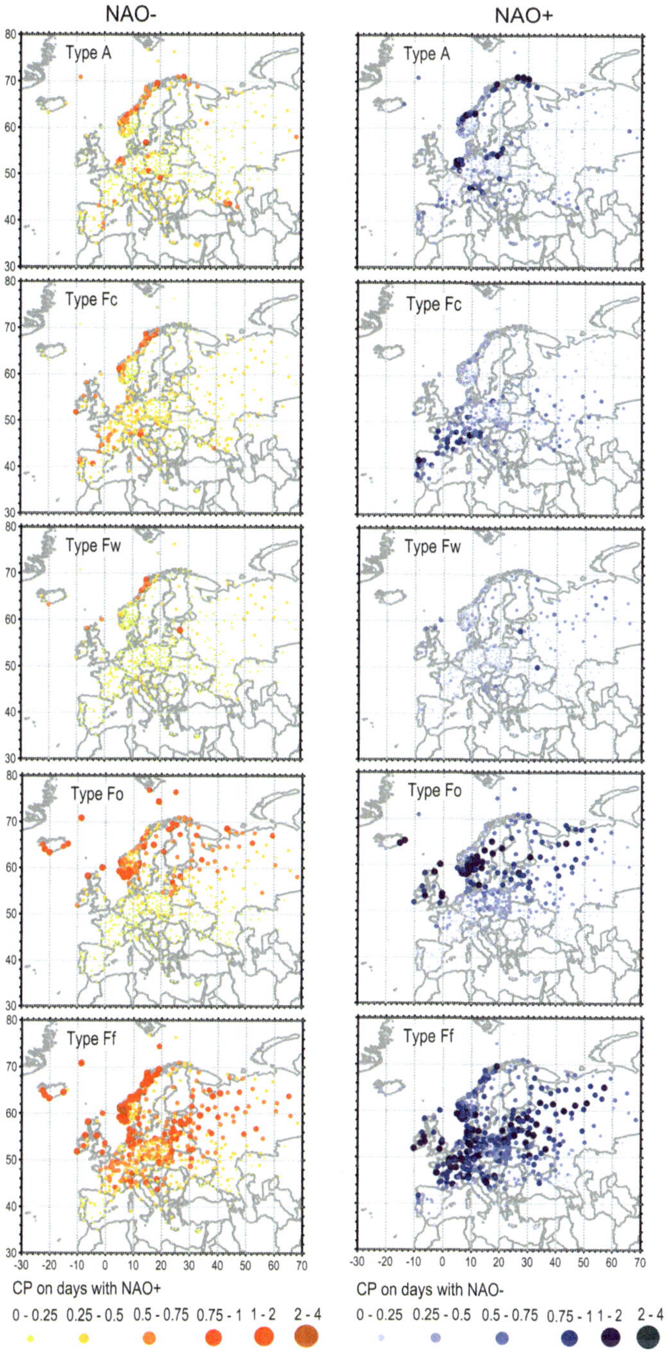

Fig. 5.8 Conditional probability (CP) of origin-based extreme precipitation types in positive (NAO+) and negative (NAO–) NAO phases, AUTUMN. *Warm colours* probability of extreme precipitation occurrence in NAO+ at least 50 % higher than in NAO–, *cold colours* opposite case

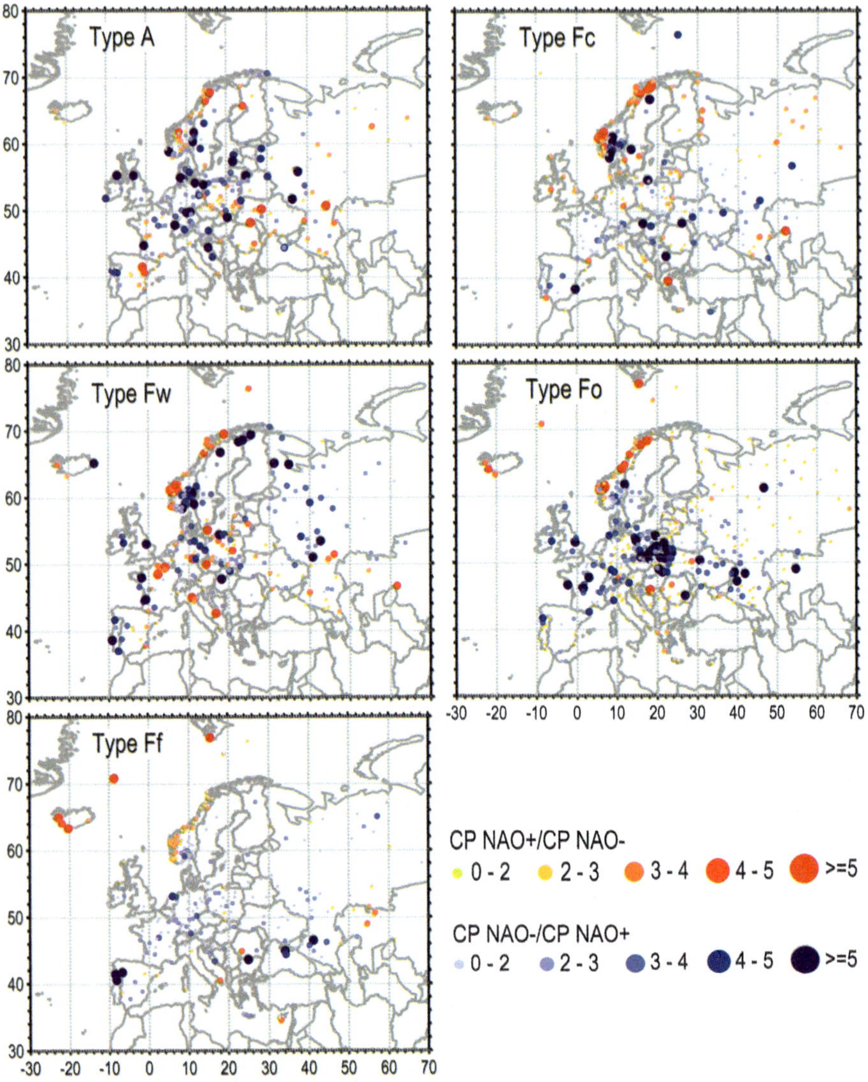

Fig. 5.9 Quotient of conditional probabilities of the occurrence of extreme precipitation types in opposite NAO phases (Quotient CP), AUTUMN. *Warm colours* probability of extreme precipitation occurrence in NAO+ at least 50 % higher than in NAO−, *cold colours* opposite case

tionships between NAO and ExP are poor in Eastern Europe. The negative NAO phase increases the frequency of weather fronts in southern Europe (Tramblay et al. 2013), as a result of which that part of Europe observes an increase in the likelihood during NAO−, especially in the western part of the Iberian Peninsula, which lies the closest to the cyclogenesis area in the Atlantic Ocean. Thus, in the south of Europe, the negative NAO phase leads to an increase in the frequency of precipitation regard-

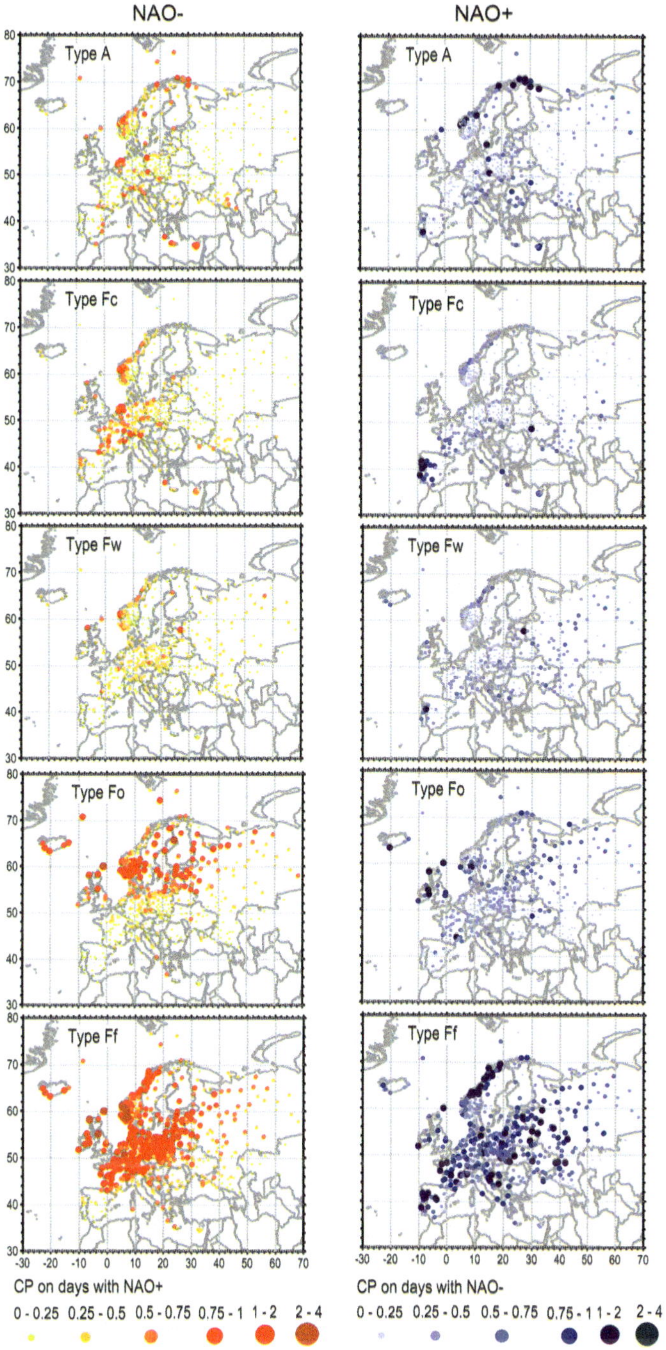

Fig. 5.10 Conditional probability (CP) of origin-based extreme precipitation types in positive (NAO+) and negative (NAO−) NAO phases, WINTER. *Warm colours* probability of extreme precipitation occurrence in NAO+ at least 50 % higher than in NAO−, *cold colours* opposite case

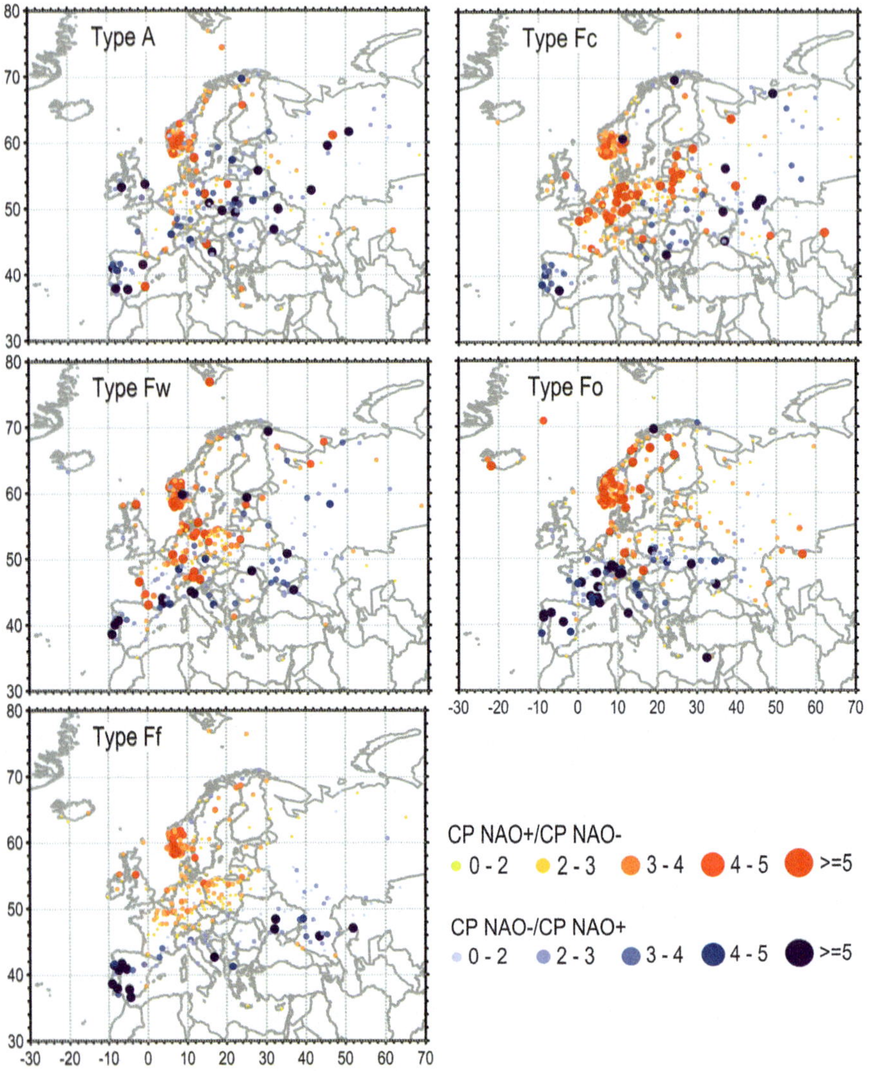

Fig. 5.11 Quotient of conditional probabilities of the occurrence of extreme precipitation types in opposite NAO phases (Quotient CP), WINTER. *Warm colours* probability of extreme precipitation occurrence in NAO+ at least 50 % higher than in NAO–, *cold colours* opposite case

less of its extremity (Houssos and Bartzokas 2006; Trigo 2006; Lima 2015) as well as extreme precipitation, notably in the western part of the area, as was confirmed by research by Krichak et al. (2014) and Lima et al. (2015). In the south of Europe, the negative NAO phase favoures the occurrence of type Ff, Fo, and Fw precipitation. The declining trends in extreme precipitation observable since 1970 are linked to the growing frequency of the positive NAO phase (Lima 2015). The positive NAO phase in winter fosters extreme precipitation in the southernmost part of the

Scandinavian Peninsula and in Western and Central Europe. Here, the NAO phase is related, in the first place, to cold and warm front precipitation. In winter, the NAO Index is the key determinant of extreme precipitation trends across Europe; its effect on the occurrence of precipitation extremes is greater than that of local factors (Haylock and Goodess 2004). In the summer, the importance of local factors strongly increases.

5.3 Conclusions

The frequency of extreme precipitation in Europe depends significantly on macroscale and mesoscale atmospheric circulation. In many parts of Europe the occurrence of extreme precipitation is linked to the intensity of the zonal circulation measured with the NAO index. This index was analysed by comparing empirical distribution functions of NAO on days with ExP and on days without ExP. Both the strength and the spatial extent of these relationships are seasonal in nature, with the peak in winter, and vary depending on the origin-based type of precipitation in question. In autumn, the influence of NAO on ExP occurrence is weaker than in winter, but clearly stronger than in summer and in spring. In winter, across most of Europe (including Northern, Western, and Central Europe), a majority of extreme precipitation coincided with the positive NAO phase and with an inflow of moist air-masses associated with it over the parts of Europe involved. The majority of precipitation events recorded at that time in Western Europe are associated with a cold front (Fc), although in the south of the Scandinavian Peninsula with an occluded front (type Fo.). In the southern part of the continent the air advection from over the Atlantic Ocean and the occurrence of extreme precipitation are associated primarily with a negative NAO phase. This extreme precipitation is associated with the passage of various fronts (type Ff) or with an occluded front (type Fo). In autumn, the nature of the relationship between ExP and NAO changes: most of the extreme precipitation recorded in Central and Western Europe, as well as in the western part of the Iberian Peninsula, is observed during NAO–, whereas that along the western Scandinavian coast and on the Atlantic islands is related to NAO+. The negative NAO phase in Western and Central Europe, as well as in the southern part of Eastern Europe, mostly coincides with type Ff and type Fo precipitation. NAO+ favours the occurrence of these two precipitation types in the northwestern part of Europe, including the Atlantic islands and the western Scandinavian coast.

References

Alexandersson H, Schmith T, Iden K, Tuomenvirta H (1998) Long-term variations of the storm climate over NW Europe. The Glob Ocean Atmos Syst 6:97–120

Casanueva A, Rodríguez-Puebla C, Frías MD, González-Reviriego N (2014) Variability of extreme precipitation over Europe and its relationships with teleconnection patterns. Hydrol Earth Syst Sci 18:709–725
Climate Prediction Center (2006) National Centers for Environmental Prediction, National Weather Service NOAA US Department of Commerce. http://www.cpc.ncep.noaa.gov
Deser C, Walsh JE, Timlin MS (2000) Arctic sea ice variability in the context of recent atmospheric circulation trends. J Climate 13:617–633
Dickson RR, Osborn TJ, Hurrell JW, Meincke J, Blindheim J, Adlandsvik B, Vigne T, Alekseev G, Maslowski W (2000) The Arctic Ocean response to the North Atlantic Oscillation. J Climate 13:2671–2696
Francuz P, Mackiewicz R (2007) Liczby nie wiedzą skąd pochodzą. Przewodnik po meteorologii i statystyce nie tylko dla psychologów. Wydawnictwo KUL, Lublin
Haylock MR, Goodess CM (2004) Interannual variability of European extreme winter rainfall and links with mean large-scale circulation. Int J Climatol 24:759–776. doi:10.1002/joc.1033
Houssos EE, Bartzokas A (2006) Extreme precipitation events in NW Greece. Adv Geosci 7:91–96
Hurrell JW (1995) Decadal trends in the North Atlantic Oscillation: regional temperatures and precipitation. Science 269:676–679
Hurrell JW, Deser C (2010) North Atlantic climate variability: the role of the North Atlantic Oscillation. J Mar Syst 79:231–244. doi:10.1016/j.jmarsys.2009.11.002
Hurrell JW, Van Loon H (1997) Decadal variations in climate associated with the North Atlantic Oscillation. Clim Change 36:301–326
Hurrell JW, Kushnir Y, Visbeck M, Ottersen G (2003) An overview of the North Atlantic Oscillation. In: Hurrell JW, Kushnir Y, Ottersen G, Visbeck M (eds) The North Atlantic oscillation. Climatic significance and environmental impact, Geophysical monograph series 134. American Geophysical Union, Washington, DC, pp 1–35
Jones PD, Jonsson T, Wheeler D (1997) Extension to the North Atlantic Oscillation using early instrumental pressure observations from Gibraltar and south-west Iceland. Int J Climatol 17:1433–1450
Kolmogorov A (1941) Confidence limits for an unknown distribution function. Ann Math Stat 12:461–463
Koronacki J, Mielniczuk J (2006) Statystyka dla studentów kierunków technicznych i przyrodniczych (Statistics for technical and environmental purposes). Wydawnictwa naukowo-Techniczne, Warszawa
Krichak SO, Breitgand JS, Gualdi S, Feldstein SB (2014) Teleconnection: extreme precipitation relationships over the Mediterranean region. Theor Appl Climatol 117:679–692. doi:10.1007/s00704-013-1036-4
Lima MIP, Espírito Santo F, Ramos AM, Trigo RM (2015) Trends and correlations in annual extreme precipitation indices for mainland Portugal, 1941–2007. Theor Appl Climatol 119:55–75
Nikiforov AM (1994) Algorithm AS 288: exact two-sample Smirnov test for arbitrary distributions. Appl Stat 43(1):265–270
Rodríguez-Puebla C, Encinas AH, Sáenz J (2001) Winter precipitation over the Iberian Peninsula and its relationship to circulation indices. Hydrol Earth Syst Sci 5:233–244
Rogers JC (1984) The association between the North Atlantic Oscillation and the Southern Oscillation in the Northern Hemisphere. Mon Weather Rev 112:1999–2015
Rogers JC (1990) Patterns of low-frequency monthly sea level pressure variability (1899–1986) and associated wave cyclone frequencies. J Climate 3:1364–1379
Rogers JC (1997) North Atlantic storm track variability and its association to the North Atlantic Oscillation and climate variability of northern Europe. J Climate 10:1635–1647
Serreze MC, Carse F, Barry RG, Rogers JC (1997) Icelandic low cyclone activity: climatological features, linkages with the NAO, and relationships with recent changes in the Northern Hemisphere circulation. J Climate 10:453–464

Smirnov NV (1939) On the estimation of the discrepancy between empirical curves of distribution for two independent samples. Bull Mosc Univ 2:3–14

Stanisz A (2005) Biostatystyka (Biostatistics). Wydawnictwo Uniwersytetu Jagiellońskiego, Kraków

Stanisz A (2006) Przystępny kurs statystyki z zastosowaniem STATISTICA PL na przykładach z medycyny (Course of Statistics using medical examples). Tom 1. Statystyki podstawowe. StatSoft Polska, Kraków

Tramblay Y, Adlouni SE, Servat E (2013) Trends and variability in extreme precipitation indices over Maghreb countries. Nat Hazards Earth Syst Sci 13:3235–3248. doi:10.5194/nhess-13-3235-2013

Trigo IF (2006) Climatology and inter-annual variability of Storm-Tracks in the EuroAtlantic sector: a comparison between ERA40 and NCEP/NCAR reanalyses. Clim Dyn 26:127–143

Trigo RM, Pozo-Vazquez D, Osborn TJ, Castro-Diez Y, Gamiz-Fortis S, Esteban-Parra MJ (2004) North Atlantic Oscillation influence on precipitation, river flow and water resources in the Iberian Peninsula. Int J Climatol 24:925–944

van Loon H, Rogers JC (1978) The seesaw in winter temperatures between Greenland and Northern Europe. Part I: general description. Mon Weather Rev 106:296–310

Vicente-Serrano S, Beguería S, López-Moreno J, El Kenawy A, Angulo M (2009) Daily atmospheric circulation events and extreme precipitation risk in Northeast Spain: the role of the North Atlantic Oscillation, Western Mediterranean Oscillation, and Mediterranean Oscillation. J Geophys Res-Atmos 114:D08106. doi:10.1029/2008JD011492

Walker GT, Bliss EW (1932) World weather. V. Mem R Meteorol Soc 4:53–84

Wallace JM, Gutzler DS (1981) Teleconnections in the geopotential height field during the Northern Hemisphere winter. Mon Weather Rev 109:784–812

Wilks DS (2006) Statistical methods in the atmospheric sciences, International geophysics series. Elsevier, Amsterdam

Zorita E, Kharin V, von Storch H (1992) The atmospheric circulation and sea surface temperature in the North Atlantic area in winter: their interaction and relevance for Iberian precipitation. J Climate 5:1097–1108

Chapter 6
Air-Mass and Frontal Extreme Precipitation Occurrence in Synoptic Situations

Abstract The frequency of origin-based types of extreme precipitation in Europe displays a regionally varied relationship with the direction of the advection and the type of the pressure system. This chapter discusses relationships between origin-based types of precipitation and mesoscale circulation. For the purpose of this chapter, circulation types were identified for each grid point between 30°N and 80°N and 30°W and 70°E at the interval of 2.5° of latitude and longitude. The direction of air advection was identified with the geostrophic wind direction. The type of the pressure system was identified using the vorticity formula. The study also identified the conditional probability of the occurrence of each origin-based type of extreme precipitation in each of these circulation types.

All origin-based types of extreme precipitation occur in both cyclonic and anticyclonic situations, but the occurrence frequency of the former is higher. Dependencies between the occurrence of frontal precipitation and atmospheric circulation are characterised by a greater degree of regional variability than these between air-mass precipitation and atmospheric circulation. Extreme air-mass precipitation has the strongest association with the circulation in Southern Europe, in mountainous areas, on the southeastern North Sea coast, and on the western slopes of the Scandinavian Mountains. Extreme precipitation of the frontal type tends to be linked with both the direction of air advection and the type of the pressure system. In cyclones these relationships display strong seasonality, whereas in anticyclones they remain stable throughout the year. The breakdown of frontal precipitation into the types of associated fronts helps better define the relationship between its occurrence and atmospheric circulation, especially in cyclonic synoptic situations.

Keywords Circulation types • Precipitation probability • Cyclone • Anticyclone • Synoptic situation

Atmospheric circulation, which is a key determinant of weather and climatic conditions, is among the most frequently considered factors in studies aiming to explain both the temporal and the spatial variability of many climate elements, including precipitation (Barry and Perry 1973; Harmann and Winkler 1991; Yarnal 2000). The occurrence of weather fronts, which are noncontinuity zones originating in strongly baroclinic areas of the atmosphere (Djurić 1994), is strictly related to synoptic situations, which determine the direction of air-mass advection, as a result of

E. Łupikasza, *The Climatology of Air-Mass and Frontal Extreme Precipitation*, Springer Atmospheric Sciences, DOI 10.1007/978-3-319-31478-5_6

which air-masses having different thermodynamic properties (temperature and humidity) come into contact. Thus, the same types of weather fronts generating heavy precipitation may be associated with the inflow of air from different directions depending on the location on the continent.

This chapter describes the relationships between the occurrence of origin-based extreme precipitation types (air-mass and frontal precipitation) and atmospheric circulation by seasons. The direction of air-mass flow is not equivalent to the direction of weather front movement. As is observed by Wilby (1998), including weather fronts in studies of the relationship between precipitation and atmospheric circulation provides much better results than research based exclusively on the distribution of atmospheric pressure.

6.1 Classification of Circulation Types for Europe

To determine the relationships between air advection and the type of pressure system and the occurrence of origin-based precipitation types (air-mass and frontal), separate calendars of synoptic situations were created for each node point every 2.5° in longitude and latitude between 30°N and 80°N and between 30°W and 70°E. To this end, use was made of data on the daily mean pressure at sea level in the years 1951–2008 obtained from the NCEP Reanalysis (National Center for Environmental Prediction) database, from NOAA/OAR/ESRL PSD, Boulder, Colorado, USA, from the website http://www.esrl.noaa.gov/psd (Kalnay et al. 1996).

The direction of air advection was determined based on the direction of geostrophic wind (Ustrnul 1997). The speed of geostrophic wind was used to select nonadvective situations for which the boundary value $v \leq 2$ m*s^{-1} was adopted. The type of the pressure system was determined on the basis of the vorticity equation (Piotrowski 2009). Eighteen model synoptic situations were identified (Table 6.1). A similar approach to investigating the relationship between extreme precipitation

Table 6.1 Circulation types (synoptic situations) used for analysis of the relationships between origin-based extreme precipitation types types and atmospheric circulation

Number	Symbol	Description	Number	Symbol	Description
1	Na	Northern anticyclonic	10	Nc	Northern cyclonic
2	NEa	Northeastern anticyclonic	11	NEc	Northeastern cyclonic
3	Ea	Eastern anticyclonic	12	Ec	Eastern cyclonic
4	SEa	Southeastern anticyclonic	13	SEc	Southeastern cyclonic
5	Sa	Southern anticyclonic	14	Sc	Southern cyclonic
6	SWa	Southwestern anticyclonic	15	SWc	Southwestern cyclonic
7	Wa	Western anticyclonic	16	Wc	Western cyclonic
8	NWa	Northwestern anticyclonic	17	NWc	Northwestern cyclonic
9	Ba	Nonadvective anticyclonic	18	Bc	Nonadvective cyclonic

and atmospheric circulation, involving the creation of separate catalogues of synoptic situations for each station based on grid data, but only for three weather stations, was used by Tolik et al. (2007). An automatic typology prepared on the basis of data for node points every 2.5° in latitude and longitude may, in some cases, reflect the characteristics of local circulation, especially in areas with complex landform.

The names of the types of synoptic situations distinguished correspond to the typology of Z. Ustrnul (1997), which is based, in turn, on the typology proposed by T. Niedźwiedź (1981). The types of atmospheric circulation, which are, according to Yarnal (2000), its climatology record, were determined by an objective (automatic) method. Automatic (objective) circulation classification methods are also characterised by a large dose of subjectivity given the multiple decisions to be taken when constructing classification algorithms. However, in contrast to manual methods, automatic methods offer the possibility of reproducing the classification procedure in a way that guarantees achieving identical results (Yarnal 2000).

The occurrence of precipitation depends not only on the state of the atmosphere near the surface, but also in its higher layers. However, it was proven that the synoptic situation observed near the surface is strongly related with the occurrence of precipitation and explains the precipitation regimes in a very good way (Sumner 1996; Saaroni et al. 2010). The usefulness of sea-level pressure charts for determining the geostrophic flow that reflects well the actual movements of the atmosphere in the extratropical zone was also observed by Hare (1964). Weather fronts, that is, the central element in the present study, are only marked in surface charts, and as a result this study relies on synoptic situations identified on the basis of sea-level pressure. The frequency of the occurrence of the individual origin-based extreme precipitation types, which is a rare phenomenon by definition, is low, and therefore when analysing the relationships between its occurrence and the circulation types at the individual weather stations, use was made of a simplified classification, consisting of the following ten synoptic types: N+NEa, E+SEa, S+SWa, W+NWa, Ba and N+NEc, E+SEc, S+SWc, W+NWc, Bc.

6.2 Frequency of Air-Mass and Frontal Extreme Precipitation Occurrence in Synoptic Situations

The relationships between origin-based extreme precipitation types (ExPT) and atmospheric circulation were examined by analysing the relative frequency of their occurrence in high-pressure systems (combined frequency in all anticyclonic situations) and low-pressure systems (combined frequency in all cyclonic situations) and for the ten synoptic situations types (shortened version), as described in Sect. 6.1, Classification of Circulation Types for Europe. The frequency is expressed as a percentage of all the extreme precipitation (ExP) that occurred at a given stations in the season concerned. The results of the research are presented in the form of climatic charts of the relative frequency of ExPT in each synoptic situation. As in the previous chapters, histograms were prepared to show the percentage of the stations where the relative frequency falls within specific brackets. The research was conducted by seasons. In this chapter, precipitation linked to stationary front and discontinuity line was excluded from the analysis because of its low frequency.

6.2.1 *Extreme Precipitation of All Origins (ExP)*

6.2.1.1 Frequency of Extreme Precipitation (ExP) in Anticyclones (ACT) and Cyclones (CCT)

In Europe, on average in the season, approximately 28 % of ExP is recorded in anticyclones (ACT) and as much as 72 % of ExP in cyclones (CCT). However, the range of variability in the frequency of ExP in ACT is wide (Table 6.2). In each season, the occurrence of extreme precipitation in high-pressure systems is favoured by the presence of land features, such as high mountain ranges, which force convection (orographic convection), leading, first, to water vapour condensation, and then, when the conditions are favourable, to precipitation. In the European continent, a high frequency of ExP in ACT is recorded by Alpine countries, the Caucasus area, the central part of the southern end of the Scandinavian Peninsula, and the Iberian Peninsula (Fig. 6.1). The high frequency of extreme precipitation in high-pressure

Table 6.2 Descriptive statistics of the frequency of extreme precipitation (ExP) [%] in anticyclones (*ACT*) and cyclones (*CCT*) in Europe, January 1951–February 2008

Pressure system	Season	Average (±SE)	Confidence intervals		Min	Max	SD	CV	ME	Quartiles	
			−95 %	+95 %						Lower	Upper
ACT	MAM	27.5 (±0.6)	26.3	28.8	1.3	80.1	14.0	50.7	25.6	17.5	35.1
	JJA	27.8 (±0.6)	26.5	29.1	0.0	95.3	14.5	52.2	25.4	19.0	33.8
	SON	28.3 (±0.7)	26.8	29.7	0.0	87.2	16.7	59.0	25.9	15.5	38.5
	DJF	28.1 (±0.9)	26.4	29.8	0.0	88.4	19.3	68.6	23.9	12.5	38.7
CCT	MAM	72.5 (±0.6)	71.2	73.7	19.9	98.7	14.0	19.3	74.4	64.9	82.5
	JJA	72.2 (±0.6)	70.9	73.5	4.7	100.0	14.5	20.1	74.6	66.2	81.0
	SON	71.7 (±0.7)	70.3	73.2	12.8	100.0	16.7	23.3	74.1	61.5	84.5
	DJF	71.9 (±0.9)	70.2	73.6	11.6	100.0	19.3	26.8	76.1	61.3	87.5

SE standard error, *Min* minimum value, *Max* maximum value, *SD* standard deviation, *CV* coefficient of variability, *ME* median

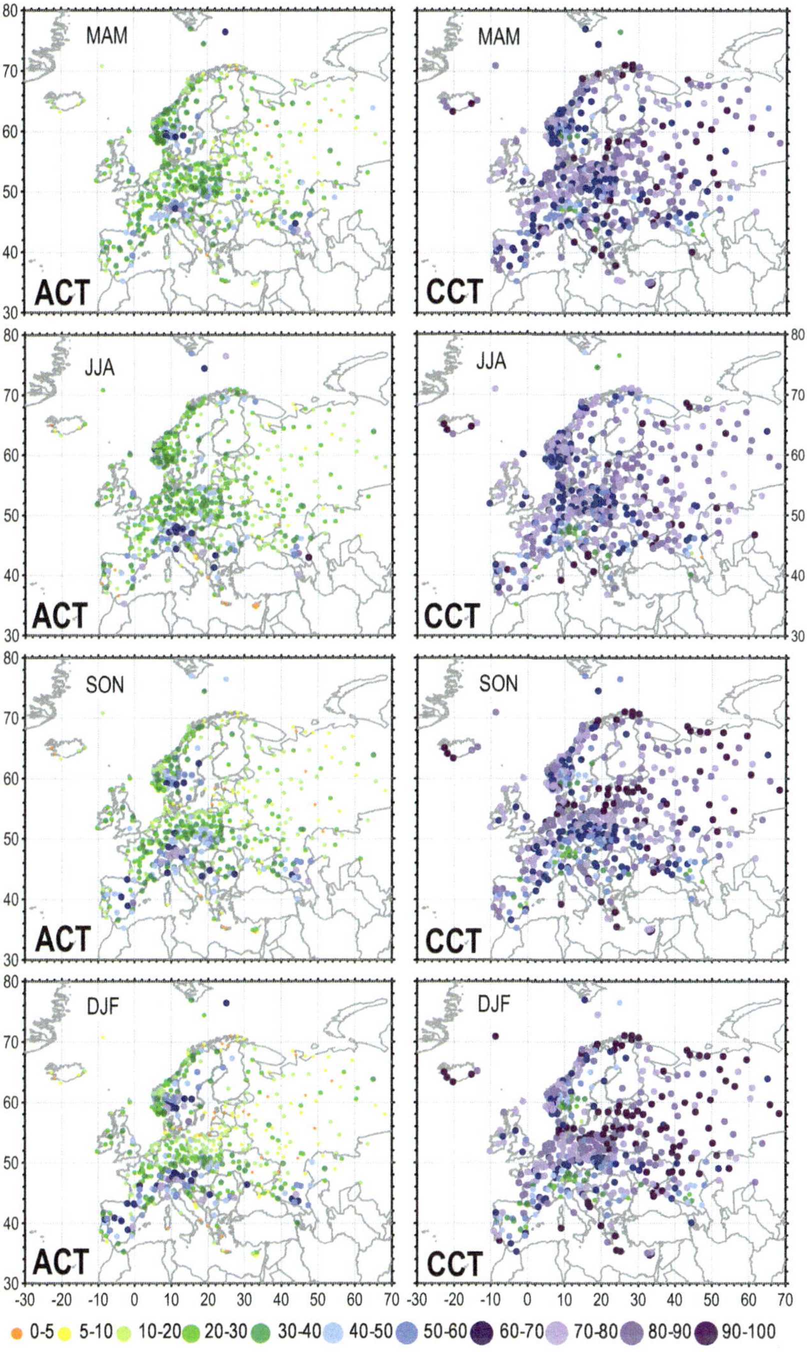

Fig. 6.1 Frequency of extreme precipitation in anticyclones (*ACT*) and cyclones (*CCT*) for January 1951–February 2008. Right closed intervals

systems in the Alps may be associated with the presence of the high-pressure ridge that stretches along that mountain range.

In the Iberian Peninsula, the frequency of ExP in ACT changes throughout the year. In autumn, the highest frequency of ExP in ACT (more than 40 % of ExP) is demonstrated by the eastern, and in winter, also by the western part of the Peninsula. In spring, the frequency of ExP in ACT in the area reaches 40 % only at isolated stations, and in summer is characterised by the highest spatial variability (between less than 5 % of ExP and more than 70 %) (Fig. 6.1).

On the continental scale, extreme precipitation in anticyclones is the least frequent in Eastern Europe and on the coasts of the Mediterranean Sea, especially in its central and eastern parts. In summer, at most stations in Eastern Europe, approximately 10–20 % of ExP is recorded in ACT. Its frequency reaches 20–30 % of ExP only at some stations. In the other seasons, especially in winter, the percentage of extreme precipitation associated with high-pressure systems in Eastern Europe is even lower, not exceeding 10 % of ExP (Fig. 6.1). In winter, a low frequency of ExP in ACT, not more than 10 %, is also seen on the southeast coast of the North Sea and the south coast of the Baltic.

In southern Poland, in the Czech Republic and Slovakia, in spring, autumn, and winter, extreme precipitation in anticyclones is more frequent (more than 30 % of ExP) than in the northern part of Central Europe (less than 30 %). In summer, in the Polish coast of the Baltic Sea and in the northern part of the Scandinavian Peninsula, more than 40 % of ExP occurs in ACT. At most stations on the British Isles, in each season, more than 20 % of ExP is attributable to anticyclones.

In Europe, the range of variability in the frequency of extreme precipitation in cyclonic situations is the highest in summer (95.3 %) and the lowest in spring (78.8 %) (Table 6.2). The spatial distribution of the frequency of ExP in CCT is, actually, a reversed version of its distribution in ACT (Fig. 6.1). In each season, at most stations (from 60 % of the stations in autumn to 64 % of the stations in summer) located in Western Europe, Central Europe, and Eastern Europe, and, except for winter, in the Scandinavian Peninsula, more than 70 % of ExP occurs in cyclones (Fig. 6.2).

As observed by Strangeways (2007), such systems, which are typical for moderate latitudes, are a source of most precipitation in those areas, not just extreme precipitation. At some stations in Eastern and Southern Europe, as well as in Iceland, and in summer and spring also on the southeast coasts of the Baltic Sea and in the northern part of the Scandinavian Peninsula, the frequency of ExP in CCT exceeds 90 % (Fig. 6.1).

Research into the relationships between precipitation and atmospheric circulation in the eastern part of the Mediterranean Basin indicates that 90 % of its annual total is represented by precipitation generated by what is referred to as the Cyprus low, which is a reactivated system inflowing to the Mediterranean area of cyclogenesis (Cyprus areas) from moderate latitudes (Goldreich 2003; Goldreich et al. 2004). Saaroni et al. (2010) estimate that the share of precipitation associated with the Cyprus low in annual precipitation is lower (75 % of the annual total). In winter, the system causes the cold air from Eastern Europe flow onto the eastern part of the

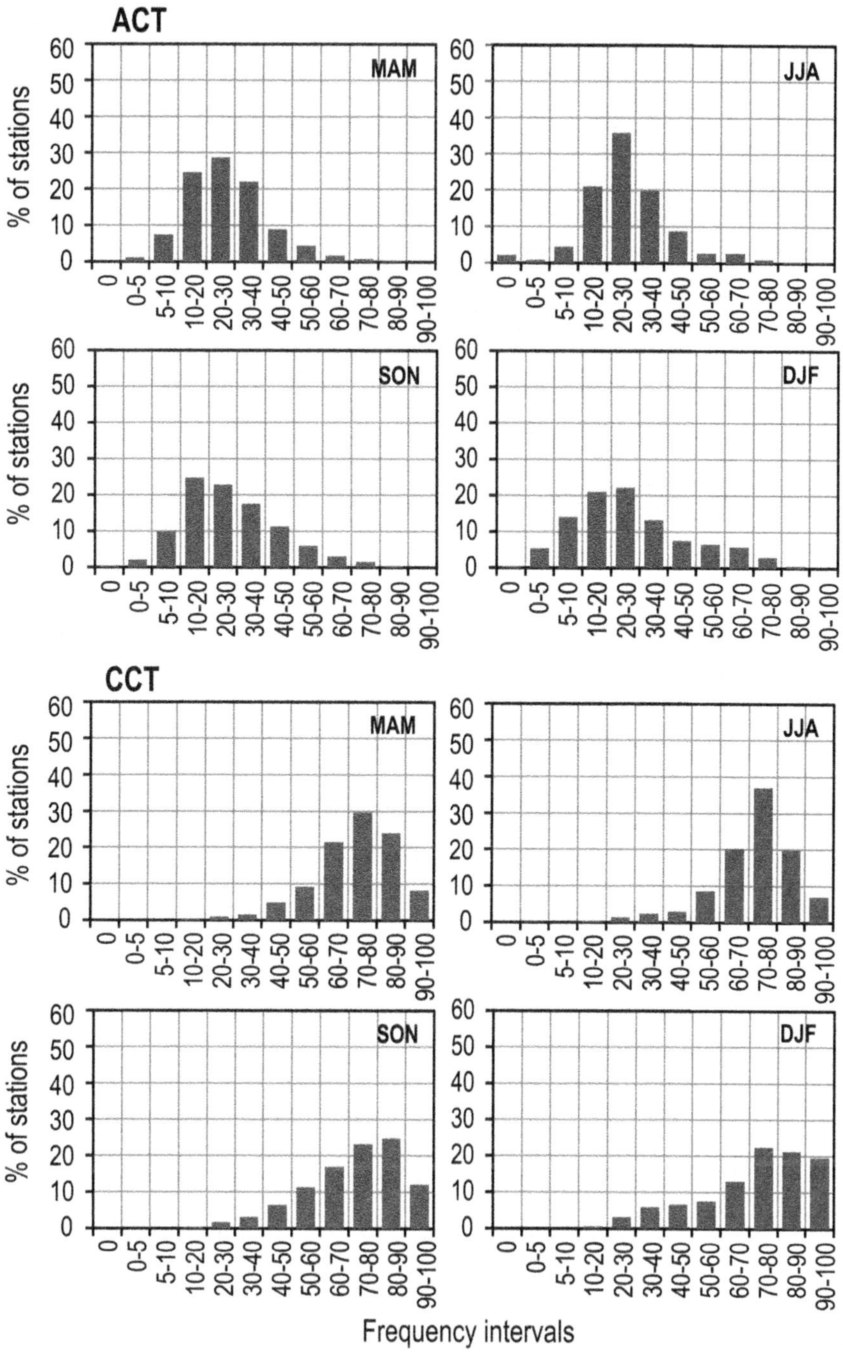

Fig. 6.2 Percentage of stations within the intervals of extreme precipitation frequency in anticyclones (*ACT*) and cyclones (*CCT*) for January 1951–February 2008. Right closed frequency intervals

Mediterranean Sea, where its humidity increases, and its stability decreases (Zangvil et al. 2003). On a continental scale, a relatively low frequency of ExP in CCT (below 60 %) is recorded by stations located in highland and mountain parts of Southern Europe (Alpine countries, Caucasus area, the Iberian Peninsula) and the southernmost part of the Scandinavian Peninsula, that is, in areas where a significant proportion of extreme precipitation is orographic precipitation, which is highly frequent in high-pressure systems. Central Europe, especially stations located in Poland, is distinguishable for a lower frequency of ExP in CCT compared to Western and Eastern Europe. In Central Europe, in the warmer half of the year, some extreme precipitation is an effect of free convection. In spring, and especially in summer, at most Polish stations, not more than 70 % of ExP is recorded in cyclonic circulation types. In winter and autumn, such a frequency is observed by stations in the southern part of the country, while in the northern part CCT is accountable for between 70 and 90 % of ExP. In winter, at some stations in the south of Europe and in the Scandinavian Peninsula (its western coasts and the southernmost part), the frequency of ExP in CCT falls below 40 % (Fig. 6.1).

6.2.1.2 Frequency of Extreme Precipitation (ExP) in Anticyclonic Synoptic Situations

The frequency of extreme precipitation in Europe is clearly related both to the direction of air inflow and to the type of the pressure system. The relationship between the occurrence of extreme precipitation and atmospheric circulation differs from region to region but shows no major seasonal changes. In Europe, in each season, the inflow of air from the N sector during an anticyclone (situations W+NWa and N+NEa) is the most frequent driver of extreme precipitation in the area of the Alps (approximately 20–30 % of ExP), and the Caucasus. In the latter area, the frequency of ExP during advection from the northern sector changes depending on the season, reaching a maximum in autumn (more than 35 % of ExP) and a relatively high frequency in winter (approximately 25–35 % of ExP). In these areas, extreme precipitation is also frequent at days with air inflow from the southeast sector (situation S+SEa). In the highland and mountain areas of Southern Poland, the highest extreme precipitation frequencies during the year are associated with the advection of air from the northeast sector (situation N+NEa), and in autumn and winter also with advection from the northwest sector (situation W+NWa) (Fig. 6.3). In the central part of the west coasts of the Iberian Peninsula and in the west coasts of the Scandinavian Peninsula, a large proportion of extreme precipitation throughout the year appears in situation W+NWa. The air inflowing from that sector in autumn and winter drives the frequency of extreme precipitation also at stations located near the coast in western Pyrenees.

Generally, in Europe, advection of air from the northern sector in anticyclonic situations generates a high frequency of ExP in mountain and highland areas, where the landform, which initiates orographic convection, is crucially important for precipitation. Extreme precipitation in the central part of the southernmost area of the

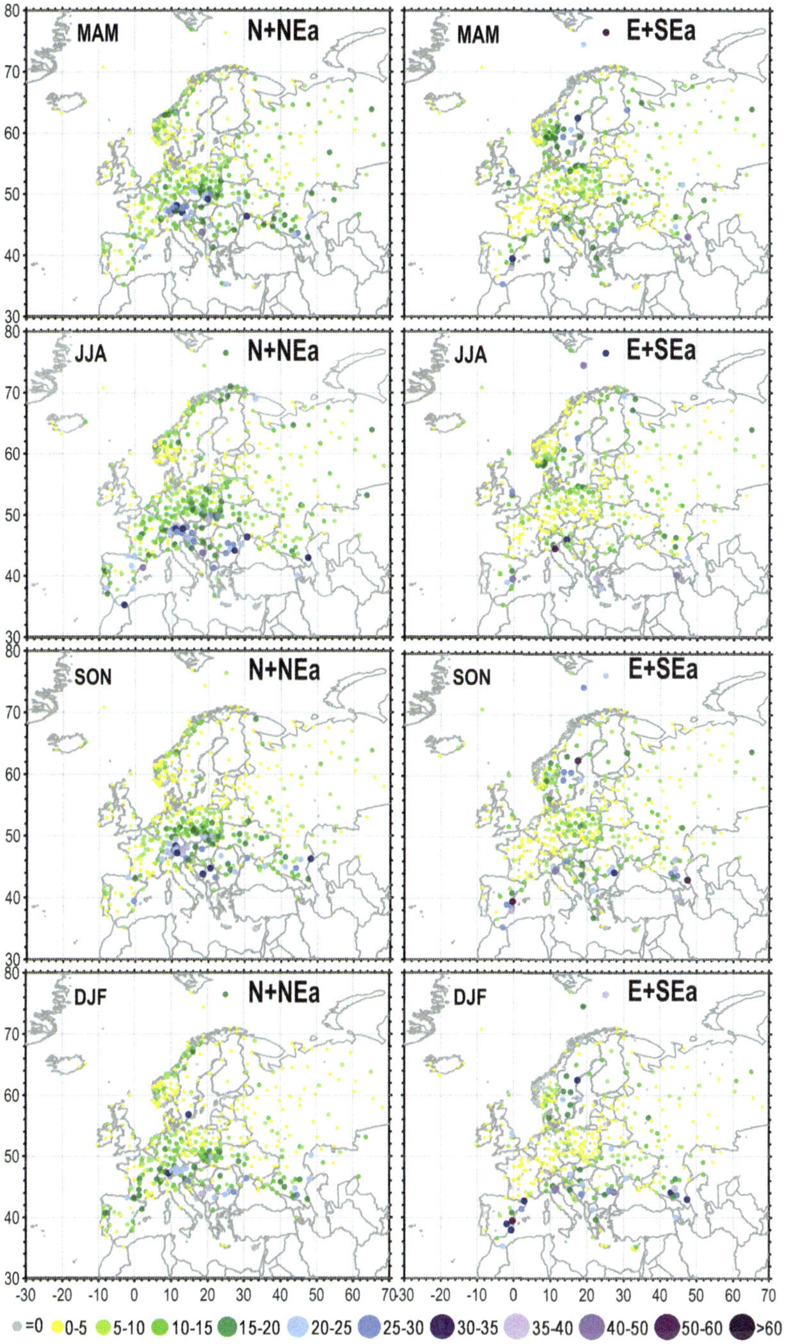

Fig. 6.3 Frequency of extreme precipitation in anticyclonic circulation types (N+NEa, E+SEa) for January 1951–February 2008. Right closed frequency intervals

Scandinavian Peninsula occurs in high-pressure systems mainly at times of air inflow from the south and southwest (situation S+SWa), with its frequency showing seasonal changes from 20 % of ExP to more than 50 % of ExP. It is an effect of dynamic orography-induced processes. In S+SWa, extreme precipitation is also frequent at stations in southwest Europe, from the Rhône to the Iberian Peninsula (more than 20 % of ExP) (Fig. 6.4).

In anticyclonic situations, with advection of air from the east and southeast (situation E+SEa), a higher frequency of extreme precipitation characterises the eastern coasts of the Iberian Peninsula and the area between the Black Sea and the Caspian Sea (e.g., in autumn and winter, from 30 % to more than 50 % of ExP).

6.2.1.3 Frequency of Extreme Precipitation (ExP) in Cyclonic Synoptic Situations

At most European stations, the frequency of extreme precipitation in cyclonic circulation types is higher than in the anticyclonic types (Figs. 6.3, 6.4, 6.5, and 6.6). The relationships between extreme precipitation and cyclonic situations change over the year and are characterised by high regional diversity. In spring, the highest frequency of extreme precipitation in Europe is associated with advection of air from the southeast sector (situation E+SEc). In E+SEc situations, the northern and southern parts of the continent record from more than 35 % of ExP to more than 50 % of ExP. At many stations, mainly in Central Europe and Eastern Europe and on the lee side (eastern side) of the Scandinavian Mountains, as well as at some stations on the British Isles, the frequency of ExP in E+SEc type exceeds in spring 20 % of ExP, whereas in the west of the continent, between the Black Sea and the Caspian Sea, and in high mountains (western coasts of the Scandinavian Mountains, the Alps, the Tatras), it is much lower (below 5 % of ExP). In Central Europe, the zone where the frequency of extreme precipitation in situation E+SEc exceeds 20 % of ExP moves eastwards during the year, in spring reaches the longitude of Jutland, and in winter, retreats beyond the eastern borders of Poland. In Southern Europe, chiefly in the central part of the Mediterranean Basin, the highest frequency of ExP during the year is associated with situation E+SEc (Fig. 6.5). In summer and transitional seasons, advection from the southeast sector is conducive to the occurrence of extreme precipitation also in the Iberian Peninsula, which was confirmed by the research of Pastor et al. (2001), according to which the heavy precipitation in the Mediterranean coast of Spain and in southern France is associated with air advection from the northeast to the southeast sector.

In general, in Europe, at times when air inflows from the southeast sector (situation E+SEc), extreme precipitation is recorded more often in Eastern Europe, and depending on the season, in Central Europe, than in its western part (Fig. 6.5).

In summer, in the southern part of Central and Eastern Europe and in the western part of the Iberian Peninsula, the highest frequency of extreme precipitation (more than 35 % of ExP) is seen in the N+NEc situation. In Poland, inflow of air from the north sector caused heavy rainfall several days long, leading to disastrous floods

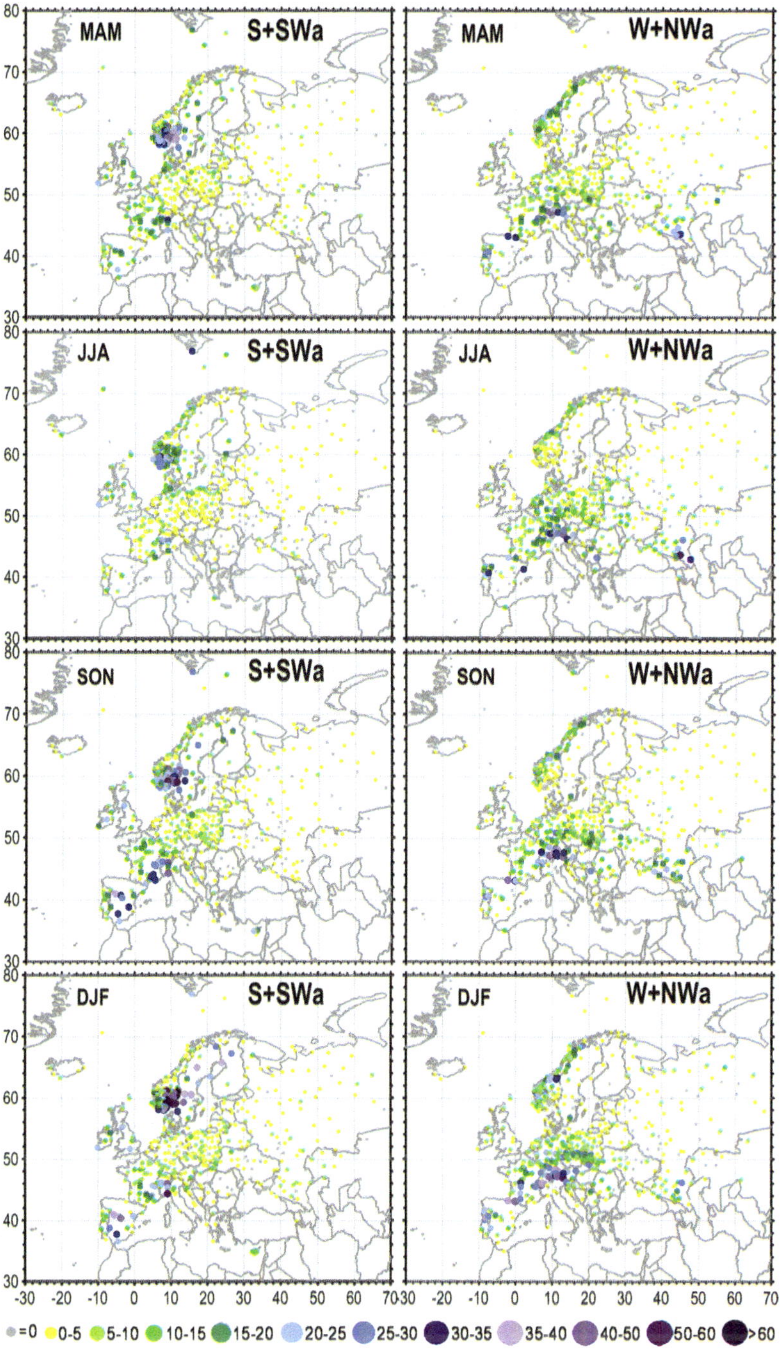

Fig. 6.4 Frequency of extreme precipitation in anticyclonic circulation types (S+SWa, W+NWa) for January 1951–February 2008. Right closed intervals

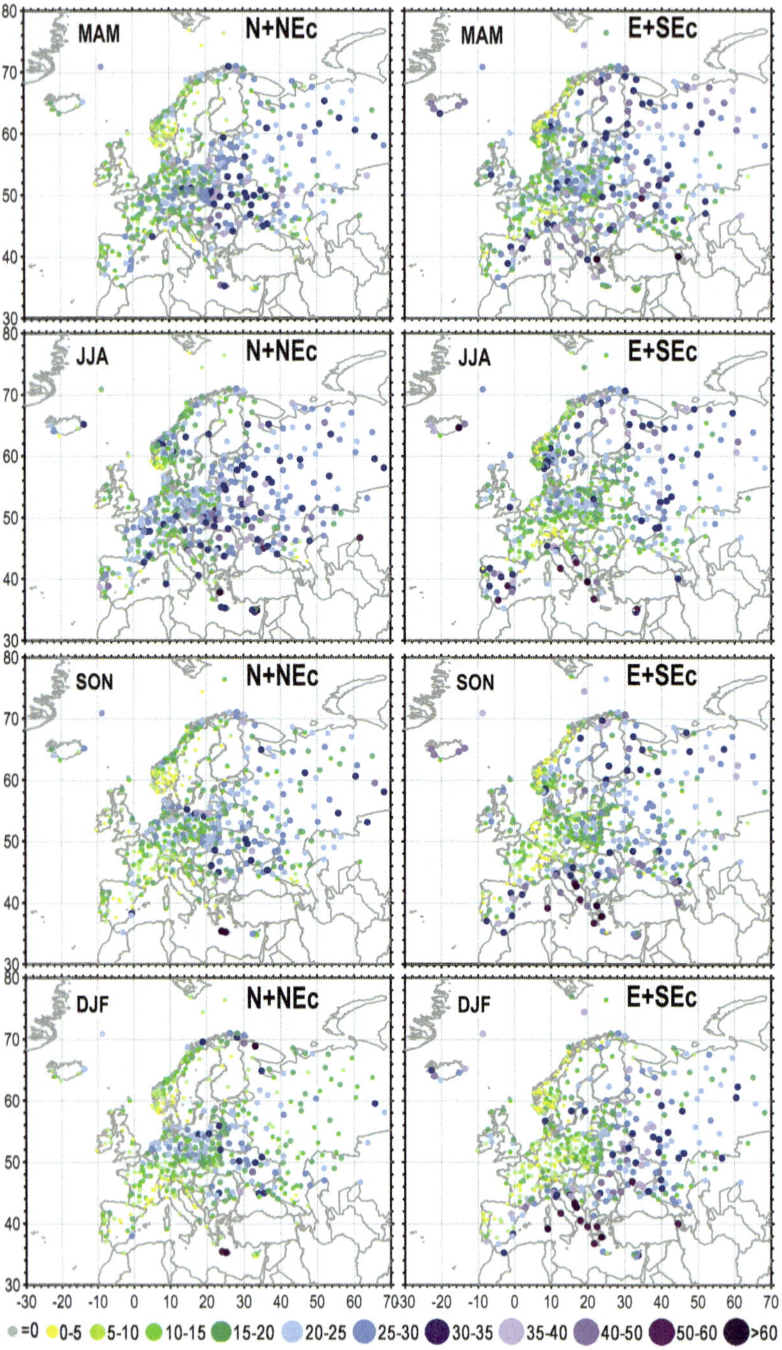

Fig. 6.5 Frequency of extreme precipitation in cyclonic circulation types (N+NEc, E+SEc) for January 1951–February 2008. Right closed intervals

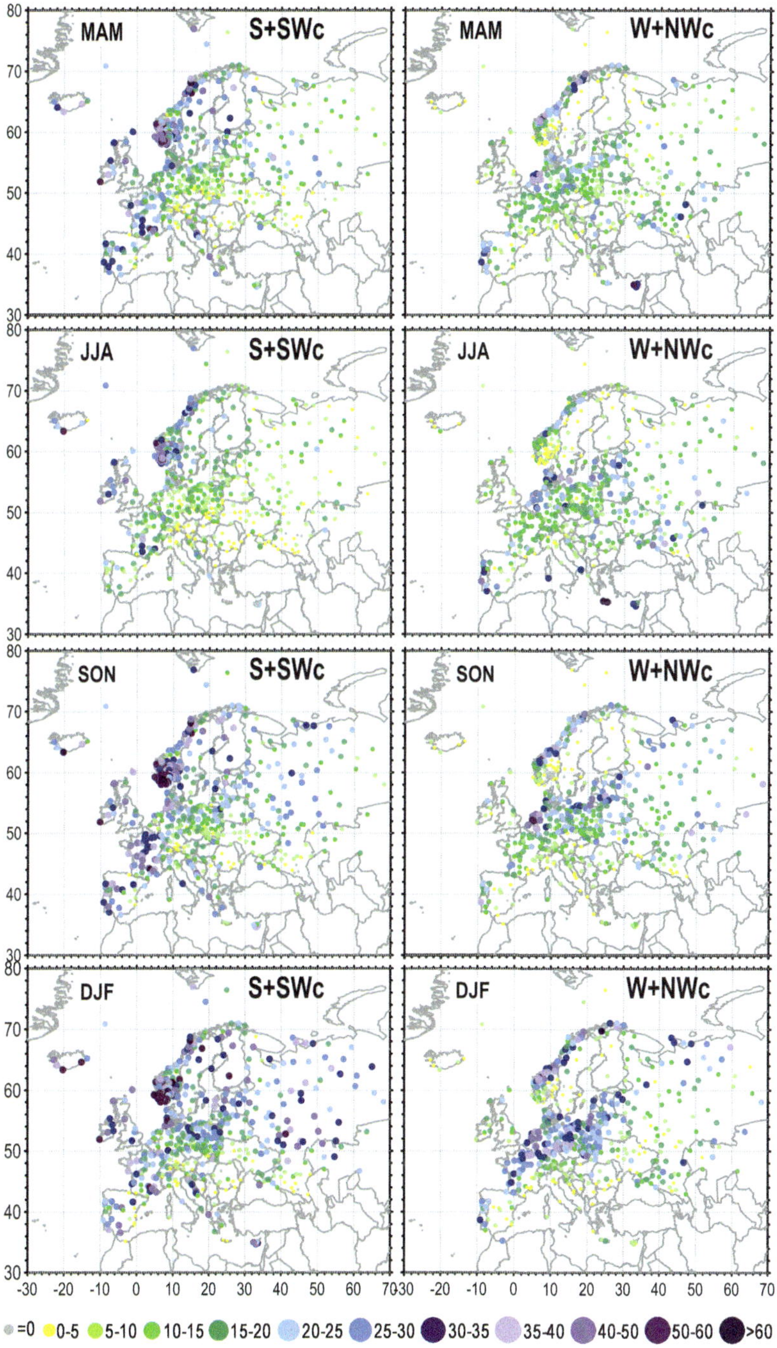

Fig. 6.6 Frequency of extreme precipitation in cyclonic circulation types (S+SWc, W+NWc) for January 1951–February 2008. Right closed intervals

(Niedźwiedź 2003a, b; Kundzewicz et al. 2012). At most stations in the rest of the continent, except for relatively small areas in the south of the Scandinavian Peninsula and some stations in Southern Europe, Poland, and Germany, the situation generates more than 20 % of ExP.

In spring, the spatial distribution of the frequency of ExP in situation N+NEc is similar to that in summer, but in spring, frequencies exceeding 35 % of ExP are rarer than in summer, and occur at stations in eastern Poland, between the Adriatic Sea and the Black Sea (areas to the north of the Balkan Peninsula), and at isolated stations in Southern Europe. In spring, in Western Europe, the frequency of ExP during air advection from the northeast sector (situation N+NEc) is lower than in summer (Fig. 6.5).

In autumn, from the Frisian Islands to Poland, as well as in Eastern Europe, approximately 20–35 % of ExP is associated with situation N+NEc. In winter, similar frequency of extreme precipitation in the same situation is found in Western Europe. In this season, most of stations in the eastern part of the continent record not more than 20 % of ExP in situation N+NEc (Fig. 6.5).

In Europe, throughout the year, during air advection from the southwest sector (situation S+SWc), the highest amount of extreme precipitation is recorded in the south of the Scandinavian Peninsula and in its west coast, in the area of Lofoten and Vesteraten, as well as on the British Isles. Compared to the rest of the continent, high ExP frequencies in situation S+SWc (more than 35 %) are also observed in the western part of Southern Europe (France, western part of the Iberian Peninsula). This regularity, even though less pronounced, is also noticeable in winter. In the latter season, a relatively high ExP frequency in situation S+SWc is recorded by many stations in Eastern Europe (Fig. 6.6). In Western and Central Europe, situation S+SWc produces more extreme precipitation in autumn and winter than in summer and spring (Fig. 6.6).

In the western coasts of the Scandinavian Peninsula and Iberian Peninsula, throughout the year, a large proportion of extreme precipitation is associated with inflow of air from the northwest sector (situation W+NWc). In winter, in situation W+NWc, clear contrasts are observable in the frequency of extreme precipitation events between Western and Central Europe (from more than 20 % to more than 50 % of ExP) and Southern and Eastern Europe (more than 15 % of ExP). In the northern and southern parts of Eastern Europe, N+NWc extreme precipitation occurs more frequently than in the central part of Eastern Europe. In the remaining seasons (spring, summer, in particular autumn) this distribution of ExP frequency in situation N+NWc maintains the key properties identified in winter, except that ExP frequency in the situation is lower at most stations in Western Europe (15–20 % ExP).

The results achieved correspond to the results of regional-scale research. Lorenzo et al. (2008) found that the most intensive precipitation in northwest Spain is associated with cyclonic types of circulation, which are characterised by inflow of air from the west and southwest, and are slightly less associated with air advection from the northwest and south.

Table 6.3 Descriptive statistics of the frequency of air mass extreme precipitation (type A) [%] in anticyclones (*ACT*) and cyclones (*CCT*) in Europe, December 1950–February 2008)

Pressure system	Season	Average (±SE)		Confidence intervals −95 %	Confidence intervals +95 %	Min	Max	SD	CV	ME	Quartiles Lower	Quartiles Upper
ACT	MAM	6.2	(±0.3)	5.6	6.7	0.0	49.2	6.5	105.8	4.1	2.1	8.0
	JJA	7.7	(±0.4)	7.0	8.4	0.0	56.1	8.1	105.4	5.4	2.9	9.4
	SON	5.7	(±0.3)	5.1	6.3	0.0	52.8	7.1	125.3	3.2	1.4	7.0
	DJF	5.1	(±0.3)	4.5	5.7	0.0	56.1	7.0	137.6	2.6	1.0	6.7
CCT	MAM	11.4	(±0.4)	10.7	12.2	1.0	67.5	14.0	19.3	9.1	5.6	14.5
	JJA	17.2	(±0.6)	15.9	18.4	0.0	100.0	14.2	82.6	13.0	8.8	20.0
	SON	10.5	(±0.4)	9.7	11.3	0.0	60.0	9.2	87.9	7.7	4.3	13.7
	DJF	8.7	(±0.4)	8.0	9.4	0.0	56.5	8.4	96.3	6.3	3.1	11.1

SE standard error, *Min* minimum value, *Max* maximum value, *SD* standard deviation, *CV* coefficient of variability, *ME* median

The analyses conducted in this chapter indicate that the occurrence of extreme precipitation in Europe is clearly related with the direction of inflowing air and the type of pressure system. These relationships are highly diversified, both in regional and in seasonal terms, which mostly applies to cyclonic synoptic situations. The relationships between the occurrence of extreme precipitation and anticyclonic situations show less seasonal diversity. The frequency of extreme precipitation in anticyclonic situations depends on the landform, which is proven by the high frequency of ExP at elevated areas; this demonstrates the high importance of the orographic factor in shaping high precipitation.

6.2.2 *Air-Mass Precipitation (Type A)*

6.2.2.1 Frequency of Air-Mass Extreme Precipitation (Type A) in Anticyclones (ACT) and Cyclones (CCT)

In Europe, the average frequency of air-mass extreme precipitation in ACT changes over the year from 5.1 % of ExP in winter to 7.7 % of ExP in summer (Table 6.3). At most stations in Europe, except those in its southern part, type A precipitation in anticyclones represents not more than 10 % of ExP. In spring, autumn, and winter the frequency of type A precipitation in ACT at more than half of the stations reaches at most 5 % of ExP, and at nearly 20 % of the stations spread over the Scandinavian Peninsula, in Central Europe, and in the south of the continent, it falls within the range of 5–10 % of ExP (Fig. 6.7). It is only in summer that the stations at which type A precipitation in ACT falls within the range 5–10 % of ExP from relatively coherent areas located at the west coasts of the Scandinavian Peninsula, on the southeast coasts of the North Sea, and in Central Europe. In those areas, the frequency of type A precipitation recorded in ACT is higher than in the east and

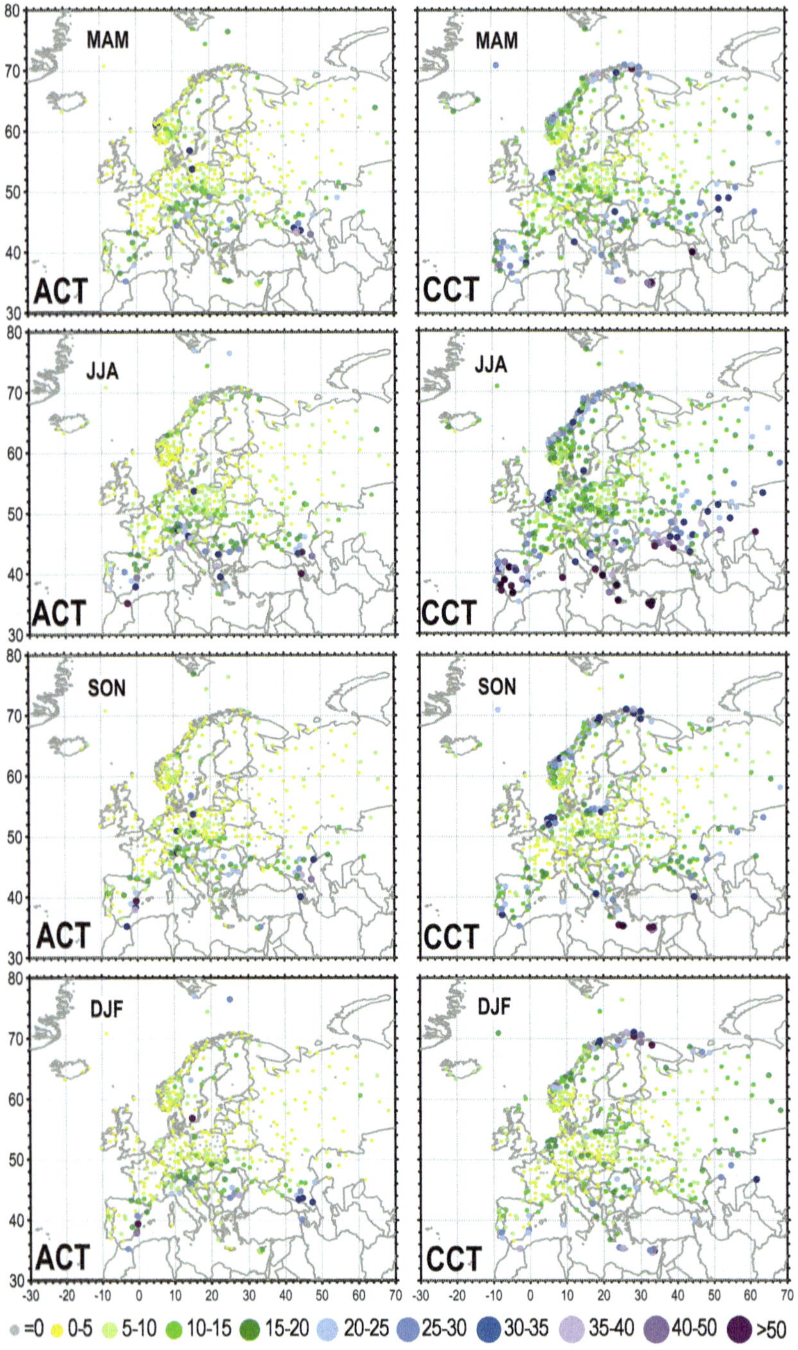

Fig. 6.7 Frequency of air mass extreme precipitation (type A) in anticyclones (*ACT*) and cyclones (*CCT*) for January 1951–February 2008. Right closed intervals

west of the continent and in the southern section of the Scandinavian Peninsula. In each season, air-mass precipitation in ACT is the most frequent (15 % of ExP and more) in Alpine areas, in Southern Europe, mainly in the eastern part of the Iberian Peninsula, in the Caucasus area, and in the Balkan Peninsula (Fig. 6.7). The mountains present in those areas enhance the formation of orographic precipitation. In the south of Europe, the highest frequency of type A extreme precipitation in ACT is recorded in summer, when the area falls under the influence of the subtropical high-pressure zone.

In Europe, the average frequency of air-mass precipitation recorded in cyclonic types ranges between approximately 9 % of ExP in winter and approximately 17 % in summer. In spring, autumn and winter, the maximum frequency of type A precipitation in CCT at individual stations reaches approximately 68 %, 60 %, and 56 %, respectively (Table 6.3). It is only in summer at the southernmost sites (Cyprus) that all air-mass precipitation occurs in cyclonic circulation types (Fig. 6.7). In spring, autumn, and winter, at most stations in the area stretching from the British Isles to the easternmost parts of the continent, the frequency of type A precipitation in low-pressure systems usually does not exceed 10 % of ExP (Fig. 6.7).

In Central and Eastern Europe, most type A precipitation in CCT occurs in summer, which results from the then prevailing conditions favouring the occurrence of free convection. In this season, the frequency of type A precipitation in CCT reaches 10–15 % of ExP at about 30 % of the stations, whereas in spring such frequency is recorded by approximately 20 % of the stations, and in autumn and winter by approximately 14–15 % of the stations (Fig. 6.8). In each season, the highest frequency of air-mass precipitation in cyclonic situations is recorded by the Atlantic coasts of the Scandinavian Peninsula, Southern Europe, and the southeast coast of the North Sea.

On the western coasts of the Scandinavian Peninsula, similarly to Central and Eastern Europe, the largest proportion of type A extreme precipitation in low-pressure systems is seen in summer (more than 20 % of ExP), whereas in the northernmost parts of Europe it occurs in winter (more than 35 % of ExP). In the southeast coast of the North Sea, in most seasons, the frequency of type A precipitation in CCT exceeds 20 % of ExP, with the highest frequency in autumn, more than 30 %, and the lowest in winter, approximately 15 %.

In the Mediterranean Basin, the highest frequency of type A precipitation in CCT, exceeding 35 % of ExP, is recorded in summer, while the lowest frequencies are recorded in winter, when the area is often crossed by weather fronts. The high frequency of air-mass precipitation in the south of the continent and in the Atlantic coasts of the Scandinavian Peninsula is also attributable to the large impact of orographic barriers present in the area on precipitation formation.

In summer, air-mass precipitation in the southern parts of Europe is fostered additionally by high air temperature, which results from the impact of the subtropical zone of high pressure on the weather conditions in the area. A high frequency of type A precipitation in CCT is also recorded by the Polish Baltic coast in autumn. Here, the increased frequency of type A ExP in CCT may be, to an extent, linked to landform.

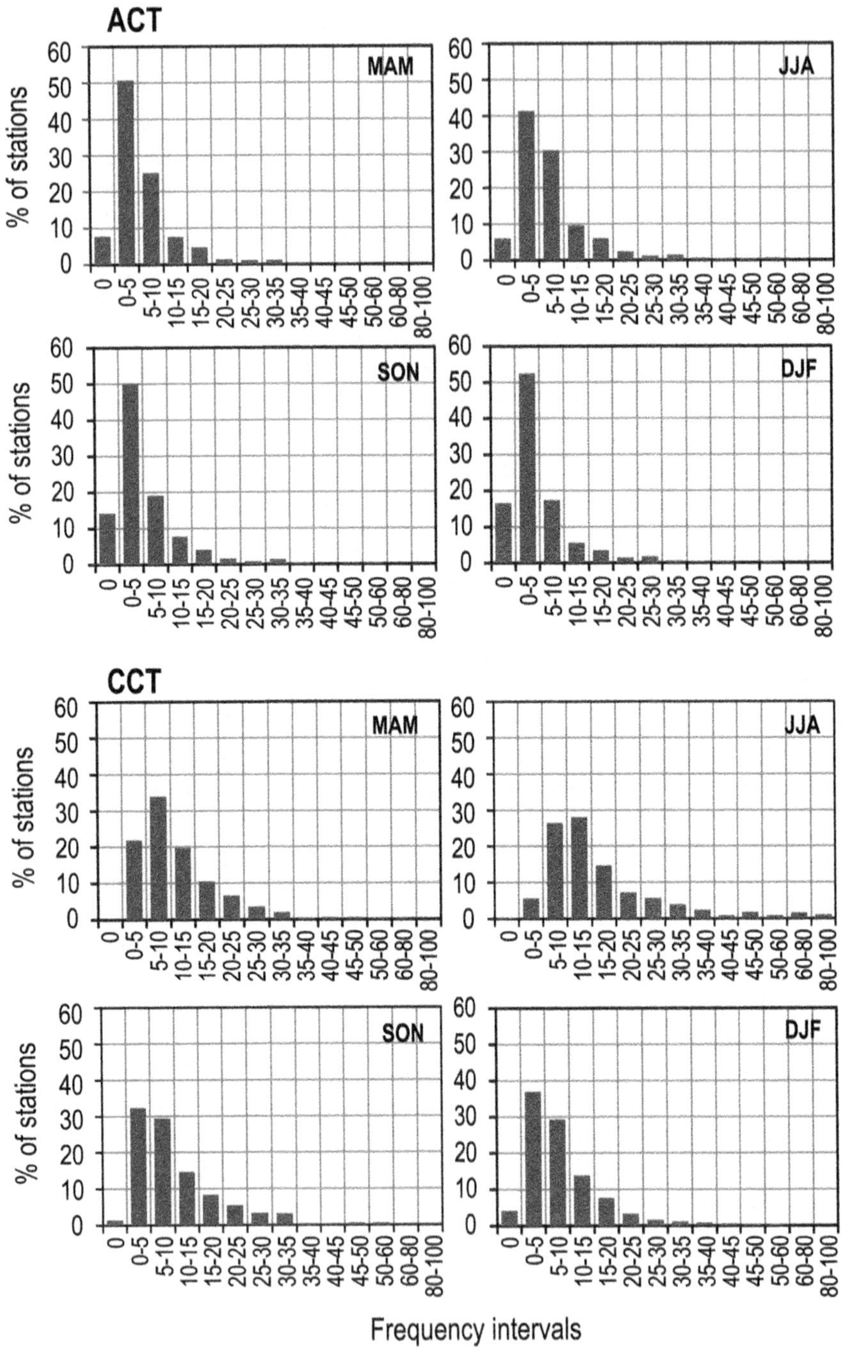

Fig. 6.8 Percent of stations within the intervals of air mass extreme precipitation (type A) frequency in anticyclones (*ACT*) and cyclones (*CCT*) in Europe for January 1951–February 2008. *Right* closed intervals

6.2.2.2 Frequency of Air-Mass Extreme Precipitation (Type A) in Anticyclonic Synoptic Situations

There is a clear relationship between the occurrence of air-mass precipitation in high-pressure systems and the direction of inflowing air in Central and Southern Europe, and in the areas between the Black Sea and the Caspian Sea (the Caucasus and surrounding areas). In spring, summer, and autumn, in the central part of the continent, covering southern Poland and Germany, the Alps, and the northern part of the Balkan Peninsula, extreme air-mass precipitation in anticyclones occurs most often during air advection from the northeast sector (situation N+NEa). In summer, in southern Poland and Germany, type A precipitation in situation N+NEa accounts for about 5–10 % of ExP, while in Alpine areas and in the central Danube basin, its frequency ranges between 10 and 25 % of ExP. In spring and autumn, as in the summer, the frequency of type A precipitation in situation N+NEa in the above areas of Central Europe reaches 5–10 % of ExP. Over the Caucasus, the frequency of type A precipitation in N+NEa situations is the highest in spring, 10–20 % of ExP, and in summer and autumn, it reaches 5–10 % of ExP and 5–15 % of ExP, respectively.

In the these areas, also in winter, air-mass precipitation in situation N+NEa is more frequent than elsewhere in Europe, except that its frequency only occasionally exceeds 5 % of ExP, mainly in the basin of the central and lower Danube and in the Caucasus (Fig. 6.9).

Throughout the year, inflow of air from the southeast sector (E+SEa) brings the highest proportions of air-mass precipitation to Southern Europe, especially to the Mediterranean coasts of Spain and the Caucasus area. In those areas, the highest frequency of type A precipitation in situations E+SEa is recorded in autumn, more than 25 % of ExP. At the Mediterranean coasts of Spain, the frequency of air-mass precipitation in situation E+SEa during the year never falls below 10 % of ExP, and in autumn it exceeds 30 % of ExP. The results achieved are consistent with the results of research by Millán et al. (2005), who found that at the coasts of Spain in the Valencia region the highest precipitation is associated with air advection from the east, caused by an extensive high-pressure system lying over Central Europe. In summer, in the area of the Caucasus, the frequency of type A ExP in S+SEa situation is the lowest, while in the central part of the Mediterranean Sea it is the highest (Fig. 6.9).

In Europe, during air advection from the southwest (situation S+SWa), the highest number of extreme precipitation in each season is recorded in the central part of the southern end of the Scandinavian Peninsula on the lee side (western side) of the Scandinavian Mountains, and in spring and winter, also in areas located at higher latitudes of the peninsula. In western areas of the Mediterranean Sea (Iberian Peninsula, southern France), in autumn, relatively high (compared to the rest of the continent) frequency of air-mass precipitation (from 5 to 10 % of ExP) is linked with situation S+SWa (Fig. 6.10).

Air inflow from the northwest (situation W+NWa) increases the frequency of type A extreme precipitation in the southern part of the western coasts of the Scandinavian Peninsula. In such a synoptic situation, a higher frequency of type A

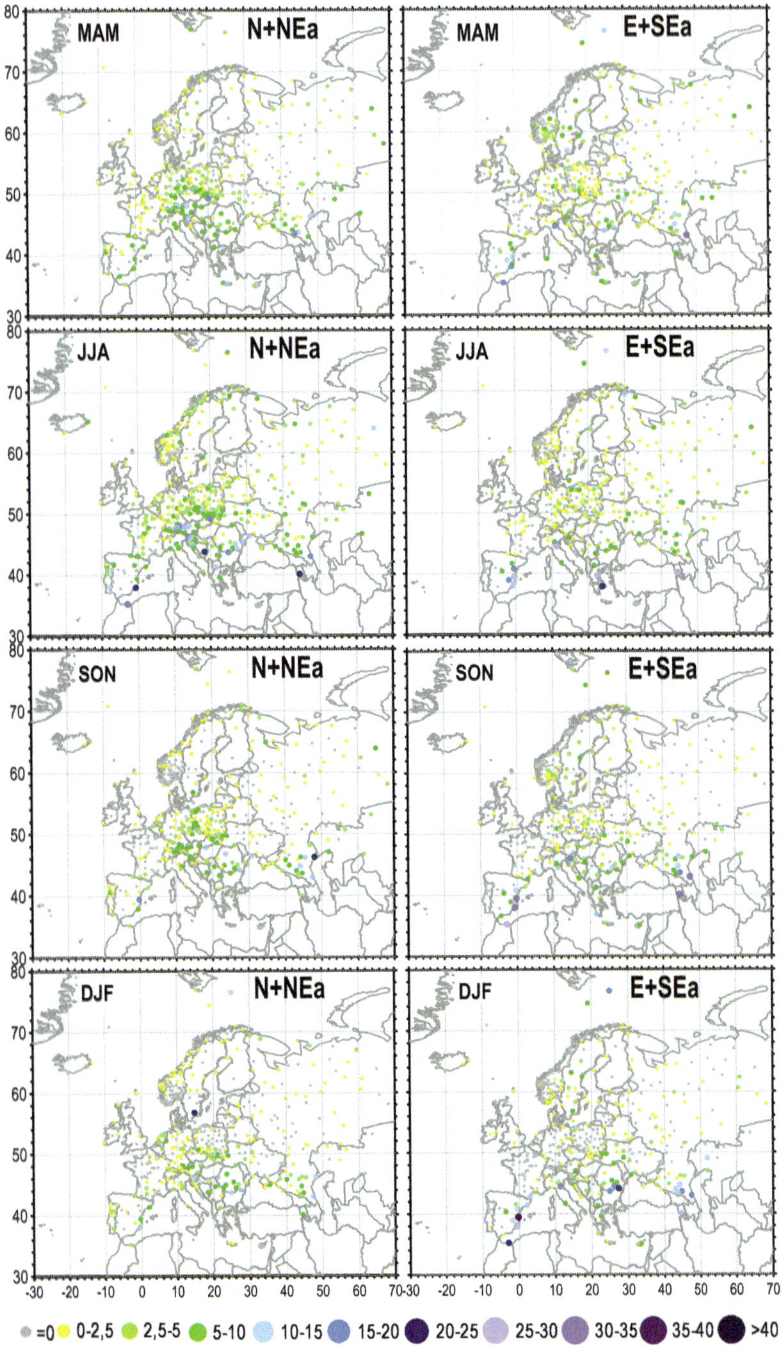

Fig. 6.9 Frequency of air-mass extreme precipitation (type A) in anticyclonic circulation types (N+NEa, E+SEa), January 1951–February 2008. Right closed frequency intervals

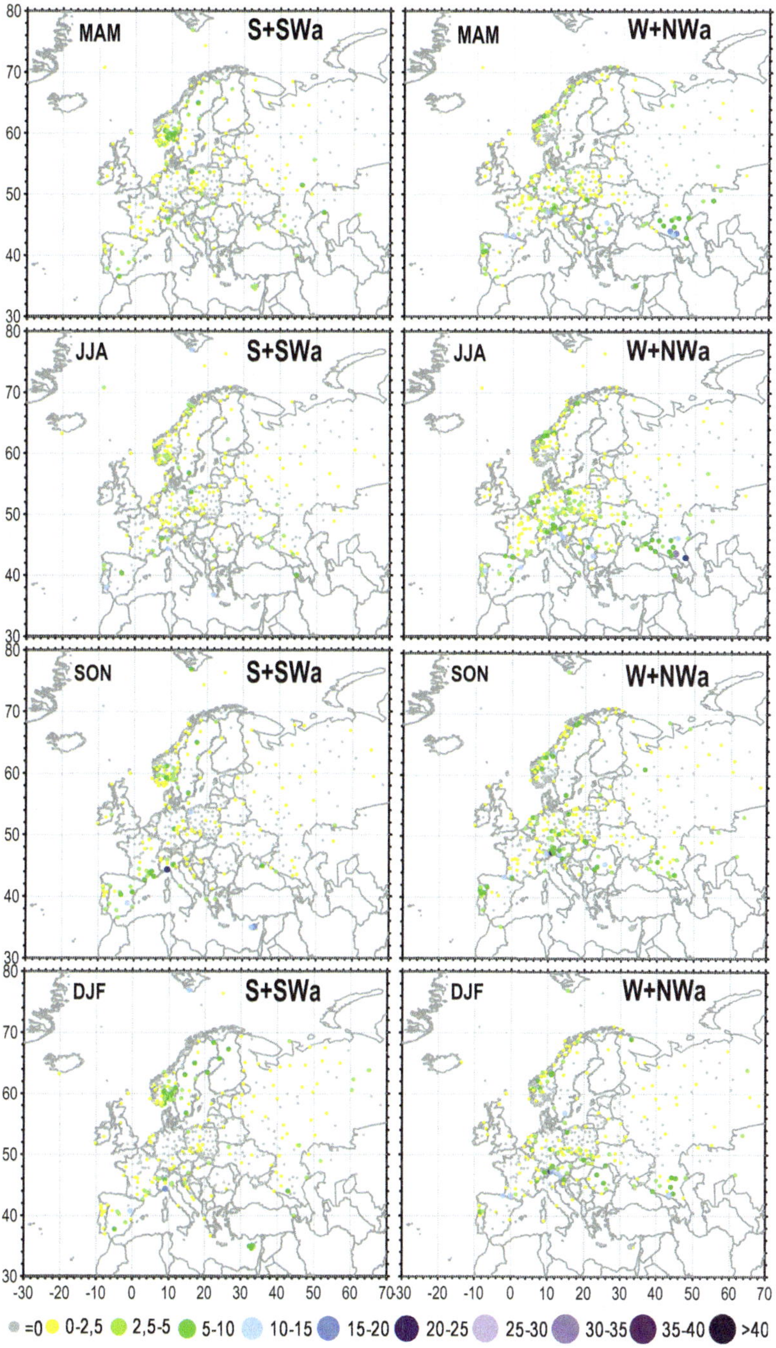

Fig. 6.10 Frequency of air-mass extreme precipitation (type A) in anticyclonic circulation types (S+SWa, W+NWa) for January 1951–February 2008. Right closed intervals

precipitation (especially in summer, more than 10 % of ExP) is recorded in the southeast coasts of the North Sea, within Alpine areas, and in the western and eastern part of Southern Europe (northern part of the Atlantic coast in the Iberian Peninsula, the Caucasus area).

6.2.2.3 Frequency of Air-Mass Extreme Precipitation (Type A) in Cyclonic Synoptic Situations

On average, in Europe, air-mass precipitation occurs most often in summer. In the season, at many stations from the west coasts of the continent to its easternmost areas, the highest frequency of air-mass precipitation is associated with advection of air from the northeast sector (situation N+NEc). In summer, type A precipitation in situation N+NEc represents between 5 and 15 % of ExP in Western Europe, in Poland, in Eastern Europe, and in the Scandinavian Peninsula. Its frequency exceeds 10 % of ExP in the southern parts of Eastern Europe and of the Mediterranean Basin, in the southeast coasts of the North Sea, as well as at isolated stations in the Carpathians and in the west coast of the Scandinavian Peninsula. In the other seasons, especially in winter, during inflow of air from N+NEc, higher frequency of air-mass precipitation occurs in the northernmost part of the Scandinavian Peninsula (in winter 35 % of ExP). In autumn, type A precipitation in situation N+NEc represents more than 10 % of ExP in the southeast coasts of the North Sea and the Polish coasts of the Baltic, while in spring in the eastern part of Southern Europe and in the southern part of Eastern Europe in the area between the Adriatic and Caspian Sea (Fig. 6.11).

In summer, the highest frequency of extreme air-mass precipitation, characterising Southern Europe (between 15 % and more than 40 % of ExP), is associated with advection of air from the southeast sector (situation E+SEc). Also, in the rest of the seasons, the frequency of air-mass precipitation in situation E+SEc (10–15 % of ExP) in the south of the continent is higher than elsewhere in Europe.

Throughout the year, northwesterly advection (situation W+NWc) generates the highest amount of air-mass precipitation (more than 10 % of ExP) in the western coasts of the Scandinavian Peninsula and in the southeast coasts of the North Sea. In the other seasons, this circulation type also involves increased ExP frequency in northern Poland, in autumn and winter, in the Mediterranean Basin, and in the Caucasus area in summer, and in spring on islands in the eastern part of the Mediterranean Sea (Fig. 6.12).

The occurrence of air-mass extreme precipitation in Europe shows the poorest relationship with situation S+SWc. The frequency of type A precipitation in this synoptic situation exceeds 10 % of ExP only in spring and summer in the southwest part of the Iberian Peninsula, and in autumn and winter on Cyprus. In each season, air-mass precipitation in situation S+SWc is slightly more frequent than elsewhere in the continent in the Scandinavian Peninsula (especially in summer and autumn) and at isolated stations in the southern part of Europe (Fig. 6.12).

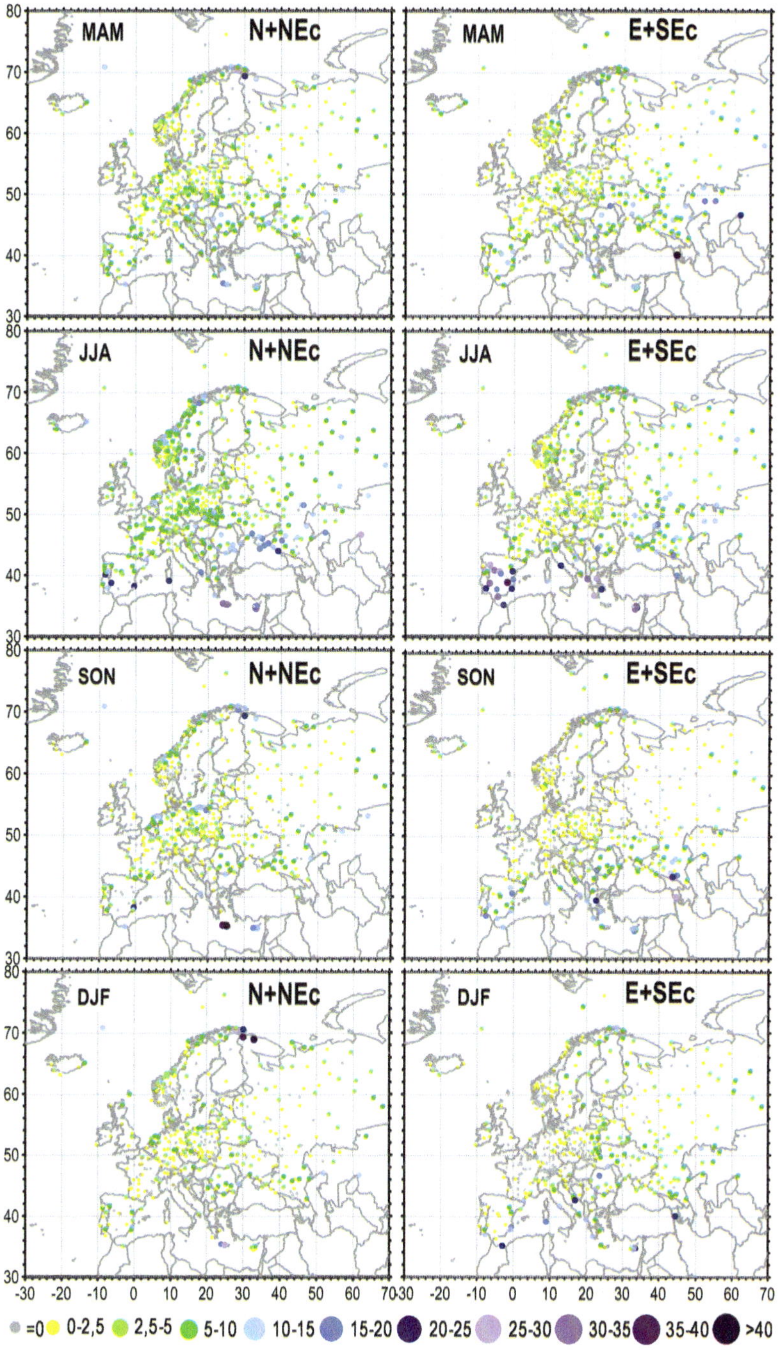

Fig. 6.11 Frequency of air-mass extreme precipitation (type A) in cyclonic circulation types (N+NEc, E+SEc) for January 1951–February 2008. Right closed frequency intervals

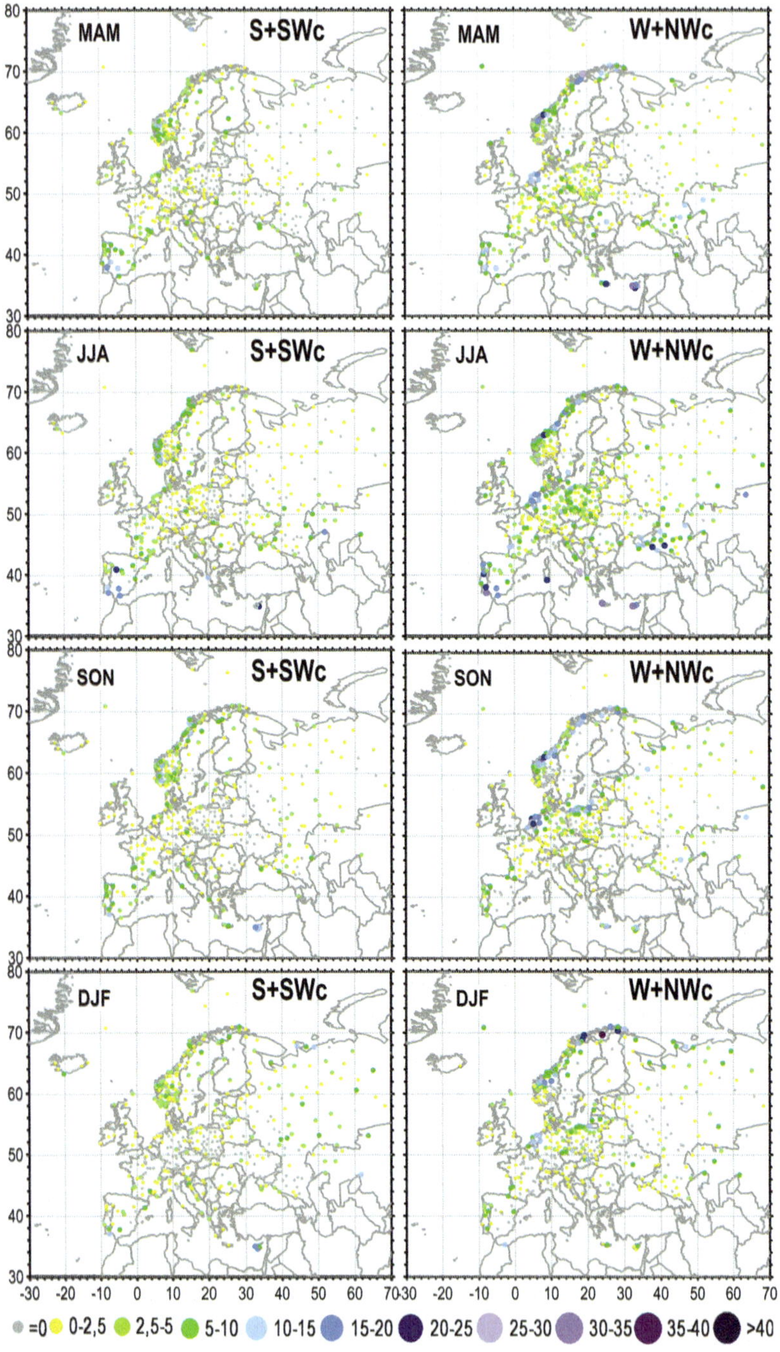

Fig. 6.12 Frequency of air-mass extreme precipitation (type A) in cyclonic circulation types (S+SWc, W+NWc) for January 1951–February 2008. Right closed intervals

Table 6.4 Descriptive statistics of the frequency [%] of frontal extreme precipitation (type F) in anticyclones (*ACT*) and cyclones (*CCT*) in Europe, January 1951–February 2008

Pressure system	Season	Average (±SE)	Confidence intervals		Min	Max	SD	CV	ME	Quartiles	
			−95 %	+95 %						Lower	Upper
ACT	MAM	21.4 (±0.5)	20.4	22.4	0.0	60.9	11.2	52.6	20.5	12.9	27.9
	JJA	20.1 (±0.5)	19.2	21.0	0.0	58.0	10.6	52.6	18.	13.0	26.5
	SON	22.6 (±0.6)	21.4	23.8	0.0	68.9	13.7	60.6	21.3	11.8	30.4
	DJF	23.0 (±0.7)	21.6	24.4	0.0	71.9	16.4	71.3	20.0	9.5	32.1
CCT	MAM	61.0 (±0.6)	59.8	62.3	13.1	92.6	14.4	23.6	62.9	51.7	71.6
	JJA	55.0 (±0.7)	53.6	56.4	0.0	89.7	16.1	29.2	57.6	46.3	65.5
	SON	61.2 (±0.7)	59.8	62.7	11.1	93.2	16.6	27.0	63.8	51.3	73.6
	DJF	63.2 (±0.8)	61.6	64.8	10.9	95.3	18.4	29.1	67.4	50.8	77.2

SE standard error, *Min* minimum value, *Max* maximum value, *SD* standard deviation, *CV* coefficient of variability, *ME* median

6.2.3 *Frontal Precipitation (Type F)*

6.2.3.1 Frequency of Frontal Extreme Precipitation (Type F) in Anticyclones (ACT) and Cyclones (CCT)

The frequency of frontal extreme precipitation in high-pressure systems is the lowest in summer (on average 20.1 % of ExP), and slightly higher in winter (23.0 %), when Europe experiences enhanced cyclonic activity (Table 6.4). The spatial distribution of the frequency of type F precipitation in ACT demonstrates seasonal variability. Its essential characteristics, which include greater frequency of type F precipitation in ACT in Western and Central Europe than in Eastern Europe, as well as increased frequency of F type precipitation in areas elevated significantly above sea level, are observable throughout the year (Fig. 6.13).

In spring, at half the weather stations, the frequency of type F precipitation in ACT falls between 12.9 and 27.9 % of ExP. In Western Europe, Central Europe, in the Atlantic coasts of the Scandinavian Peninsula, and especially in its southern part, as well as at many stations in Southern Europe, frontal precipitation in ACT represents about 20–30 % of ExP, whereas in Eastern Europe and in the central and eastern part of the Mediterranean Basin its frequency reaches not more than 20 % of ExP. Throughout the year, the amount of type F precipitation in ACT is high, approximately 35 % of ExP, in the south of the Scandinavian Peninsula. In autumn and winter, similar frequency of such precipitation in ACT is also recorded by sta-

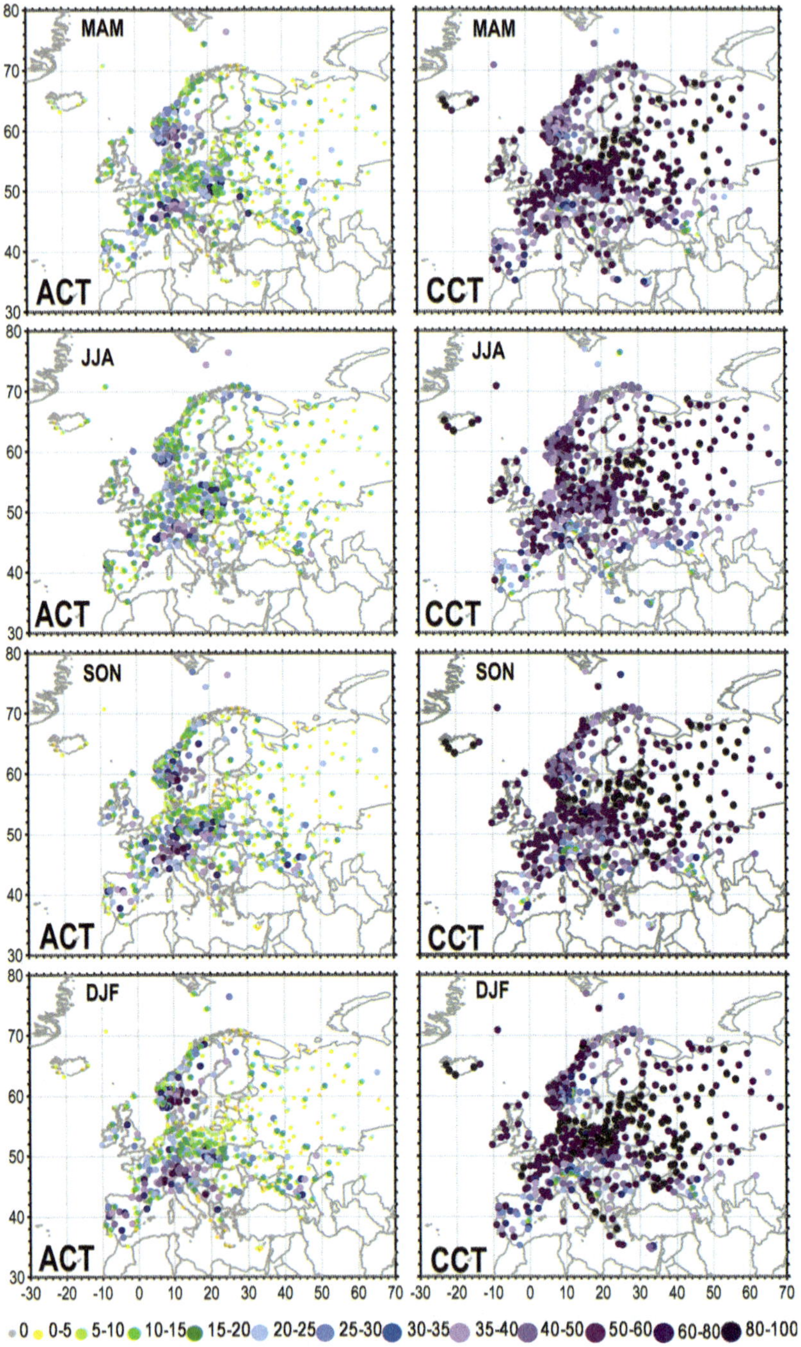

Fig. 6.13 Frequency of frontal extreme precipitation (type F) in anticyclones (*ACT*) and cyclones (*CCT*) for January 1951–February 2008. Right closed intervals

tions located on the lee side of the Scandinavian Mountains. In summer, in the Scandinavian Peninsula, extreme frontal precipitation occurs in high-pressure systems less frequently (Fig. 6.13). In each season, it represents more than 35 % of ExP also at Alpine areas.

At many stations in southern Europe, from Spain to the Alps, in the Caucasus area, and at some stations in the Balkan Peninsula frontal precipitation in ACT is most frequent in winter (Fig. 6.13). Seasonality is also noticeably in the frequency of type F in ACT in Central Europe, mainly in Poland and its vicinity, where, in autumn and winter, the frequency grows from the north where it reaches no more than 20 % of ExP, to the south, where it usually exceeds 35 % of ExP. In autumn, the frequency of extreme precipitation in anticyclones ranges between 25 and 35 % of ExP. It falls, in the border area between Poland and Germany, to 10–20 % of ExP, and exceeds 35 % of ExP at some stations in southeast Poland.

In summer, the highest frequency of type F precipitation in ACT is registered by mountain stations and individual stations in southern and northeastern Poland. As a rule, in summer, frontal precipitation in anticyclones is rare in the southernmost stations in Europe, where its frequency does not exceed 10 % (Fig. 6.13).

The average frequency of frontal precipitation in cyclones is decidedly higher. It ranges from approximately 55 % of ExP in summer and 63 % of ExP in winter, being slightly lower in the transitory seasons of the year than in winter, approximately 61 % of ExP (Table 6.4). Type F precipitation in cyclones constitutes not more than 35 % of ExP only in the southern part of the Scandinavian Peninsula (in spring, autumn, and winter), in the Alps and the Caucasus area (throughout the year), as well as in the Iberian Peninsula and the Balkan Peninsula (summer) (Fig. 6.13). Elsewhere in Europe, the frequency of type F precipitation in CCT falls between 60 and 80 % of ExP at 50 % of the stations in spring, 45 % of the stations in autumn, 46 % of the stations in winter, and only 40 % of the stations in summer (Fig. 6.14).

In Europe, the highest frequency of frontal precipitation in winter and the lowest frequency in summer are related to enhanced cyclonic activity in the cold part of the year and increased frequency of weather fronts, which are an inherent element of cyclogenesis and cyclolysis. In summer, when cyclonic activity declines, the role of free convection in the formation of rainfall increases.

Frontal precipitation (type F) represents a great majority of extreme events in Europe, hence the spatial distribution of the frequency of its occurrence in synoptic situations, as shown next, shows a strong resemblance to the spatial distribution of all extreme precipitation events, as presented in Sect. 5.2.1.

6.2.3.2 Frequency of Frontal Extreme Precipitation (Type F) in Anticyclonic Synoptic Situations

The relationships between the occurrence of extreme frontal precipitation and anticyclonic synoptic situations do not show significant seasonal changes. As a rule, a circulation type is conducive to the occurrence of type F precipitation in the same or

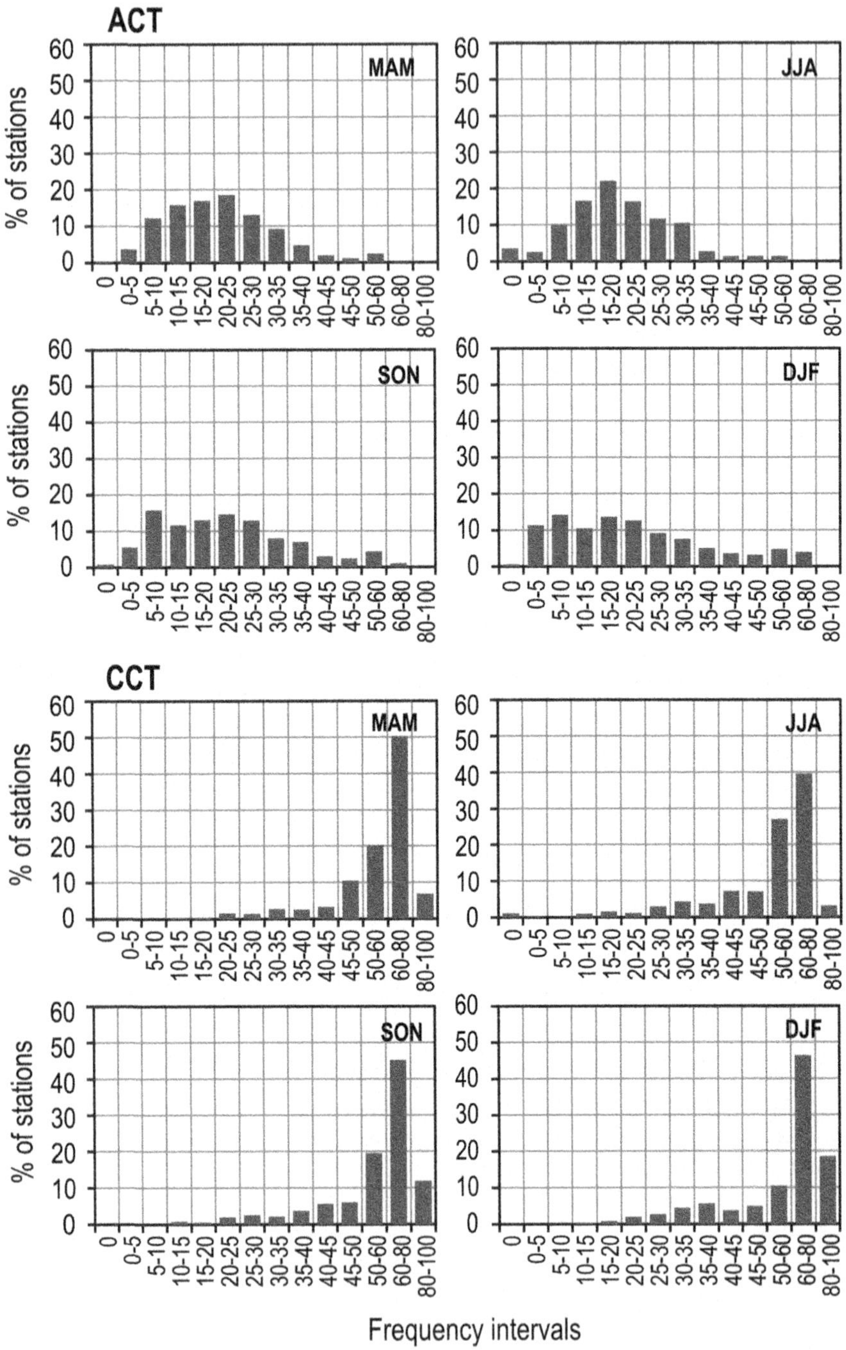

Fig. 6.14 Percentage of stations within the intervals of frontal extreme precipitation (type F) frequency in anticyclones (*ACT*) and cyclones (*CCT*) for January 1951–February 2008. Right closed intervals

similar areas of the continent throughout the year. In Europe, during inflow of air from the northeast sector (N+NEa situation), the highest proportion of type F extreme precipitation is spotted, above all, in the Alps and in southern Poland (Western Carpathians). In the Alps, in spring, summer and autumn, the frequency of F type precipitation in situation N+NEa ranges between 15 and 30 % of ExP, and drops in winter. In the Carpathians, the frequency of type F precipitation usually varies from 10 to 20 % of ExP, and in spring and summer occasionally exceeds 20 % of ExP (Fig. 6.15).

In situation N+NEa, in summer, autumn, and winter, the frequency of type F precipitation increases also in the Balkan Peninsula (20–30 % of ExP), whereas in spring, and particularly in autumn and winter, it occurs most often, compared to the rest of the continent (10–20 % of ExP), at certain stations located in the Ukrainian highlands and between the Black Sea and the Caspian Sea (the Caucasus area). However, in the Alps and the Caucasus, the highest amount of frontal precipitation is associated with situation W+NWa. In the Alps, its frequency in this situation varies from 15 to 50 % of ExP, peaking in winter. In the Caucasus area, the frequency of type F precipitation at times when the air inflows from the northwest sector is, generally, lower than in the Alps (5–35 % of ExP), with its maximum recorded in summer. Advection of air from the northwest sector (situation W+NWa) in winter brings heavy type F precipitation to southern France (Fig. 6.16), and in nearly all seasons, except for autumn, to the Atlantic coasts of the Iberian Peninsula. In each season, situation W+NWa is characterised by the highest frequency of frontal precipitation in the west coasts of the Scandinavian Peninsula, where in winter its frequency at some sites exceeds 20 % of ExP.

In the southern part of the Scandinavian Peninsula, on the eastern side of the Scandinavian Mountains, type F precipitation is the most frequent during advection of air from the south and southwest (situation S+SWa, more than 35 % of ExP). High frequencies of type F precipitation (exceeding 20 % of ExP) during inflow of air from the southeast sector also occur in autumn and winter on the British Isles, in the Iberian Peninsula, and at stations in southeastern France. These areas also are distinguished compared to the rest of the continent in terms of the high occurrence of type F precipitation in spring, except that, in this season, its frequency is lower (15–20 % of ExP) than in autumn and winter (Fig. 6.16).

The spatial diversity of the occurrence of type F precipitation in Europe is the least pronounced during advection of air from the southeast sector (situation E+SEa). In this situation, at most stations a frequency of frontal precipitation does not exceed 10 % of ExP, except for stations in the southern part of Sweden and some stations in Southern Europe, especially in autumn and winter, and in the southern coasts of the Baltic Sea in spring (Fig. 6.15). In Eastern Europe, frontal precipitation in anticyclonic synoptic situations is rare regardless of the season (Figs. 6.15 and 6.16).

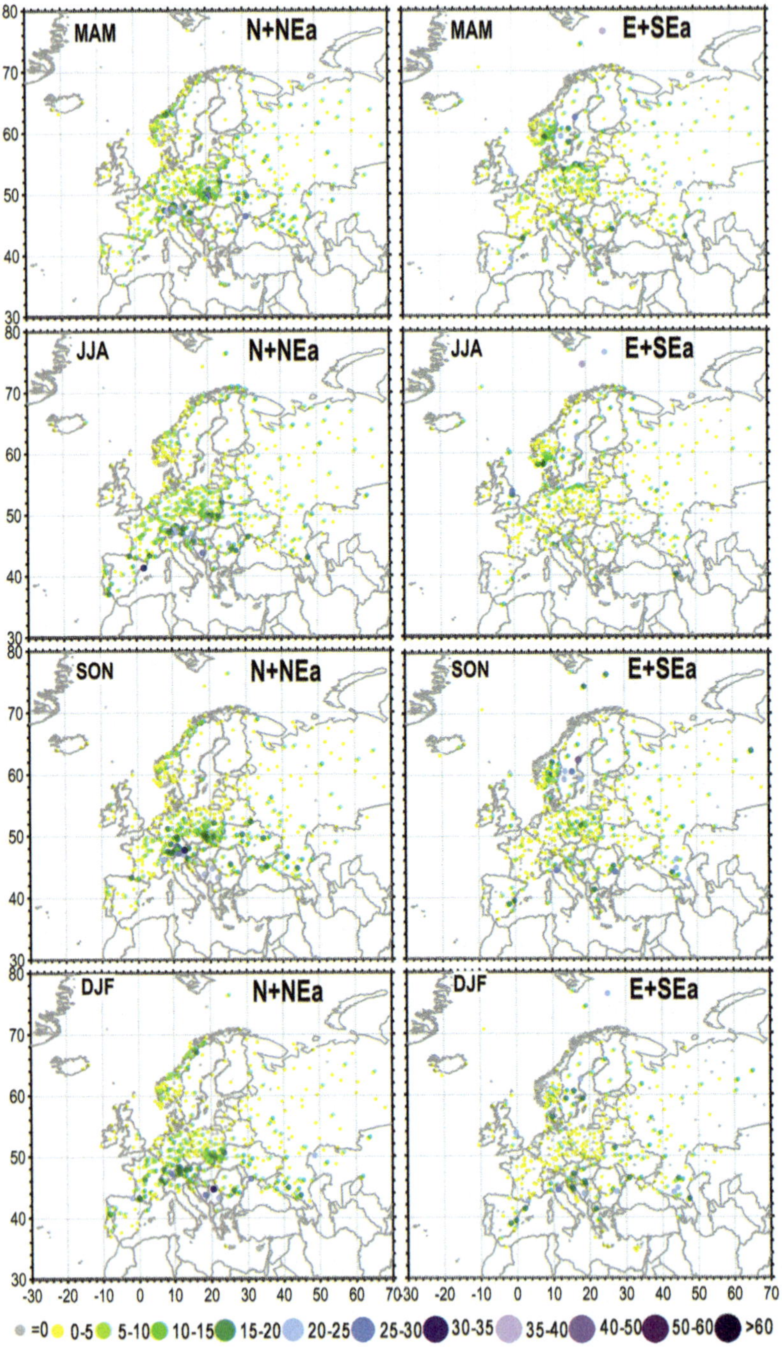

Fig. 6.15 Frequency of frontal extreme precipitation (type F) in anticyclonic circulation types (N+NEa, E+SEa) for January 1951–February 2008. Right closed intervals

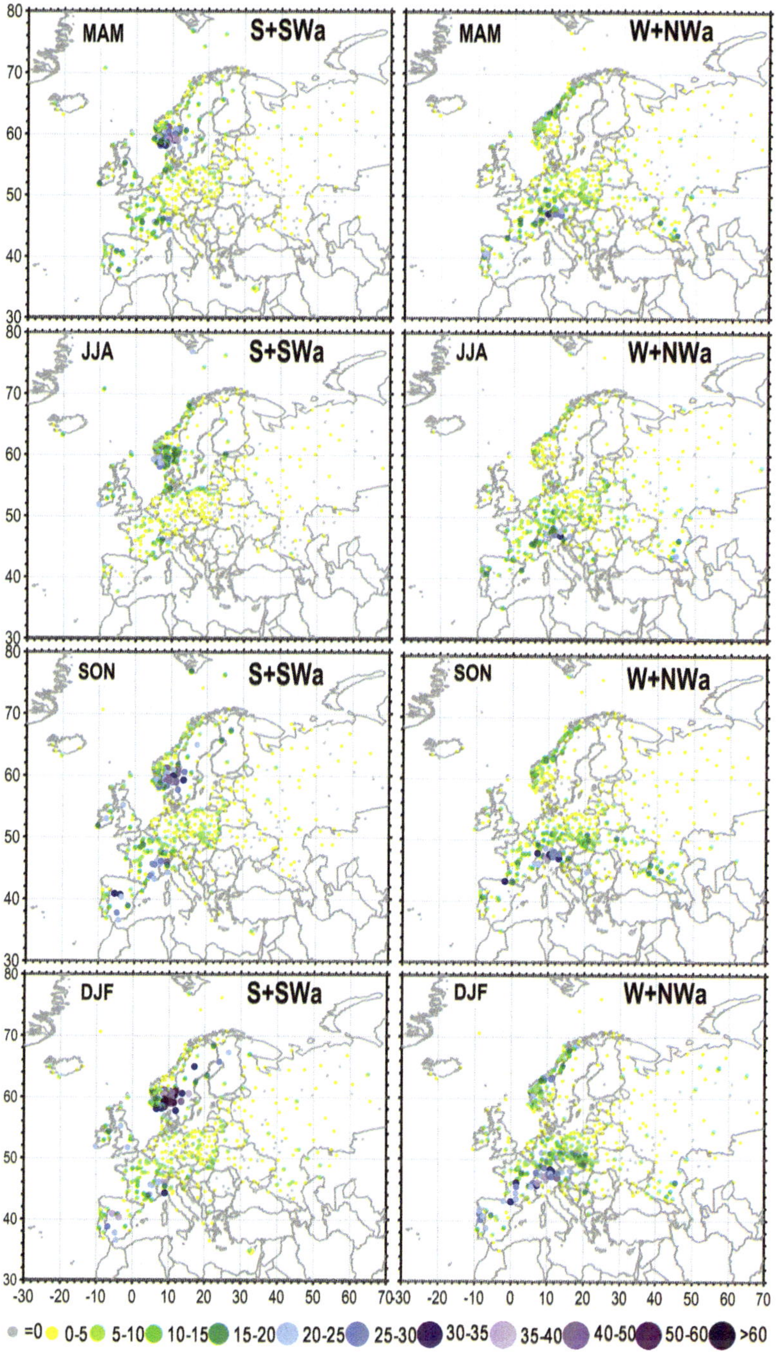

Fig. 6.16 Frequency of frontal extreme precipitation (type F) in anticyclonic circulation types (S+SWa, W+NWa) for January 1951–February 2008. Right closed frequency intervals

6.2.3.3 Frequency of Frontal Extreme Precipitation (Type F) in Cyclonic Synoptic Situations

A large majority of extreme frontal precipitation, as for all the other types of extreme precipitation, is related to cyclonic synoptic situations. The spatial distribution of the frequency of type F ExP in each synoptic situation demonstrates certain general regularities, which should be mentioned before proceeding to a detailed analysis of the regional diversity of the relationships in question. During advection of air from the northeast sector (situation N+NEc), the frequency of type F precipitation in Central Europe and in the western and southern parts of Eastern Europe is higher than in the Scandinavian Peninsula and Western and Southern Europe. Inflow of air from the southeast sector (situation E+SEc) is responsible for an increase in the frequency of type F ExP in Southern, Eastern, and Northern Europe on the lee side of the Scandinavian Mountains. In all seasons, advection of air from the northwest sector (situation W+NWc) is associated with a high frequency of the precipitation in question in Western Europe, including on the coasts of the Scandinavian Peninsula, and in winter, also in Central Europe (Figs. 6.17 and 6.18).

The spatial distribution of the frequency of type F precipitation in situation S+SWc shows seasonal diversity, with one regularity across the seasons, namely, a high frequency in the Scandinavian Peninsula, notably in southern Norway, on the British Isles, and at many French stations (Fig. 6.18). In those areas, the frequency of type F ExP in situation S+SWc exceeds 35 % of ExP at many stations. In the west coasts of the Scandinavian Peninsula, type F precipitation is also frequent at the W+NWc situation. In the southern part of Europe, from the west coasts of the Mediterranean Sea as far as the north coast of the Black Sea, the largest proportion of precipitation occurs when the air inflows from the southeast sector (situation E+SEc). In the continental coasts of the North Sea, from the Frisian Islands to Jutland, most frontal precipitation is associated with W+NWc situations (Fig. 6.18).

The areas characterised by seasonality in the relationships between the occurrence of type F precipitation and atmospheric circulation include the northern part of Europe on the lee side of the Scandinavian Mountains, where the occurrence type F precipitation in spring and summer is most strongly connected with situation E+SEc, and in autumn and winter with situation S+SWc. In spring and autumn, in the northern and southern part of Poland, the highest frequency of type F precipitation is associated with situation N+NEc, whereas in central Poland it is with situation E+SEc. In summer, the precipitation type in question is linked with N+NEc situations, and in central Poland with E+SEc situations, while in the north the frequency of type F precipitation in both these situations is similar. In winter, in central Poland, frontal precipitation occurs mainly during advection of air from S+SWc, and in the north and south of the country during northwesterly flows (situation W+NWc). In the northern part of Eastern Europe, in spring, autumn, and summer, frontal extreme precipitation is associated, in the first place, with situation E+SEc, whereas in the central part of Eastern Europe it is more frequent in situation N+NEc. Such relationships are pronounced in spring and autumn. In winter, the highest frequency of type F precipitation in that part of Europe is associated with situation S+SWc.

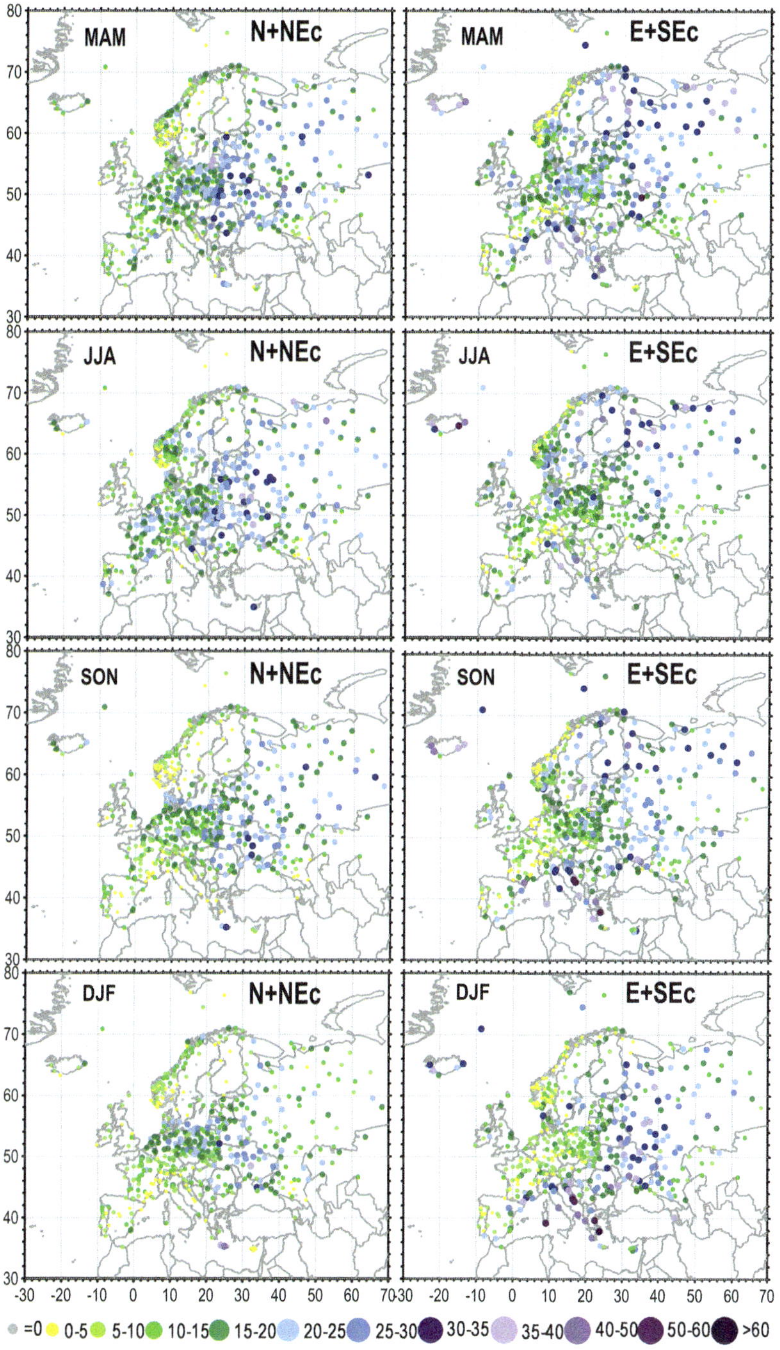

Fig. 6.17 Frequency of frontal extreme precipitation (type F) in cyclonic circulation types (N+NEc, E+SEc), January 1951–February 2008. Right closed intervals

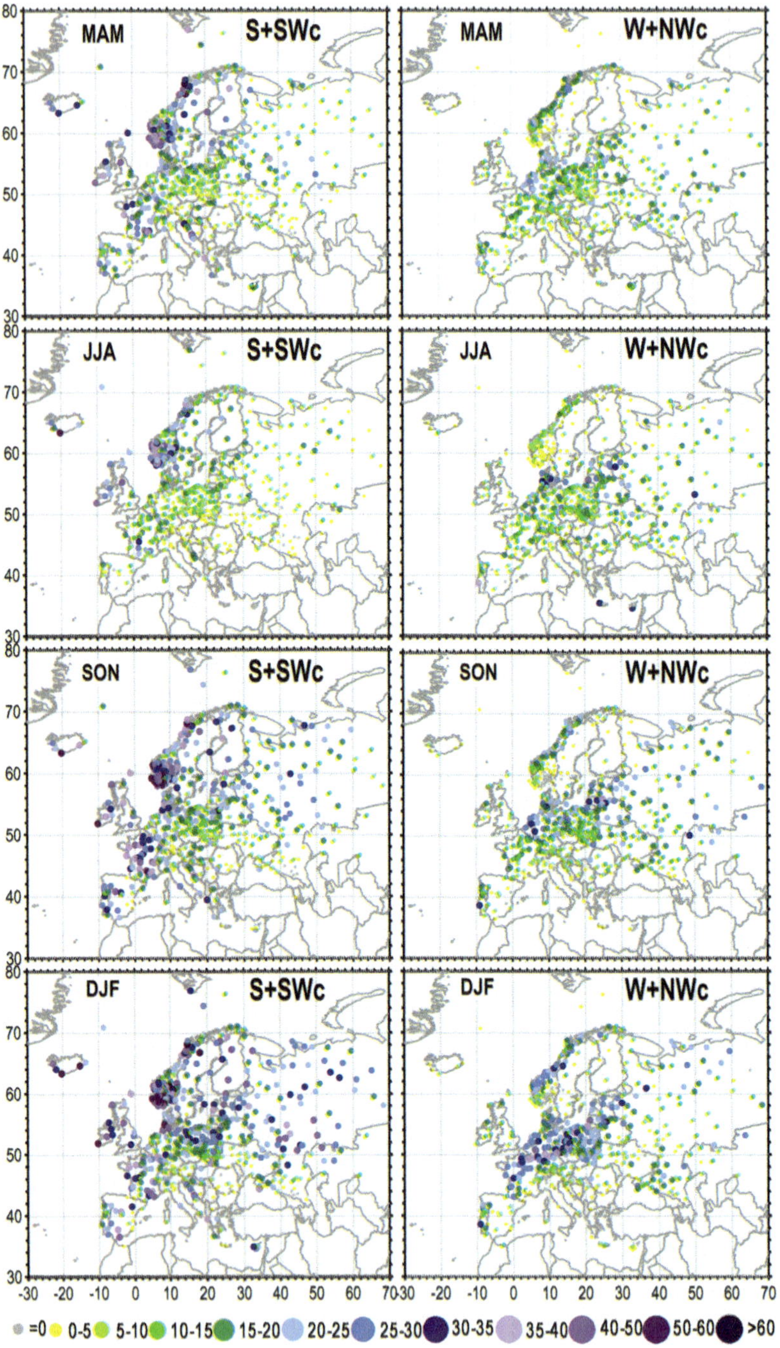

Fig. 6.18 Frequency of frontal extreme precipitation (type F) in cyclonic circulation types (S+SWc, W+NWc) for January 1951–February 2008. Right closed frequency intervals

Table 6.5 Descriptive statistics of the frequency [%] of cold front extreme precipitation (type Fc) in anticyclones (*ACT*) and cyclones (*CCT*) in Europe for January 1951–February 2008

Pressure system	Season	Average (±SE)		Confidence intervals −95 %	+95 %	Min	Max	SD	CV	ME	Quartiles Lower	Upper
ACT	MAM	4.6	(±0.2)	4.2	5.0	0.0	25.7	4.4	95.9	3.5	1.5	6.0
	JJA	6.1	(±0.2)	5.6	6.6	0.0	30.8	5.3	87.6	4.8	2.6	7.9
	SON	5.5	(±0.2)	5.0	6.0	0.0	28.0	5.5	99.2	3.8	1.7	7.3
	DJF	4.3	(±0.2)	3.9	4.8	0.0	32.8	5.0	116.3	2.9	1.1	5.4
CCT	MAM	10.7	(±0.3)	10.1	11.2	0.0	38.2	6.2	57.7	10.3	6.0	14.3
	JJA	13.2	(±0.3)	12.6	13.9	0.0	43.2	7.4	55.8	12.5	7.9	17.2
	SON	11.7	(±0.3)	11.1	12.3	0.0	36.2	6.9	59.0	10.6	6.5	15.2
	DJF	9.1	(±0.3)	8.6	9.6	0.0	44.4	6.0	66.3	8.1	4.9	12.2

SE standard error, *Min* minimum value, *Max* maximum value, *SD* standard deviation, *CV* coefficient of variability, *ME* median

In Iceland, most frontal precipitation occurs during air advection from the southern sector, whereby in spring and summer, such precipitation is more frequent in E+SEc situations, and in autumn and winter in S+SWc situations (Fig. 6.17).

6.2.4 *Cold Front Precipitation (Type Fc)*

6.2.4.1 Frequency of Cold Front Extreme Precipitation (Type Fc) in Anticyclones (ACT) and Cyclones (CCT)

In Europe, the average frequency of extreme precipitation associated with cold fronts in high-pressure systems is the highest in summer (on average, 6.1 % of ExP), and the lowest in winter (4.3 %) (Table 6.5). The essential characteristics of the spatial distribution of the frequency of type Fc precipitation in ACT do not change much during the year. In Europe, type Fc ExP in ACT is most likely to occur in the Alps, especially in summer and autumn, when type Fc in ACT accounts for more than 20 % of ExP (Fig. 6.19). In spring and winter, frequencies of more than 20 % of ExP in the area occur occasionally, with most weather stations recording type Fc in ACT at a rate of 10–20 % of ExP. Similar frequency of cold front precipitation throughout the year, notably in autumn and winter, is recorded by the area of the Caucasus and the Iberian Peninsula. In the other part of the continent, type Fc precipitation in ATC represents at most 5 % of ExP (45 % of the stations in summer, 51 % in autumn, 55 % in spring, and 56 % in winter) (Fig. 6.19).

At some stations located for the most part in Western and Central Europe, at some distance from Europe's west coasts in a stretch running between the Breton Peninsula and Jutland (except for summer) and in the Scandinavian Peninsula, the frequency of type Fc in ACT is also higher compared to its frequency elsewhere in the continent (5–10 % of ExP).

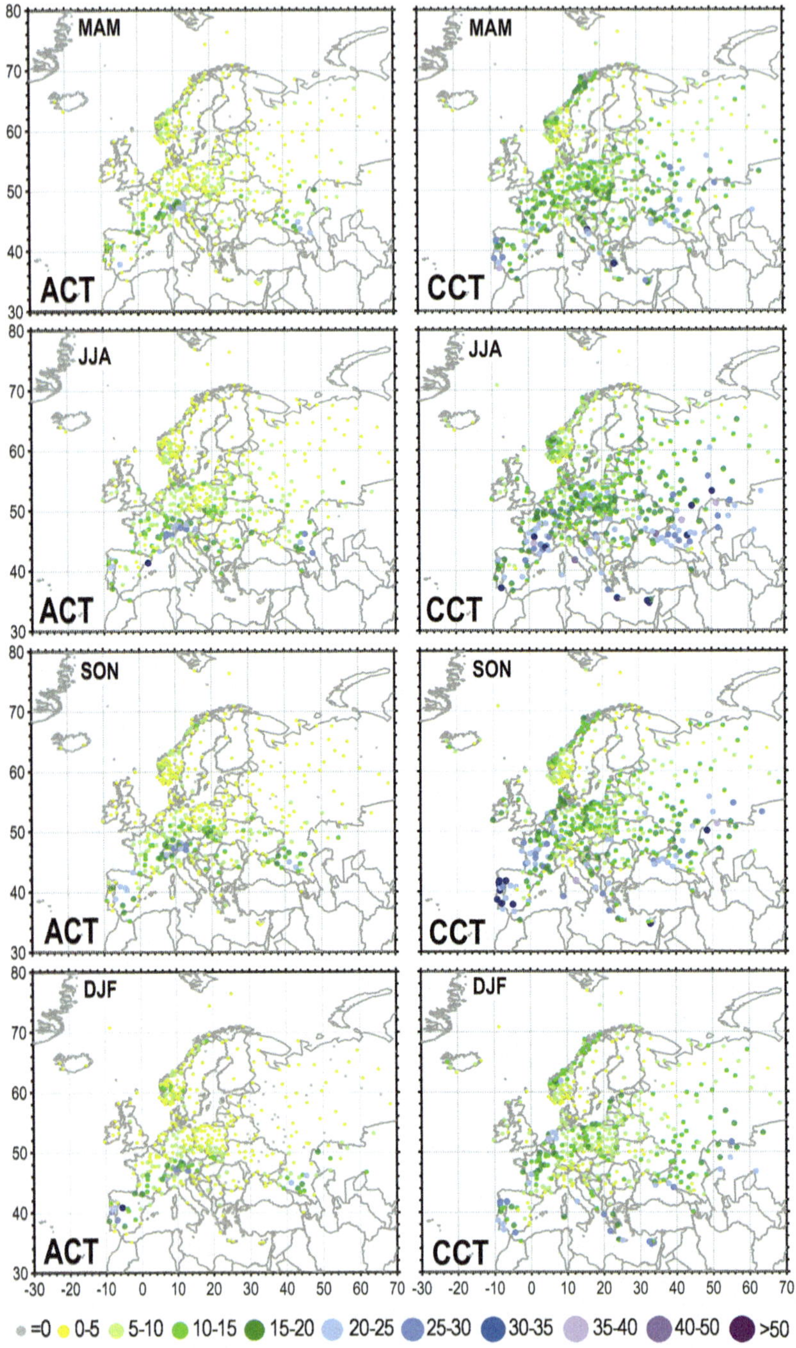

Fig. 6.19 Frequency of cold front extreme precipitation (type Fc) in anticyclones (*ACT*) and cyclones (*CCT*) for January 1951–February 2008. Right closed frequency intervals

The average frequency of extreme precipitation associated with the passage of a cold front in low-pressure systems changes during the year from approximately 9 % of ExP in winter to approximately 13 % of ExP in summer (Table 6.5), when Europe receives air from the Atlantic Ocean, which is cooler in this season of the year. The spatial distribution of the frequency of type Fc extreme precipitation in CCT varies in Europe during the year. In summer, the most frequent cold front precipitation in cyclones (more than 20 % of ExP) is spotted in the area of the Massif Central in France, as well as in the southern part of Eastern Europe, mainly in the area of the Caucasus and the Stavropol Upland, and at the southernmost stations of the Mediterranean Basin (Fig. 6.19). In this season, in the other areas of Western, Central, and Eastern Europe and in the southern part of the Atlantic coast of the Scandinavian Peninsula, Fc type in CCT represents between 10 and 20 % of ExP. In summer, the lowest frequencies of type Fc precipitation in CCT (up to 10 % of ExP) are mainly recorded in northern Europe (Fig. 6.19). Similar characteristics of the spatial distribution of type Fc ExP in CCT are observable in autumn and winter, whereby in those seasons the values of the index in question at most European stations are lower than in summer. In autumn, the frequency of type Fc precipitation in CCT exceeds 10 % of ExP at 55 % of the stations, in winter at 36 % of the stations, and in summer at 63 % of the stations (Fig. 6.20).

In autumn, cold front precipitation represents more than 20 % of ExP at stations in Western Europe, from the Iberian Peninsula to the Netherlands, in the central part of the Mediterranean Basin, and at sites located to the north of the Black Sea and the Caspian Sea (Fig. 6.19). A similar spatial distribution of the frequency of type Fc precipitation is observable in winter. Its frequency in the season exceeds 20 % of ExP at the southeast coasts of the North Sea, in the southern part of the French lowland on the coast of the Bay of Biscay, in the Iberian Peninsula, as well as in the central and eastern parts of the Mediterranean Sea and in the southern part of Eastern Europe (Fig. 6.19). In spring, Fc type precipitation in CCT represents more than 20 % of ExP in the south of the continent. In the west coasts of the Scandinavian Peninsula, throughout the year, cold front precipitation in CCT is more frequent than on the eastern side of the Scandinavian Mountains (Fig. 6.19).

6.2.4.2 Frequency of Cold Front Precipitation (Type Fc) in Anticyclonic Synoptic Situations

As shown in Sect. 3.3, on average, the highest frequency of extreme precipitation associated with the passage of cold fronts is recorded in summer. In the season, cold front precipitation in anticyclones is most frequent (10–20 %) in the Alpine region, especially in the northeast part of the Alps, when the air inflows from the northwest (situation W+NWa). A particularly high frequency of type Fc extreme precipitation in that part of the Alps may result from the reactivation of a cold front moving along the mountain range, as described by Schneider (1996).

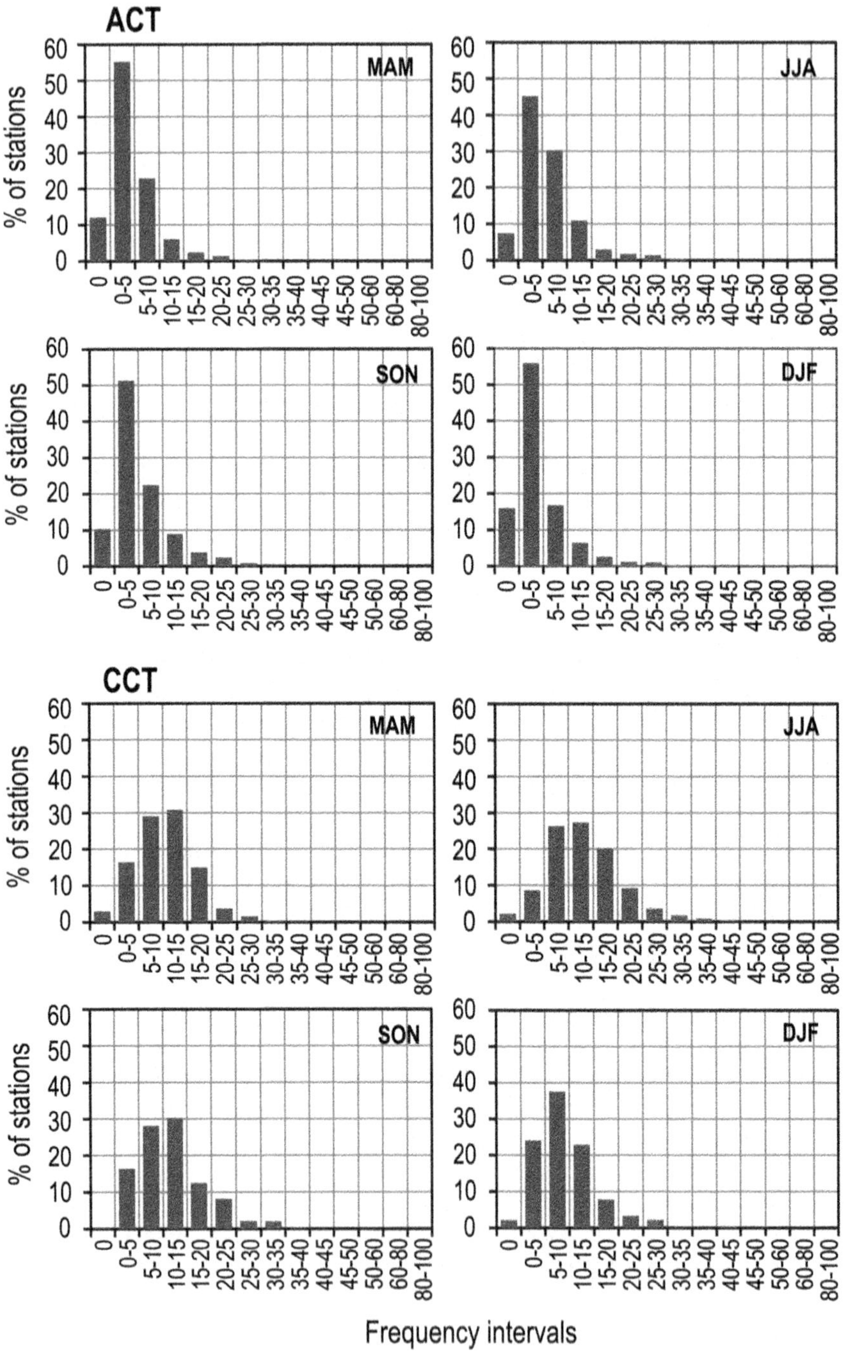

Fig. 6.20 Percentage of stations within the intervals of cold front extreme precipitation (type Fc) frequency in anticyclones (*ACT*) and cyclones (*CCT*) for January 1951–February 2008. Right closed intervals

In the other seasons (spring, autumn, winter) the foregoing advection also contributes to increased frequency of type Fc precipitation in the Alps and the area, but in those seasons, its frequency in situation W+NWa is lower (it less frequently exceeds 10 % of ExP). Outside the Alpine area, a high (compared to the rest of the continent) amount of Fc precipitation in situation W+NWa is recorded by the west coasts of the Iberian Peninsula in winter and summer (between 5 and 20 % depending on the station), and the Caucasus area in summer and autumn (usually, from 5 to 15 % of ExP, and occasionally more). In those areas in the other seasons (the Iberian Peninsula in spring and autumn, the Caucasus in winter and spring) the frequency of Fc precipitation reaches no more than 10 % of ExP; however, it is higher than at most stations in Central Europe and Eastern Europe, and in spring and winter also compared to stations in the west of the continent. In general, across the seasons, advection of air from the northwest sector causes an increase in the frequency of type Fc precipitation in southwestern Europe, from the Iberian Peninsula to the northern part of the Alps, in the Caucasus area, in the southern part of Central Europe, and in spring, autumn, and winter also on the western coast of Scandinavian Peninsula. In these areas, extreme precipitation generated at cold fronts occurs more frequently also during air inflow from the northeast (situation N+NEa), representing, usually, from approximately 5 to 10 % of ExP, and occasionally, mainly in summer, exceeding 10 % of ExP. In the other parts of the continent, during N+NEa situations, cold front precipitation constitutes at most 2.5 % of ExP (Figs. 6.21 and 6.22).

Inflow of air from the southeast sector (situation S+SWa) is favourable for the formation of type Fc precipitation in the Iberian Peninsula in autumn (usually between 10 and 20 % of ExP) and in winter (between 5 and 15 %), and at stations located in the western part of the Po Plain in summer (from 10 to 15 % of ExP) and autumn (up to 25 % of ExP). Slightly lower, but also distinguishable, compared to the rest of the continent, frequency of Fc precipitation in S+SWa occurs, throughout the year, at some sites in France and in the southern part of the Scandinavian Peninsula. This situation also generates a large proportion of type Fc precipitation in Southern Poland in autumn, and in Ireland, as well as in the northern part of Germany in summer (Fig. 6.22).

Of all the advective anticyclonic circulation types, the poorest relationship was observed between the occurrence of cold front precipitation and situation E+SEa. At many stations in Europe, during inflow of air from the southwest sector cold front extreme precipitation does not occur at all or its frequency does not exceed 2.5 % of ExP. It is only at isolated stations located, for the most part, in the south of Europe in autumn and winter, as well as in Central Europe in spring and summer, that the frequency of type Fc precipitation in situation E+SEa exceeds 10 % of ExP (Fig. 6.21).

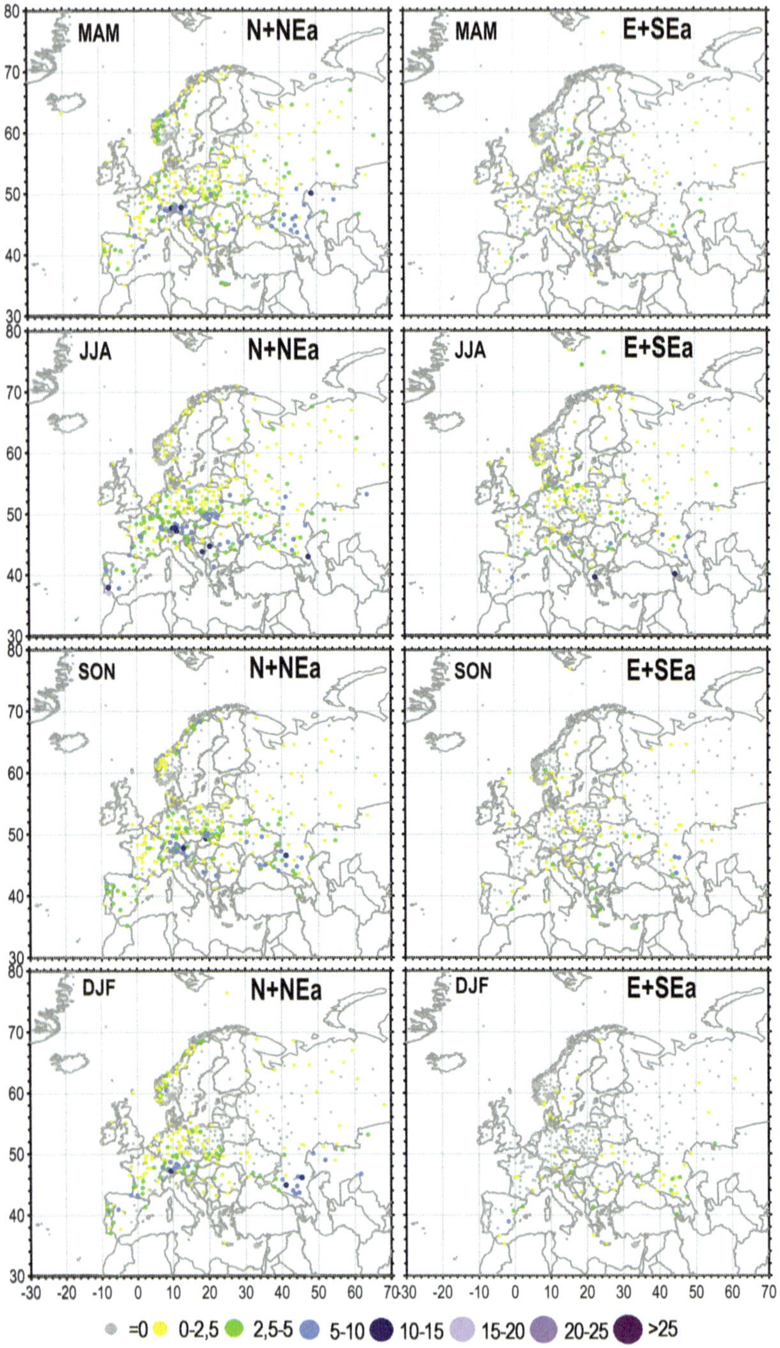

Fig. 6.21 Frequency of cold front extreme precipitation (type Fc) in anticyclonic circulation types (N+NEa, E+SEa) for January 1951–February 2008. Right closed intervals

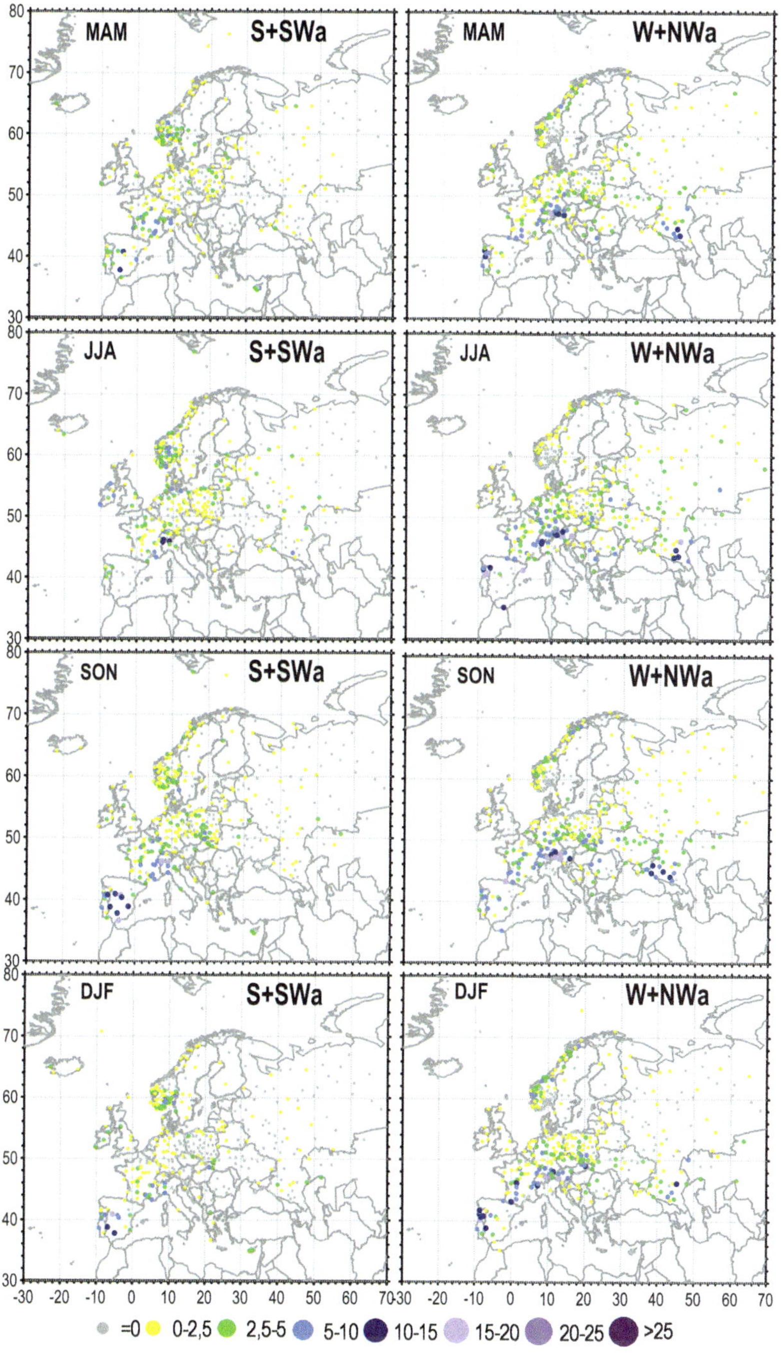

Fig. 6.22 Frequency of cold front extreme precipitation (type Fc) in anticyclonic circulation types (S+SWa, W+NWa) for January 1951–February 2008. Right closed frequency intervals

6.2.4.3 Frequency of Cold Front Extreme Precipitation (Type Fc) in Cyclonic Synoptic Situations

The distribution of the cold front extreme precipitation frequency in cyclonic synoptic situations is characterised by a less ordered spatial pattern than its distribution in anticyclonic situations and shows seasonal variability. The highest frequency of cold front precipitation in cyclonic situations is recorded by stations located in the south of Europe and near the west coasts of the continent (Figs. 6.23 and 6.24). In those areas, the relationships between the occurrence of Fc precipitation and the individual synoptic situations are the strongest. Inflow of air from the northeast sector (situation N+NEc) determines the occurrence of cold front precipitation in summer. In that season, the highest frequencies of type Fc precipitation in situation N+NEc, in excess of 10 % of ExP, occur mainly at stations located in the area stretching from the Adriatic Sea (northern part) to the southern part of Eastern Europe, as well as in the southwest part of the Iberian Peninsula. The lowest spatial variability in the type Fc precipitation frequency during inflow of air from the northeast sector is observable in winter (Fig. 6.23).

Similar to the earlier discussed precipitation types, the occurrence of type Fc is least related to the air advection from the southeast sector (situation E+SEc), chiefly in winter. In this situation, in autumn and winter, cold front extreme precipitation occurs only occasionally or does not occur at all at most of the European stations. Its frequency exceeds 5 % of ExP only at some stations, mainly in the southeast part of the study area, and 10 % of ExP at isolated stations in the central part of the Mediterranean Basin. The highest frequency of type Fc precipitation in situation E+SEc, exceeding at most 25 % of ExP, is recorded in summer at stations located in the Mediterranean Basin and at isolated stations between the Black Sea and the Caspian Sea. In summer, a higher (compared to most of Europe, 5–10 % of ExP) frequency of type Fc precipitation in E+SEc situation is also observed at stations located in the lowlands of Central Europe.

During air advection from the southwest sector in summer the frequency of type Fc precipitation increases in southwest Europe (the Iberian Peninsula and France), in Ireland, and in the central part of the Mediterranean Sea, where it constitutes 10–25 % of ExP. In summer, similar frequency of type Fc precipitation in situation S+SWc is also observed in the southern part of the Scandinavian Peninsula (Norway). In those areas, the frequency of type Fc precipitation under S+SWc conditions is higher compared to the rest of the continent also in the other seasons, but such precipitation represents a smaller percentage, that is, approximately 5–10 % of ExP. Type Fc extreme precipitation occurs at a similar frequency (5–10 % of ExP) also in summer and autumn in Western Europe and Central Europe, and across the year in the southern part of Eastern Europe.

Westerly and northwesterly airflow (situation W+NWc) leads, throughout the year, to an increase in the number of days with extreme cold front precipitation in the southern part of Eastern Europe, in autumn and winter also in the west coasts of Europe from the Iberian Peninsula to the Frisian Islands, and in spring and summer on islands in the eastern area of the Mediterranean Basin (Fig. 6.24).

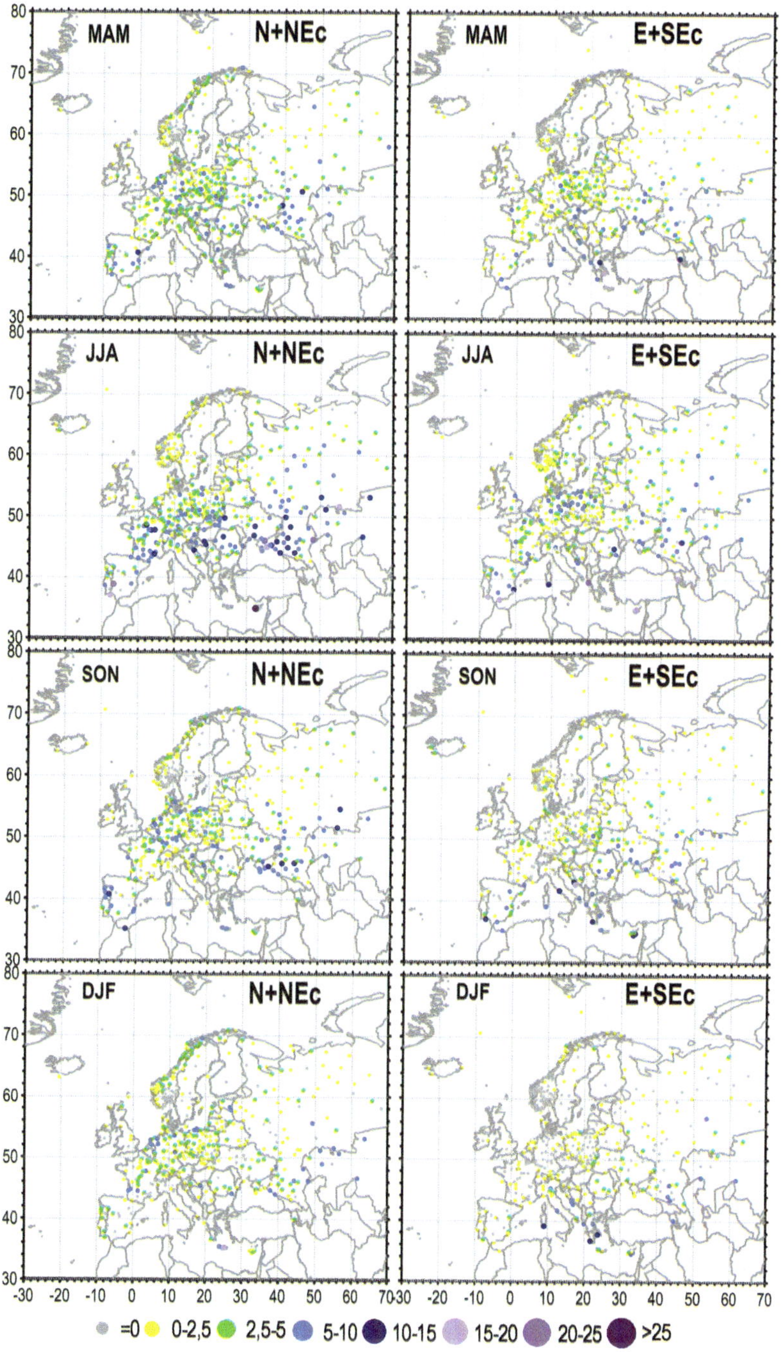

Fig. 6.23 Frequency of cold front extreme precipitation (type Fc) in cyclonic circulation types (N+NEc, E+SEc) for January 1951–February 2008. Right closed intervals

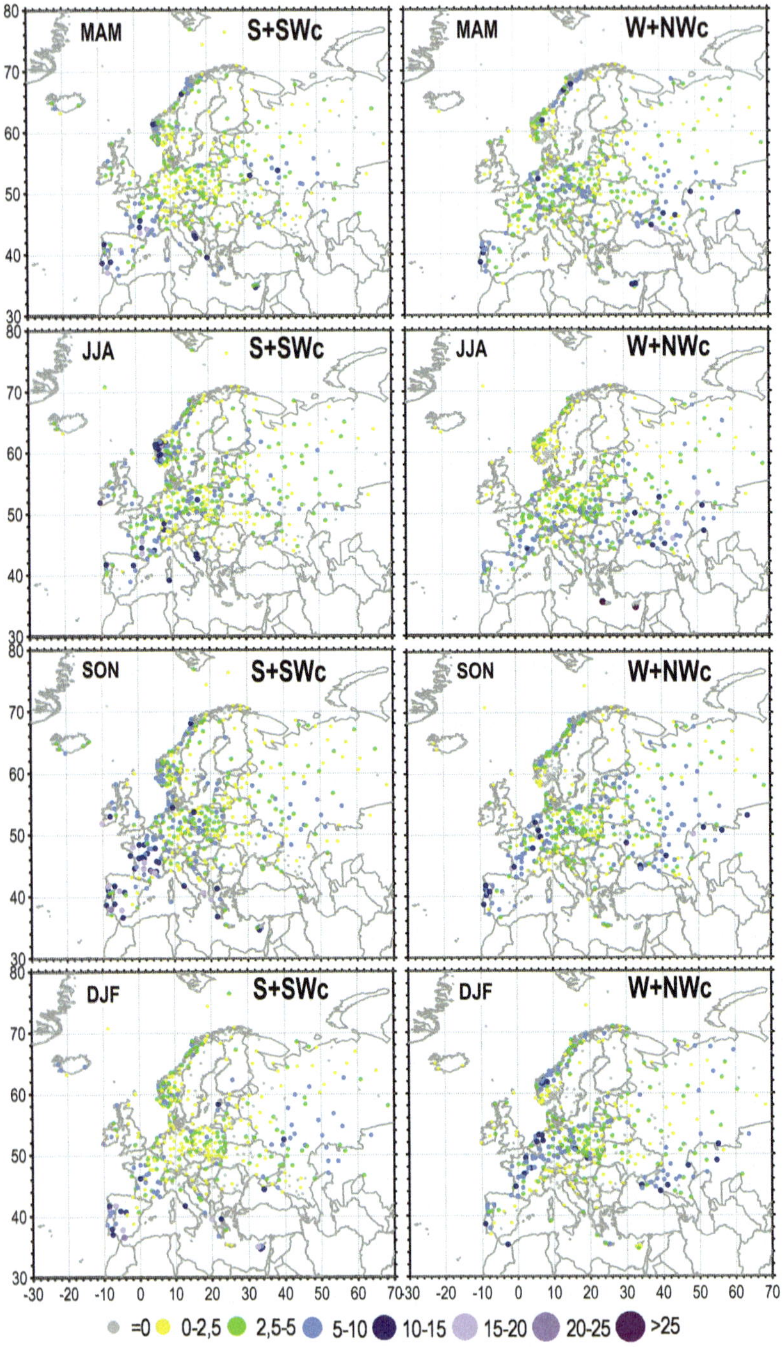

Fig. 6.24 Frequency of cold frontal extreme precipitation (type Fc) in cyclonic circulation types (S+SWc, W+NWc) for January 1951–February 2008. Right closed intervals

6.2.5 *Warm Front Precipitation (Type Fw)*

6.2.5.1 Frequency of Warm Front Extreme Precipitation (Type Fw) in Anticyclones (ACT) and Cyclones (CCT)

In Europe, the average frequency of extreme precipitation associated with the passage of a warm front in anticyclonic types of circulation is low; it ranges, annually, between 2.7 % of ExP in summer and 4.3 % of ExP in winter. At most weather stations (59 % of stations in spring, 63 % in summer, 60 % in autumn, 52 % in winter) type Fw precipitation in ACT does not represent more than 5 % of ExP (Figs. 6.25 and 6.26); the maximum frequencies fluctuate between 14.5 % of ExP in spring and summer and 18.9 % of ExP in winter (Table 6.6)

In winter, the frequency of warm front precipitation in anticyclonic situations exceeds 10 % of ExP at a few stations in the Alpine area, in the southern part of Poland and Germany, as well as in the northern part of the Balkans. A distinguishable (on a continental scale) frequency of type Fw precipitation (5–10 %) is found in areas located in the borderland between Central Europe and Southern Europe, in the area of the Caucasus and in the northern part of the west coast of the Iberian Peninsula, as well as on the lee side of the Scandinavian Mountains. In other areas of Europe, type Fw precipitation in ACT represents not more than 5 % of ExP (Fig. 6.25).

In autumn, the spatial distribution of the index resembles its distribution in winter, with the frequency of type Fw precipitation in ACT being slightly lower than in winter. The Fc precipitation in ACT constitutes 5–10 % of ExP at 29 % of the stations in winter and at 25 % of the stations in autumn. Its frequency reaches more than 10 % of ExP at 7 % of the stations in winter and at 2 % of the stations in autumn. In spring, the frequency of type Fc precipitation in ACT does not actually exceed 10 % of ExP. At most stations, the type represents less than 5 % of ExP, exceeding this percentage only in east Poland, in the Alps, in the northern part of the Iberian Peninsula, in the southern part of the Scandinavian Peninsula, and at some sites in Eastern Europe. In summer, at 63 % of the stations, the frequency of type Fw precipitation in ACT does not exceed 5 % of ExP, and at 22 % of the sites, warm fronts do not generate heavy precipitation during anticyclones at all (Fig. 6.25).

The spatial distribution of the frequency of type Fw precipitation in low-pressure systems is characterised by a higher seasonal diversity than in high-pressure systems. The frequency of type Fw precipitation in CCT is also low compared to the other origin-based types of precipitation, ranging from 4.7 % of ExP in summer to 8.2 % of ExP in winter, and is lower in autumn than in spring (Table 6.6).

In general, irrespective of the season, in Eastern Europe warm front precipitation in CCT is more frequent than in Western and Central Europe (Fig. 6.25). This property of the spatial distribution mainly characterises spring and winter. In those seasons, in Western Europe and Central Europe and in the Scandinavian Peninsula type Fw precipitation in CCT accounts for not more than 10 % of ExP. At most sites in Eastern Europe, the frequency of type Fw precipitation exceeds 10 % of ExP, and in

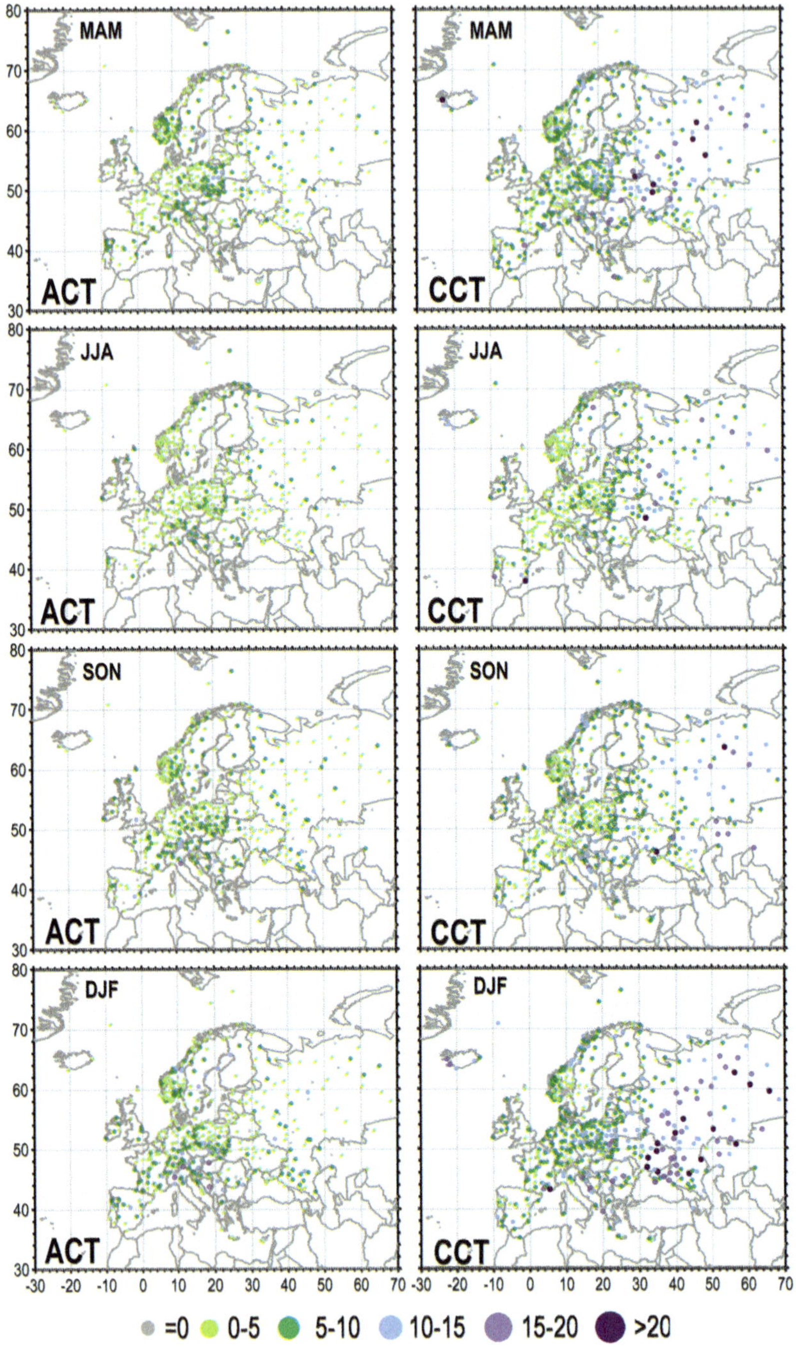

Fig. 6.25 Frequency of warm front extreme precipitation (type Fw) in anticyclones (*ACT*) and cyclones (*CCT*), January 1951–February 2008. Right closed intervals

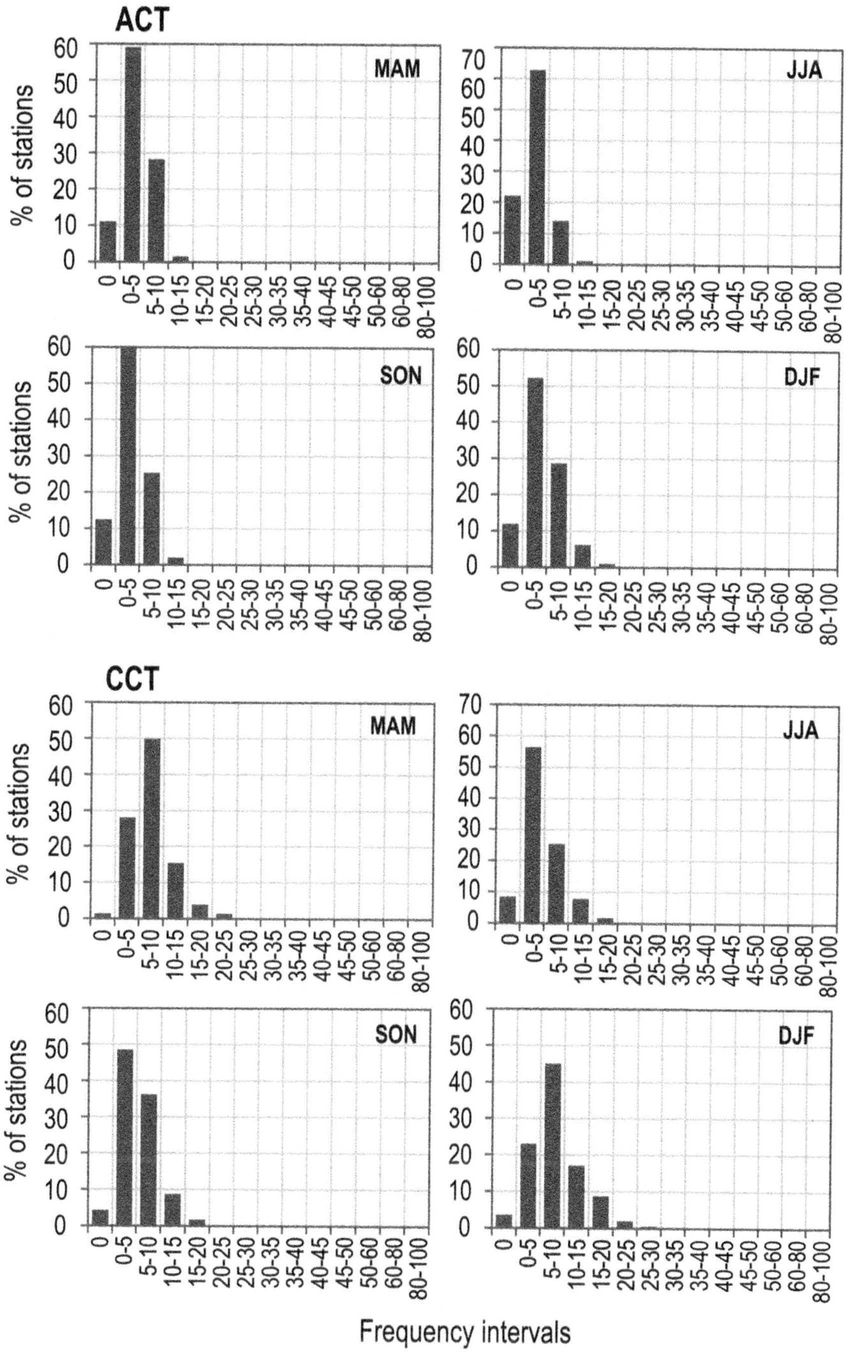

Fig. 6.26 Percentage of stations within the intervals of warm front extreme precipitation (type Fw) frequency in anticyclones (*ACT*) and cyclones (*CCT*) for January 1951–February 2008. Right closed frequency intervals

Table 6.6 Descriptive statistics of the frequency [%] of warm front extreme precipitation (type Fw) in anticyclones (*ACT*) and cyclones (*CCT*) Europe for January 1951–February 2008. Right closed intervals

Pressure system	Season	Average (±SE)	Confidence intervals		Min	Max	SD	CV	ME	Quartiles	
			−95 %	+95 %						Lower	Upper
ACT	MAM	3.8 (±0.1)	3.6	4.1	0.0	14.5	2.7	69.3	3.5	1.9	5.7
	JJA	2.7 (±0.1)	2.5	2.9	0.0	14.5	2.4	88.0	2.4	1.0	4.1
	SON	3.6 (±0.1)	3.4	3.8	0.0	16.4	2.8	76.3	3.2	1.4	5.3
	DJF	4.3 (±0.2)	4.0	4.6	0.0	18.9	3.4	78.9	3.8	1.7	6.3
CCT	MAM	7.4 (±0.2)	7.0	7.7	0.0	22.4	4.1	55.3	6.7	4.7	9.6
	JJA	4.7 (±0.2)	4.4	5.1	0.0	22.2	3.7	77.4	3.8	2.3	6.5
	SON	5.6 (±0.2)	5.3	5.9	0.0	26.3	3.8	68.4	4.8	2.8	7.8
	DJF	8.2 (±0.2)	7.7	8.6	0.0	27.8	5.1	62.6	7.2	4.9	10.8

SE standard error, *Min* minimum value, *Max* maximum value, *SD* standard deviation, *CV* coefficient of variability, *ME* median

winter, in the southern and easternmost parts of the area, type Fw precipitation in CCT represents even more than 20 % of ExP (Fig. 6.25). Nearly throughout the year, except for summer, heavy precipitation associated with a warm front in low-pressure systems is likely to occur on the coasts of the Scandinavian Peninsula in the area of Lofoten and Vesteraten.

6.2.5.2 Frequency of Warm Front Precipitation (Type Fw) in Anticyclonic Synoptic Situations

In Europe, the frequency of extreme precipitation associated with the passage of a warm front in anticyclonic situations rarely reaches 10 % of ExP in spring, autumn, and winter, and is exceptionally low in summer, not exceeding 2.5 % of ExP in all the anticyclonic situations. The relationships between the occurrence of type Fc extreme precipitation and atmospheric circulation can be noticed in winter, when the type of precipitation occurs more frequently than in other seasons of the year. In the remaining seasons, because of the low frequency of extreme warm front precipitation, the relationships between its occurrence and atmospheric circulation are even less pronounced.

In winter, the occurrence of type Fw precipitation is associated with situations W+NWa and S+SWa. During inflow of air from the northwest sector, the frequency of type Fw is slightly higher compared to the continent as a whole (up to 10 % of ExP) at stations in the northern part of the Iberian Peninsula, in southern France, in the Alps, in Southern Poland, and in the south of Germany, as well as at isolated stations on the west coasts of the Scandinavian Peninsula. At the other stations in Western Europe and Central Europe, the frequency of warm front precipitation accounts for at most 2.5 % of ExP; in Eastern Europe, such precipitation does not occur in W+NWa situations.

Southerly and southwesterly airflow in a high-pressure system (situation S+SWa) causes, in winter, an increase in the number of days with type Fw precipitation mainly in the central part of the southernmost areas of the Scandinavian Peninsula and on the lee side of the Scandinavian Mountains, as well as at single stations in southwest Europe. Although weaker, this are also observable in spring and autumn (Fig. 6.27).

During advection of air from the southeast sector (E+SEa), in winter, the frequency of warm front precipitation in Southern Europe and at many stations in Eastern Europe is slightly higher compared to elsewhere in the continent (not more than 10 % of ExP). Air inflowing from the north and northeast (situation N+NEa), in spring and autumn, fosters the frequency of warm front extreme precipitation in the central part of the study area, which is also observable, although to a lesser extent, in winter (Fig. 6.28).

6.2.5.3 Frequency of Warm Front Extreme Precipitation (Type Fw) in Cyclonic Synoptic Situations

Extreme warm front precipitation in cyclonic situations occurs mainly in the southern and eastern part of Europe. The pattern of its frequency in the individual synoptic situations undergoes seasonal changes. In summer, in cyclonic situations, as in anticyclonic situations, type Fw extreme precipitation is also very rare. In this season, during air advection from northwest to south (situations W+NWc and S+SWc), many stations recorded no warm front precipitation during the study period at all (Fig. 6.30). In the other circulation types such precipitation constitutes not more than 2.5 % of ExP.

In winter, when the average frequency of type Fw ExP in Europe is the highest, its occurrence is mainly related to inflow of air from the southeast sector (situation E+SEc). In the situation, at many stations in Eastern Europe and in the central and eastern part of Southern Europe, type Fw precipitation represents between 5 and 15 % of ExP. In situation E+SEc, also in spring the frequency of Fw precipitation exceeds 5 % at stations in Eastern Europe and Southern Europe (Fig. 6.29).

In winter, the frequency of type Fw precipitation depends on the air advection from the southwest sector in Eastern Europe and at isolated coastal stations in various parts of the continent, including the Scandinavian Peninsula. At these stations, type Fw precipitation constitutes 5–10 % of ExP. In the other seasons (spring, autumn), precipitation of Fw type in situation S+SWc represents 10 % of ExP in the western coasts of the Scandinavian Peninsula and at several stations in Eastern Europe (the latter in autumn only) (Fig. 6.30).

The relationship between the occurrence of type Fw precipitation and situation W+NWc exists in winter. Then, a relatively high frequency of this precipitation type (2.5–10 % of ExP) is noticeable in the west coast of the Scandinavian Peninsula and the area between the Frisian Islands and Poland. In the other seasons (spring and autumn), this relationship is limited to the coasts of the Scandinavian Peninsula in

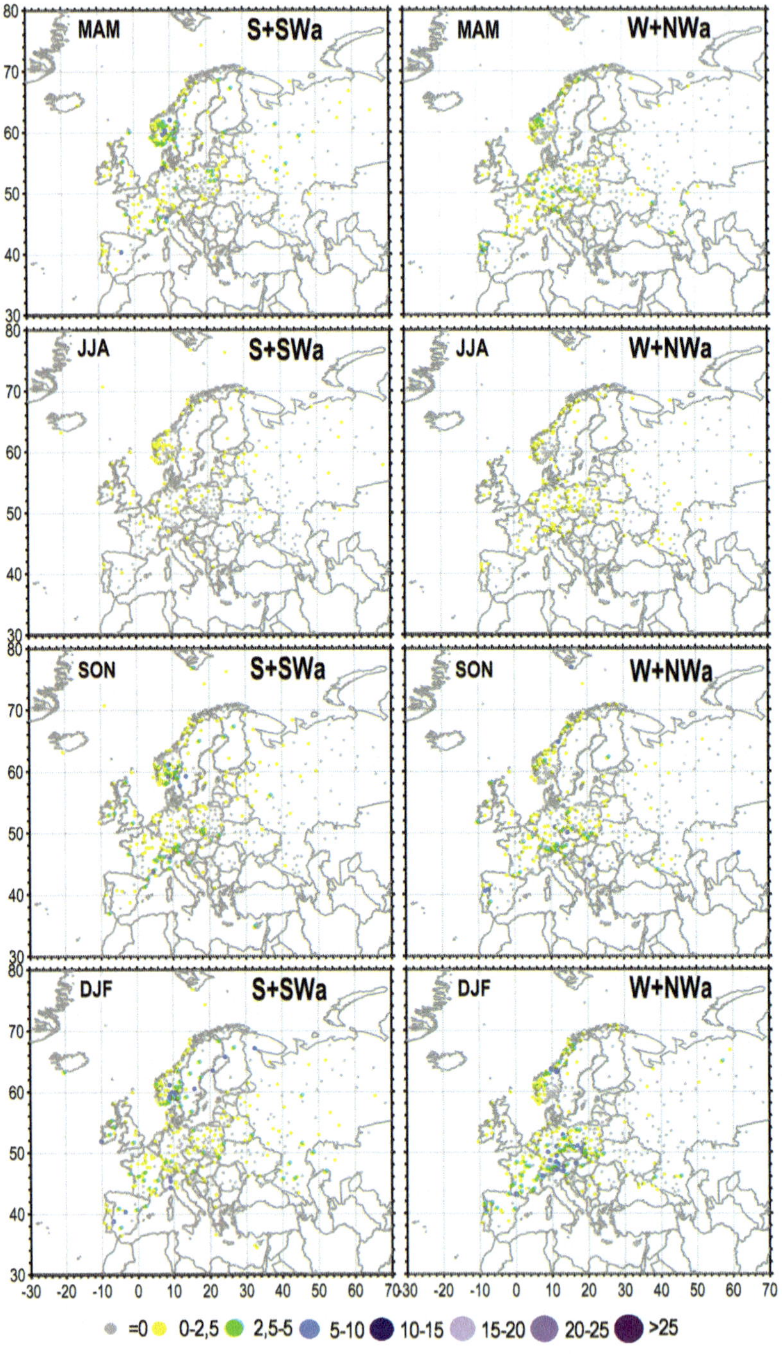

Fig. 6.27 Frequency of warm front extreme precipitation (type Fw) in anticyclonic circulation types (S+SWa, W+NWa) for January 1951–February 2008. Right closed intervals

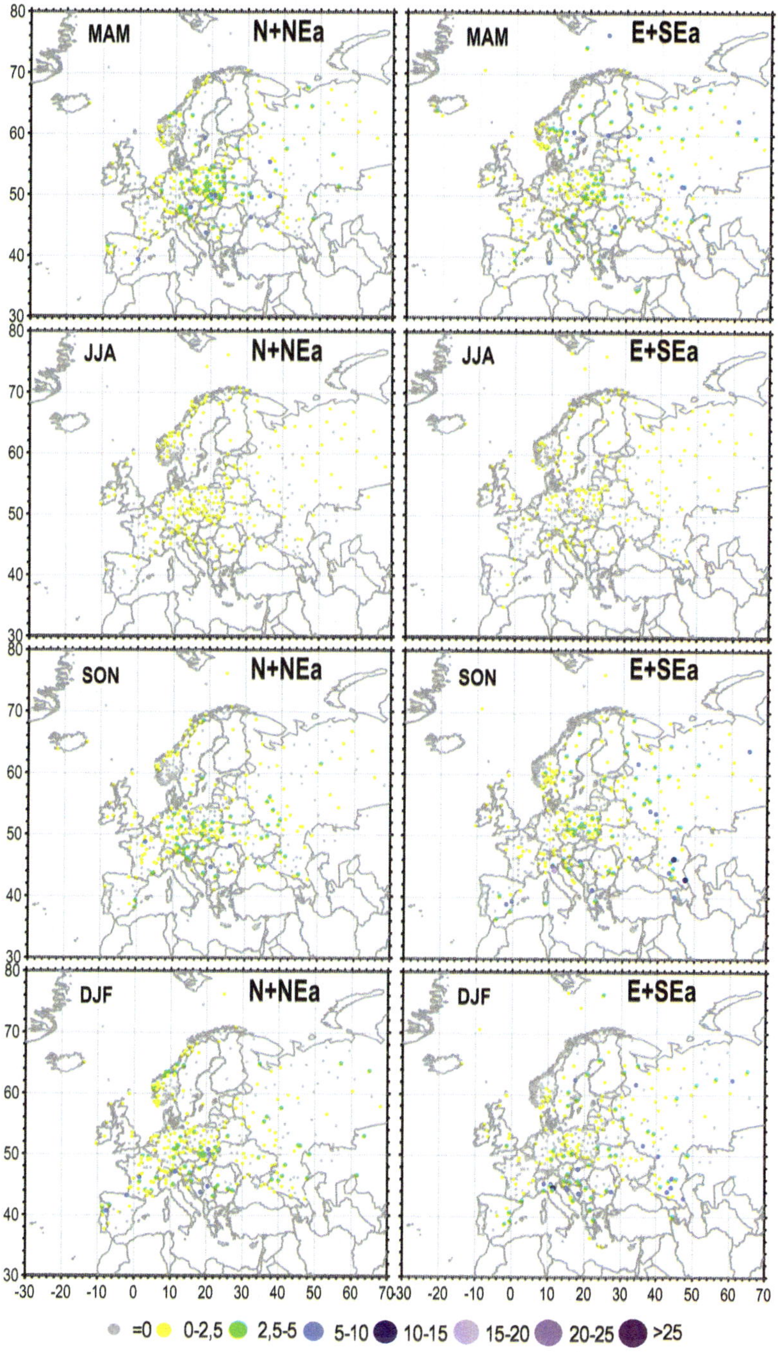

Fig. 6.28 Frequency of warm front extreme precipitation (type Fw) in anticyclonic circulation types (N+NEa, E+SEa) for January 1951–February 2008. Right closed frequency intervals

the area of Lofoten and Vesteraten. In spring, there is also a clear linkage between the occurrence of type Fw precipitation and advection of air from the northeast sector. At the time, at many stations in Central Europe and southeast Europe, the frequency of type Fw in situation N+NEc reaches 5–10 % of ExP (Fig. 6.29).

6.2.6 *Precipitation Linked to the Passage of Different Fronts (Type Ff)*

6.2.6.1 Frequency of Extreme Precipitation Linked to the Passage of Different Fronts (Type Ff) in Anticyclones (ACT) and Cyclones (CCT)

In Europe, extreme precipitation linked to the passage of several fronts in anticyclones is most frequent in winter (approximately 9 % of ExP) and nearly as frequent in autumn (8 % of ExP). It is the least frequent in summer (approximately 6 % of ExP). Throughout the year, the frequency of type Ff precipitation in ATC at most stations in Eastern Europe does not exceed 5 % of ExP, and is higher in Western Europe and Central Europe, showing seasonal variability. In winter, precipitation associated with the passage of several fronts in ACT occurs most frequently (more than 20 % of ExP) in the Alps, in the southern part of the Scandinavian Peninsula, in the Balkan Peninsula, and at some stations in northwest Europe. In the southwest Europe and Western Europe, as well as in southern Poland, type Ff precipitation in ACT constitutes between 10 and 20 % of ExP. A similar frequency of this precipitation type is also recorded on the Atlantic coasts of the Scandinavian Peninsula and on the British Isles (Fig. 6.31). Relatively infrequently, the passage of several fronts in winter generates high precipitation in Iceland and at the southernmost Mediterranean stations, as well as in Eastern Europe. In Poland, in winter, the frequency of type Ff precipitation in ACT declines from north to south (Fig. 6.31). Similar spatial distribution of the frequency of type Ff ExP in ACT is observable in autumn, except that, in some areas of the continent, that is, in eastern Poland and Iceland, this type is more frequent in autumn than in winter. In the Iberian Peninsula type Ff is less frequent in autumn than in winter. In spring, the passage of several weather fronts in anticyclonic situations leads to extreme precipitation mainly in the Alps and in the central part of the western coasts of the Scandinavian Peninsula (more than 20 % of ExP). At many stations in Central Europe (chiefly in southern Poland), in Western Europe, as well as in the southernmost areas of the Scandinavian Peninsula, the frequency of type Ff ExP in ACT exceeds 10 % of ExP in that season. In summer, when the frequency of precipitation associated with the passage of several fronts is the lowest, it constitutes at most 20 % of ExP only in the southernmost part of the Scandinavian Peninsula, in the northern part of Central Europe (Poland and Denmark), and in the Alps. In summer, many stations in southern Europe recorded no extreme precipitation of type Ff in anticyclones in the study period.

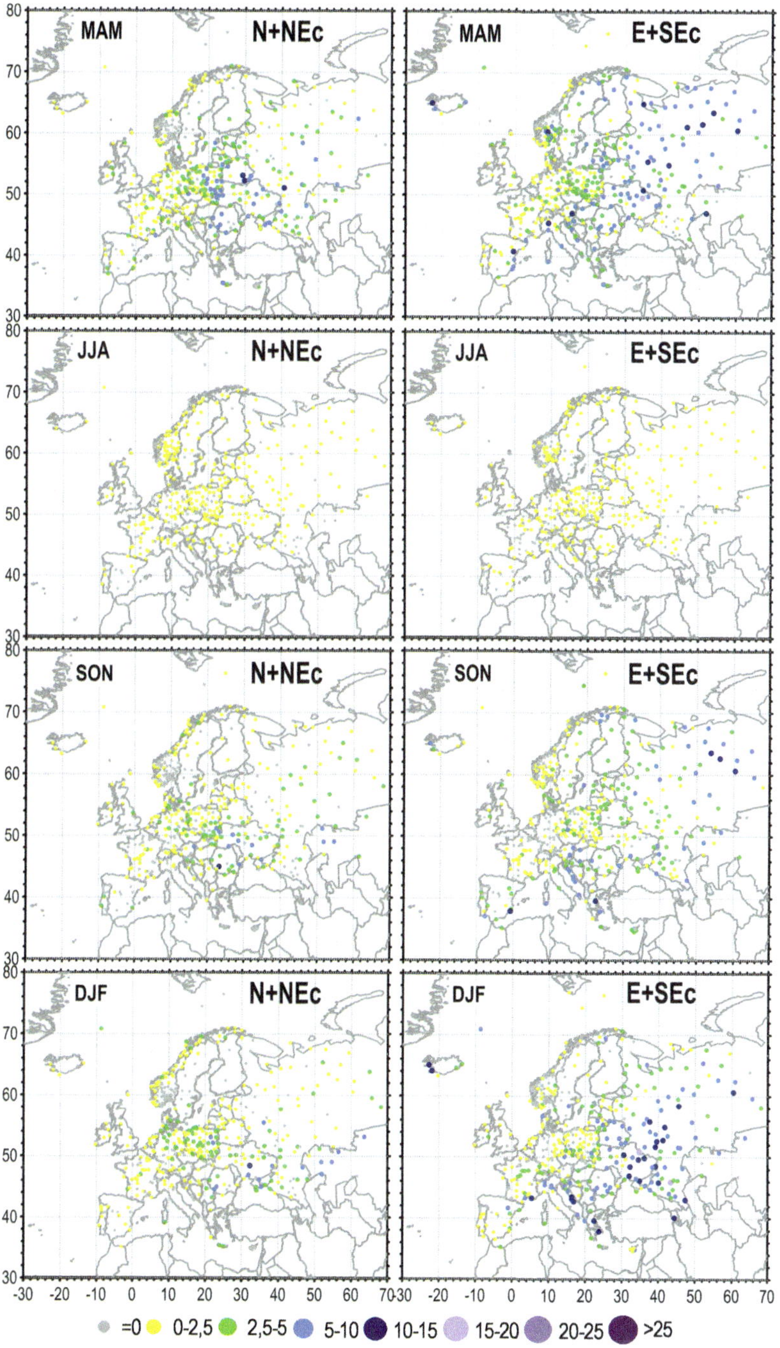

Fig. 6.29 Frequency of warm front extreme precipitation (type Fw) in cyclonic circulation types (N+NEc, E+SEc) for January 1951–February 2008. Right closed intervals

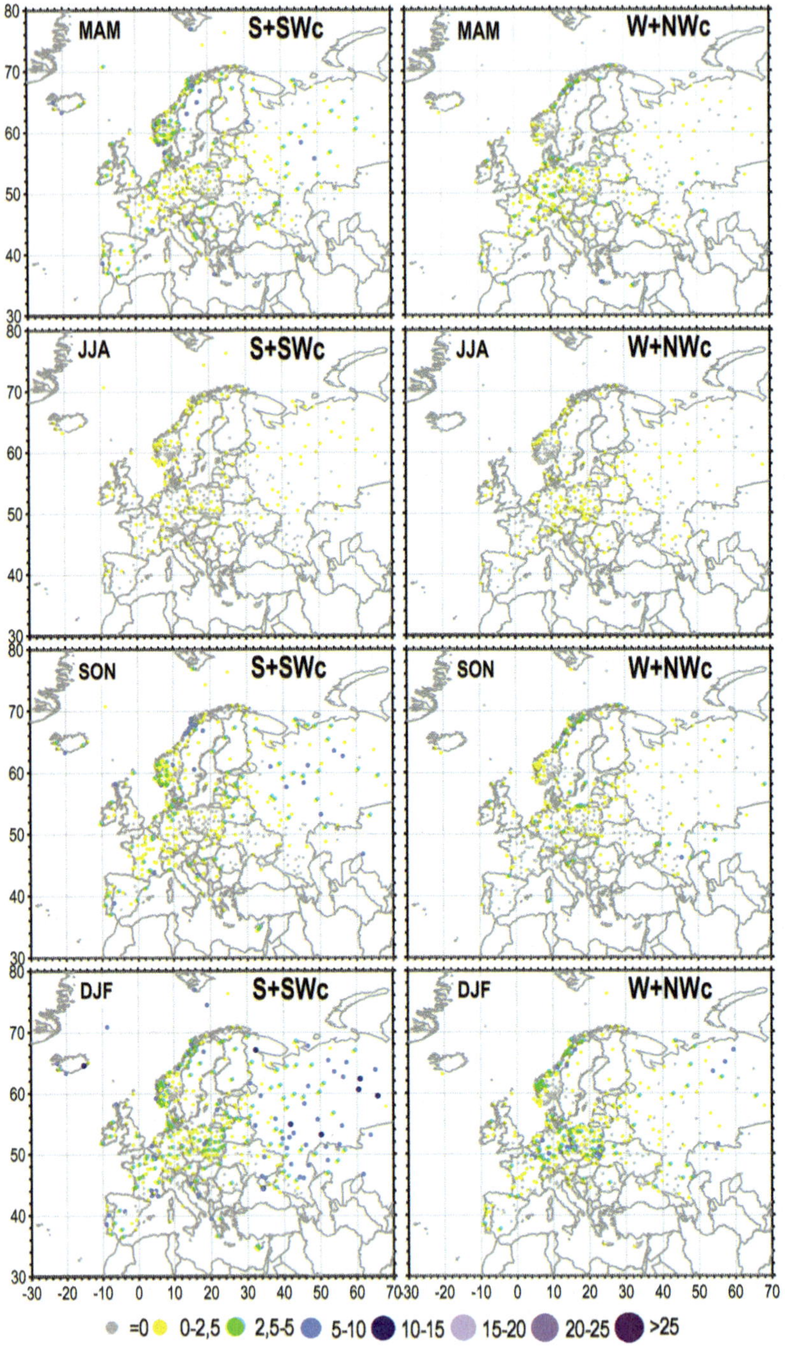

Fig. 6.30 Frequency of warm frontal extreme precipitation (type Fw) in cyclonic circulation types (S+SWc, W+NWc) for January 1951–February 2008. Right closed intervals

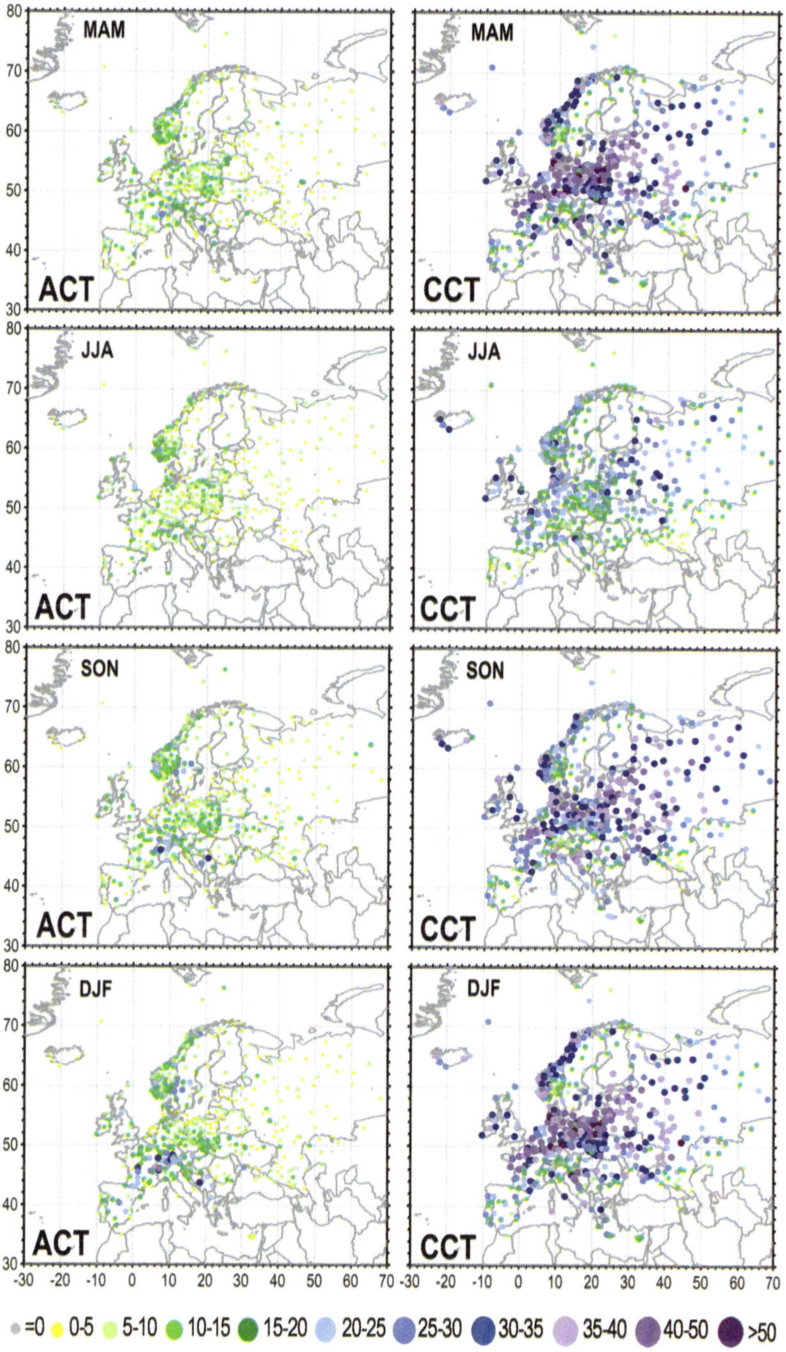

Fig. 6.31 Frequency of extreme precipitation associated with the passage of different fronts (type Ff) in anticyclones (*ACT*) and cyclones (*CCT*) for January 1951–February 2008. Right closed frequency intervals

Table 6.7 Descriptive statistics of the frequency [%] of extreme precipitation associated with the passage of different fronts (type Ff) in anticyclones (*ACT*) and cyclones (*CCT*) in Europe for January 1951–February 2008

Pressure system	Season	Average (±SE)		Confidence intervals −95 %	+95 %	Min	Max	SD	CV	ME	Quartiles Lower	Upper
ACT	MAM	7.5	(±0.2)	7.1	8.0	0.0	27.5	5.0	66.1	7.0	3.6	10.4
	JJA	6.4	(±0.2)	6.0	6.8	0.0	25.0	4.6	73.1	5.7	2.7	9.3
	SON	8.4	(±0.3)	7.9	9.0	0.0	33.3	6.1	72.0	7.3	3.8	12.3
	DJF	9.3	(±0.3)	8.6	9.9	0.0	46.1	7.4	79.9	7.6	3.2	13.7
CCT	MAM	24.0	(±0.4)	23.2	24.8	0.0	47.7	9.5	39.7	24.6	17.0	31.0
	JJA	18.0	(±0.3)	17.4	18.7	0.0	37.7	7.9	43.5	18.6	13.3	23.3
	SON	26.1	(±0.4)	25.3	27.0	0.0	49.1	10.1	38.8	27.4	18.7	33.8
	DJF	28.2	(±0.5)	27.2	29.2	0.0	54.3	11.6	41.0	29.2	19.3	37.1

SE standard error, *Min* minimum value, *Max* maximum value, *SD* standard deviation, *CV* coefficient of variability, *ME* median

The frequency of type Ff extreme precipitation in cyclones is much higher than in anticyclones, showing annual changes from 18 % of ExP in summer to 28 % of ExP in winter (Table 6.7). Precipitation extremes related to the passage of several fronts in cyclones are frequent, exceeding 30 % of ExP at 47 % of the stations in winter, at 38 % of the stations in autumn, at 30 % of the stations in spring, and only 5 % of the stations in summer (Fig. 6.32).

In winter, the highest frequency of type Ff precipitation in CCT (more than 35 % of ExP) occurs in areas from the Bay of Biscay to northern and central Poland, in Iceland, in Eastern Europe as far as around 50°E and at single stations in the south of Europe and the southern part of the Scandinavian Peninsula (Fig. 6.31). In autumn, the spatial distribution of the occurrence of type Ff precipitation in CCT is similar, except that the stations recording the highest values are more dispersed in the eastern part of the continent, whereas in Central Europe and Western Europe such precipitation is less frequent than in winter (Fig. 6.31). In summer, the changes in the frequency of type Ff extreme precipitation in CCT reflect its distribution in winter, but the values recorded in that season do not, usually, exceed 35 % of ExP, while in winter they reach as much as 60 % of ExP (2 % osstations only) (Fig. 6.32). In spring, high frequencies of type Ff precipitation (more than 35 % of ExP) are limited to stations in an area stretching from the Frisian Islands, across Poland, to the Gulf of Finland, and appear at isolated stations in the northern part of Eastern Poland.

The lowest frequency of type Ff ExP in CCT during the entire year occurs in the southern part of the continent, notably at stations located in the Iberian Peninsula, between the Black Sea and the Caspian Sea (Caucasus area) and in the Alps, and in most of the seasons (except summer) also on the lee side of the Scandinavian Mountains. On the British Isles, at most stations, the frequency of type Ff precipitation in CCT (approximately 20–30 % of ExP) shows no major annual changes (Fig. 6.31). Type Ff extreme precipitation is the most frequent type, therefore the

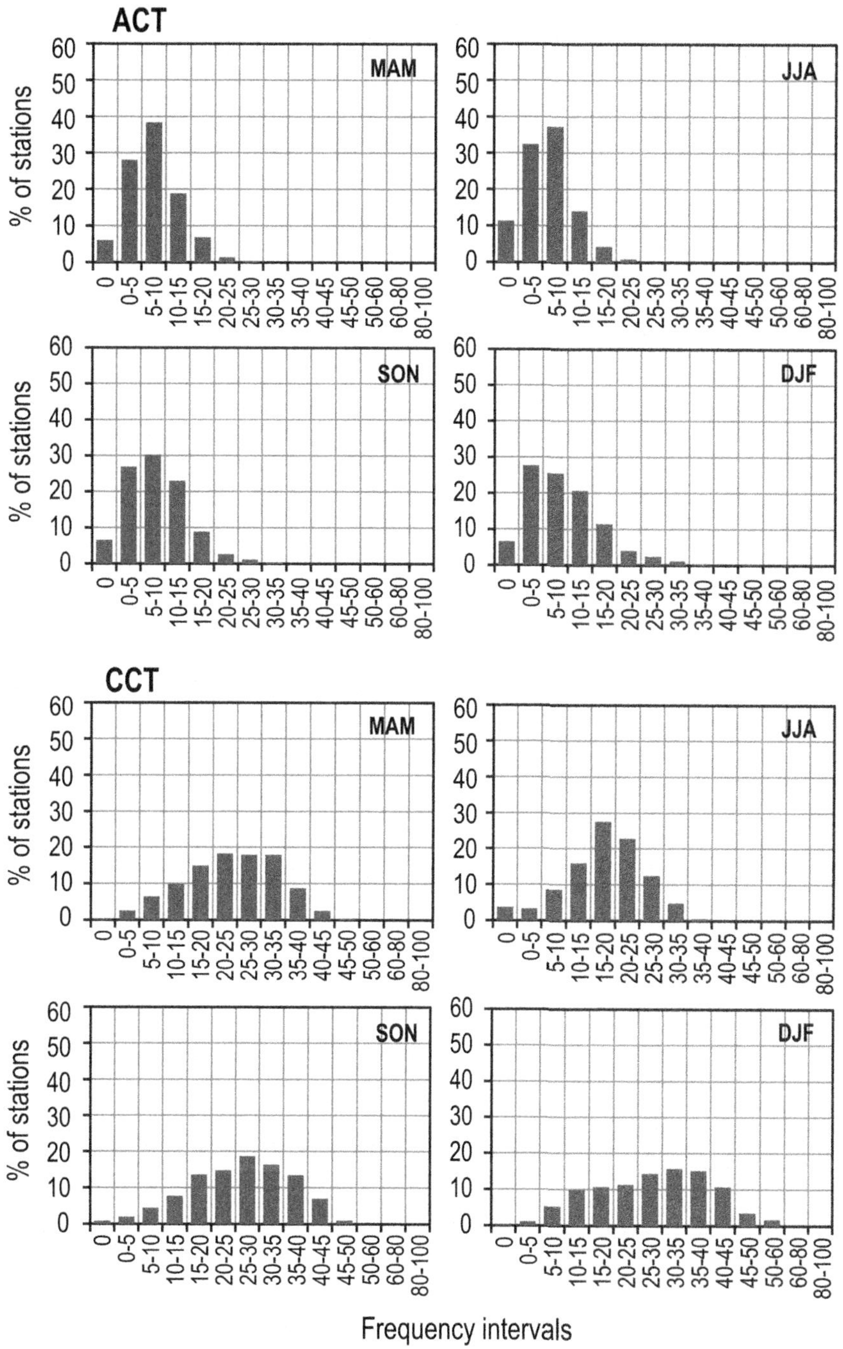

Fig. 6.32 Percentage of stations within the intervals of the frequency of extreme precipitation associated with the passage of different fronts (type Ff) in anticyclones (*ACT*) and cyclones (*CCT*) for January 1951–February 2008. Right closed intervals

relationships between its occurrence and the individual circulation types are similar to analogous relations applying to frontal precipitation in general (type F).

6.2.6.2 Frequency of Extreme Precipitation Associated with the Passage of Different Fronts (Type Ff) in Anticyclonic Synoptic Situations

The spatial distribution of the frequency of type Ff extreme precipitation changes considerably from one synoptic situation to another. It also demonstrates seasonal variability, although less prominent. On a regional scale, the occurrence of type Ff ExP in anticyclones is related most strongly to the S+SWa and W+NWa situations, especially in autumn and winter, when such precipitation is the most frequent in Europe. During inflow of air from the southeast sector (situation S+SWa), the highest frequencies of type Ff precipitation in Europe throughout the year occur in the southernmost part of the Scandinavian Peninsula and on the British Isles. In those areas, the precipitation type in question represents between 5 and 15 % of ExP. In winter, its frequency there reaches as much as 20 % of ExP. In situation S+SWa, type Ff precipitation is also more frequent, compared to the rest of the continent, in the Iberian Peninsula (autumn and winter) and in France (spring, autumn, winter) (Fig. 6.33). The latter synoptic situation (W+NWa) fosters type Ff precipitation in the west coasts of the Scandinavian Peninsula and in the Alps, with its importance being relatively lower in summer. In spring, and even more so in winter, during inflow of air from W+NWa, the frequency of type Ff precipitation is higher in the area extending from the Iberian Peninsula to southern Poland, whereas in autumn, the frequency of type Ff precipitation in Western and Central Europe is greater than in the southern and eastern parts of the continent (Fig. 6.33).

The influence of air advection from the northeast (situation N+NEa) on the occurrence of extreme precipitation associated with the passage of different fronts is lesser than that of S+SWa and W+NWa situations already discussed, as is proven by the lower number of stations where type Ff precipitation in those situations constitutes 10 % of ExP. However, the area under the influence of situation N+NEa, marked by the enhanced number of days with type Ff ExP, is vaster (especially in winter) and demonstrates the most pronounced seasonal changes compared to the other synoptic situations. In winter, frequencies of type Ff precipitation in situation N+NEa falling between 5 and 10 % of ExP are recorded at many stations located in the area stretching from France to Poland, including the Alps and the northern part of the west coast of the Scandinavian Peninsula in the area of Lofoten and Vesteraten. In the season, higher frequency, that is, exceeding 10 % of ExP, of type Ff precipitation is recorded also in the Alps and at many stations in the Balkan Peninsula (Fig. 6.34).

In autumn and spring, the highest frequency of type Ff precipitation in situation N+NEa is recorded at Alpine stations, in southern Germany and Poland, and at some stations in the Balkan Peninsula, as well as, but only in autumn, in the Transnistria region (to the north of the Black Sea). In the summer, the relationship between the occurrence of type Ff precipitation and situation N+NEa is the least pronounced: the number of stations where the frequency of type Ff precipitation exceeds 5 % of ExP is low, and its occurrence is limited to areas located in Alpine regions and in the northern part of the Balkan Peninsula, as well as individual stations in the other parts of Europe. Extreme precipitation associated with the passage of several fronts in high-pressure systems is the least frequent during inflow of air from the east and southeast (Fig. 6.34).

6.2.6.3 Frequency of Extreme Precipitation Associated with the Passage of Different Fronts (Type Ff) in Cyclonic Synoptic Situations

Most of the type Ff precipitation occurring in Europe is associated with low-pressure systems. As for anticyclonic situations, the spatial distribution of the frequency of this type of precipitation differs depending on the synoptic situation, yet with lower seasonal changeability of the frequency in each situation.

During advection of air from the northeast sector (situation N+NEc), the highest frequency of type Ff precipitation in the continent (5–10 % of ExP) is recorded in Central Europe and, depending on the season, in various parts of Eastern Europe. Only in summer does type Ff precipitation in situation N+NEc exceed 10 % of ExP at relatively few (compared to the other seasons) stations located in the border area between France and Belgium and in the western part of Eastern Europe. In Central Europe (mainly in Poland), the frequency of type Ff precipitation in situation N+NEc in summer is lower than in any other seasons, but it remains the highest of all the cyclonic advective types (Fig. 6.35).

Advection of air from the southeast sector (situation E+SEc) in autumn and winter causes an increase in the frequency of Ff precipitation mainly in Southern Europe and in the west of Eastern Europe. At some stations located in the south of Europe, the frequency of type Ff precipitation in situation E+SEc in these seasons exceeds as much as 20 % of ExP. In Central Europe and on the western coasts of the Scandinavian Peninsula, the occurrence of type Ff precipitation is associated, above all, with advection of air from the northwest sector (situation W+NWc), particularily in winter (Fig. 6.36).

Air masses inflowing over Europe from the south and southwest throughout the year bring the highest amount of type Ff extreme precipitation to the west coast and southern part of the Scandinavian Peninsula, to the British Isles and Western Europe. In these areas, type Ff precipitation represents more than 10 % of ExP, and occasionally even more than 25 % of ExP. In winter, the zone in which the frequency of type Ff precipitation in situation S+SWc exceeds 10 % of ExP reaches Poland's eastern borders (Fig. 6.36). This change results from the faster movement of low-pressure systems in Europe in the cooler seasons, when occlusion begins in areas more distanced from the coasts of the continent, than in warmer ones, when lower thermal gradients in the Northern Hemisphere lead to reduced wind speeds.

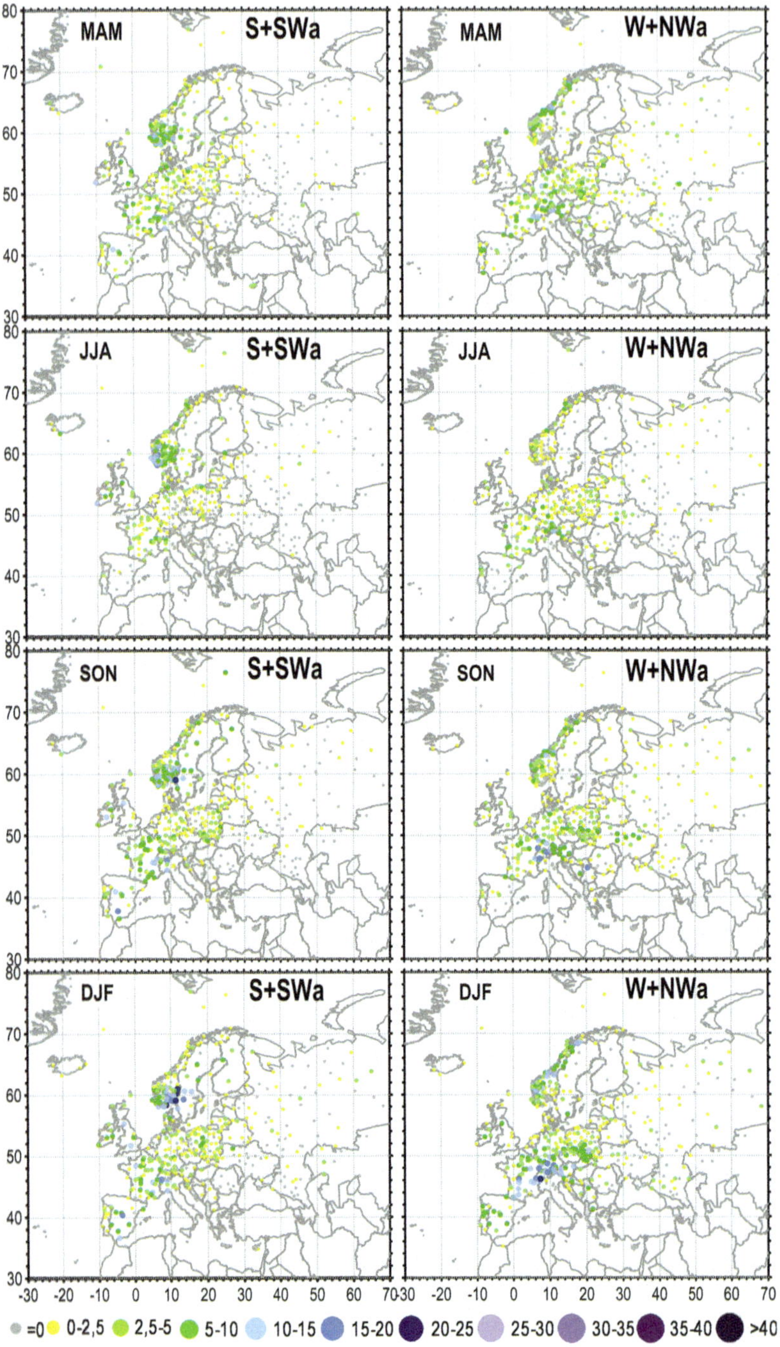

Fig. 6.33 Frequency of extreme precipitation associated with the passage of different fronts (type Ff) in anticyclonic circulation types (S+SWa, W+NWa) for January 1951–February 2008. Right closed intervals

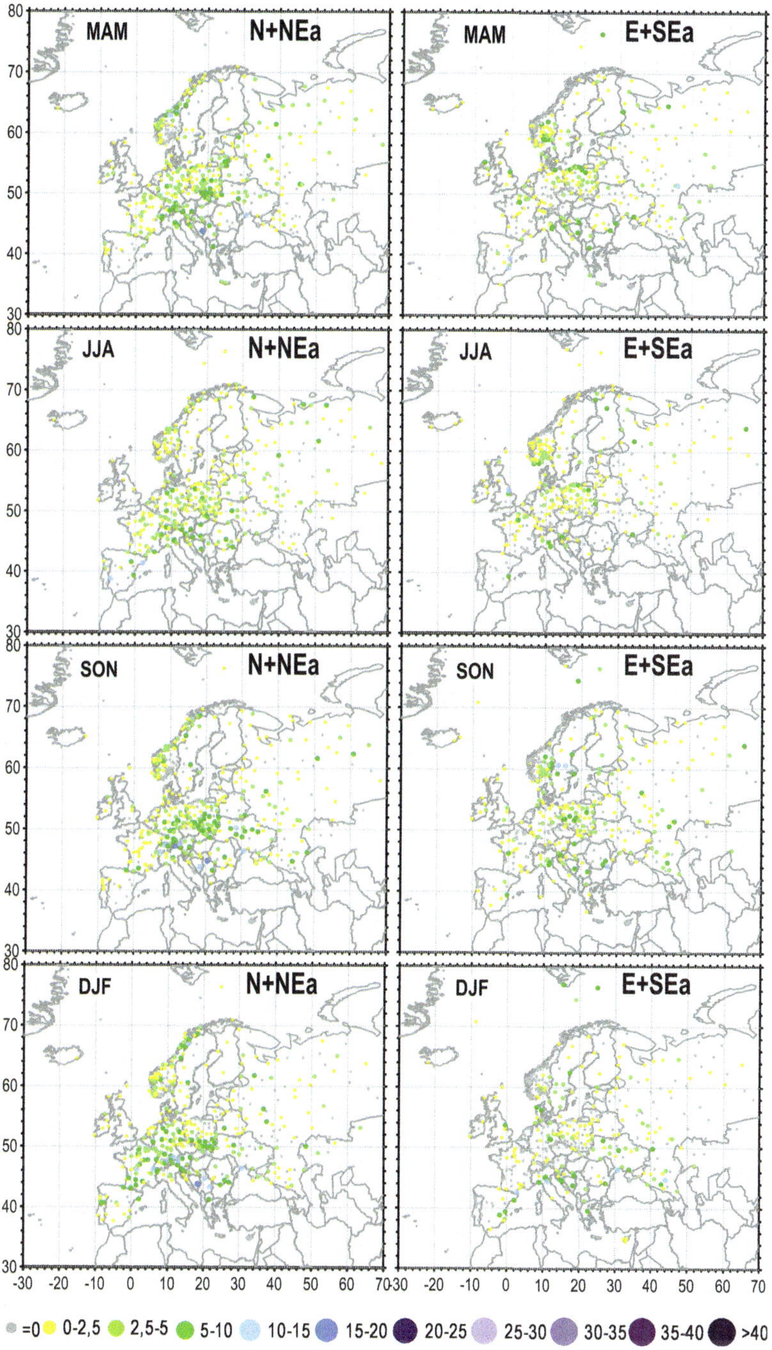

Fig. 6.34 Frequency of extreme precipitation associated with the passage of different fronts (type Ff) in anticyclonic circulation types (N+NEa, E+SEa) for January 1951–February 2008. Right closed intervals

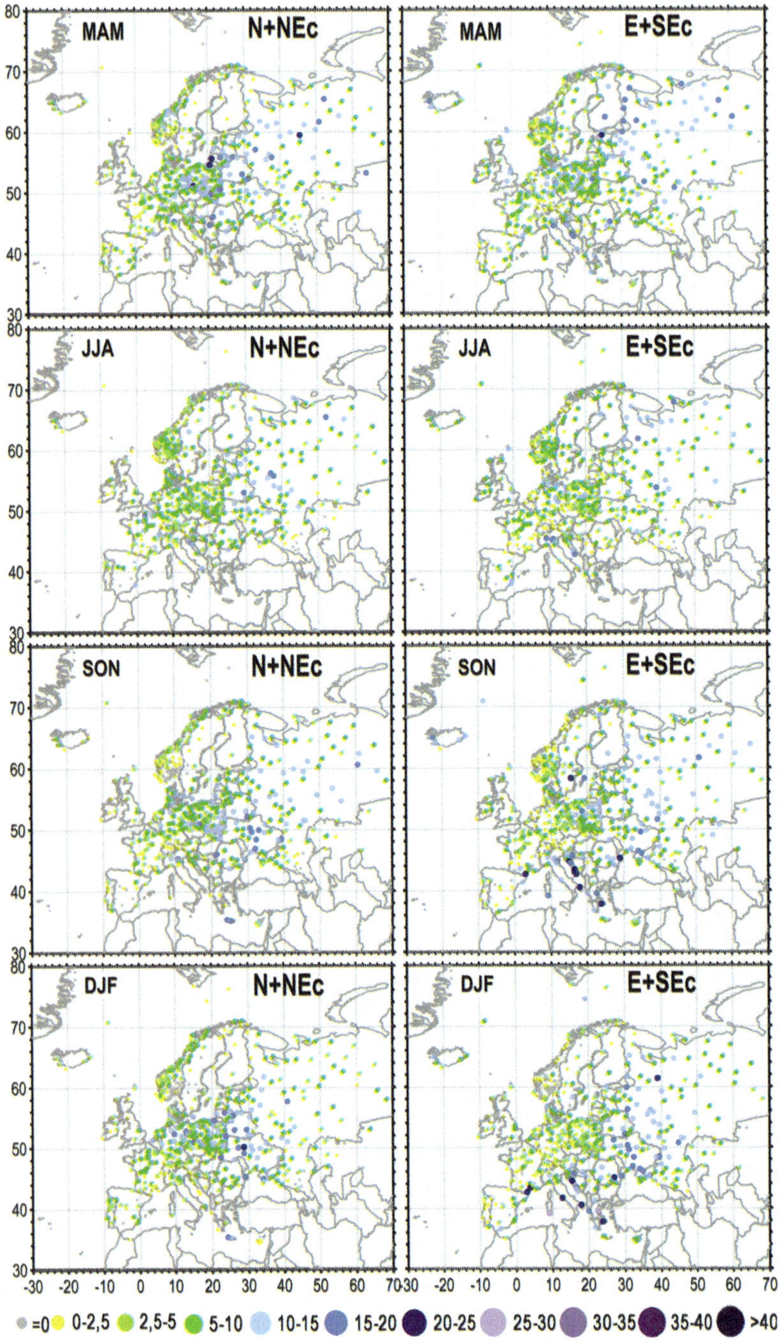

Fig. 6.35 Frequency of extreme precipitation associated with the passage of different fronts (type Ff) in cyclonic circulation types (N+NEc, E+SEc) for January 1951–February 2008. Right closed intervals

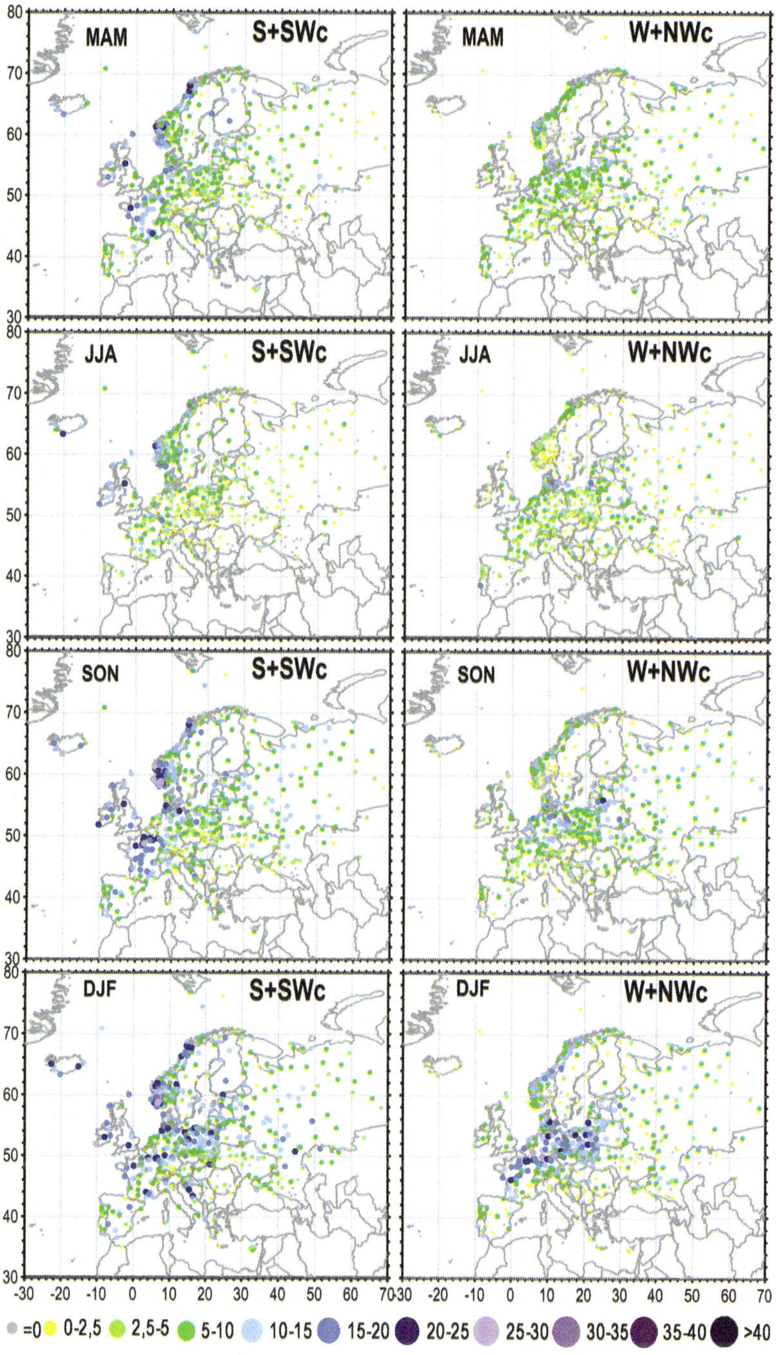

Fig. 6.36 Frequency of extreme precipitation associated with the passage of different fronts (type Ff) in cyclonic circulation types (S+SWc, W+NWc) for January 1951–February 2008. Right closed frequency intervals

6.2.7 *Occluded Front Precipitation (Type Fo)*

6.2.7.1 Frequency of Occluded Front Extreme Precipitation (Type Fo) in Anticyclones (ACT) and Cyclones (CCT)

The range of seasonal changes in the occurrence of occluded front extreme precipitation, in both high-pressure and low-pressure systems, is small and amounts to 0.8 % of ExP in ACT (from 3.5 % of ExP in summer to 4.3 % of ExP in winter) and 1.1 % of ExP in CCT (from 15.6 % of ExP in autumn to 16.7 % of ExP in spring) (Table 6.8). In all seasons, the frequency of type Fo precipitation in anticyclones does not exceed 5 % of ExP in at least half of the stations under study, which are dispersed across the continent (59 % of the stations in spring, 50 % in summer, 54 % in autumn, 59 % in winter) (Fig. 6.37). The frequency of type Fo precipitation in ACT demonstrates slightly higher values (more than 5 % of ExP) at 13 % of the stations, located mainly in Southern Europe in winter, at 15 % of the stations, located in the northern part of Poland and Germany in summer, and at 12 % of the stations in spring and 13 % of the stations in autumn, which are dispersed in both cases across the continent except for most of the area of Eastern Europe (Figs. 6.37 and 6.38). The share of type Fo extreme precipitation in ACT in the southern part of the Scandinavian Peninsula exceeds 50 % of ExP and is lower only in summer. In the period under investigation, many stations (from 19 % in spring and winter to 27 % in summer) recorded no type Fo precipitation in anticyclones (Figs. 6.37 and 6.38).

The spatial distribution of the frequency of type Fo ExP in cyclonic situations is exceptionally regular. The frequency of type Fo precipitation in CCT falls from the north to the south of Europe, from approximately 30 % to less than 5 % of ExP. At some stations in the Scandinavian Peninsula and its vicinity, as well as on Iceland, type Fo precipitation in CCT represents as much as 35–50 % of ExP. Generally, in the southern part of the continent the frequency of type Fo precipitation in CCT at

Table 6.8 Descriptive statistics of the frequency [%] extreme precipitation associated with occlusion (type Fo) in anticyclones (*ACT*) and cyclones (*CCT*) in Europe (January 1951–February 2008)

Pressure system	Season	Average (±SE)		Confidence intervals		Min	Max	SD	CV	ME	Quartiles	
				−95 %	+95 %						Lower	Upper
ACT	MAM	4.2	(±0.3)	3.7	4.8	0.0	40.6	6.2	146.4	2.4	1.1	4.6
	JJA	3.5	(±0.2)	3.1	3.9	0.0	33.7	4.3	124.3	2.3	0.0	4.7
	SON	3.8	(±0.3)	3.3	4.3	0.0	36.6	5.7	150.8	2.2	0.0	4.5
	DJF	4.3	(±0.3)	3.7	4.9	0.0	45.0	6.9	160.7	2.1	0.9	4.4
CCT	MAM	16.7	(±0.4)	15.9	17.4	0.0	49.4	8.9	53.4	14.7	10.0	22.4
	JJA	15.7	(±0.5)	14.7	16.6	0.0	58.2	10.8	68.7	13.6	7.5	23.5
	SON	15.6	(±0.4)	14.8	16.5	0.0	52.0	10.0	64.0	13.6	8.0	22.8
	DJF	15.7	(±0.4)	14.9	16.5	0.0	48.9	9.4	59.9	13.6	8.6	21.6

SE standard error, *Min* minimum value, *Max* maximum value, *SD* standard deviation, *CV* coefficient of variability, *ME* median

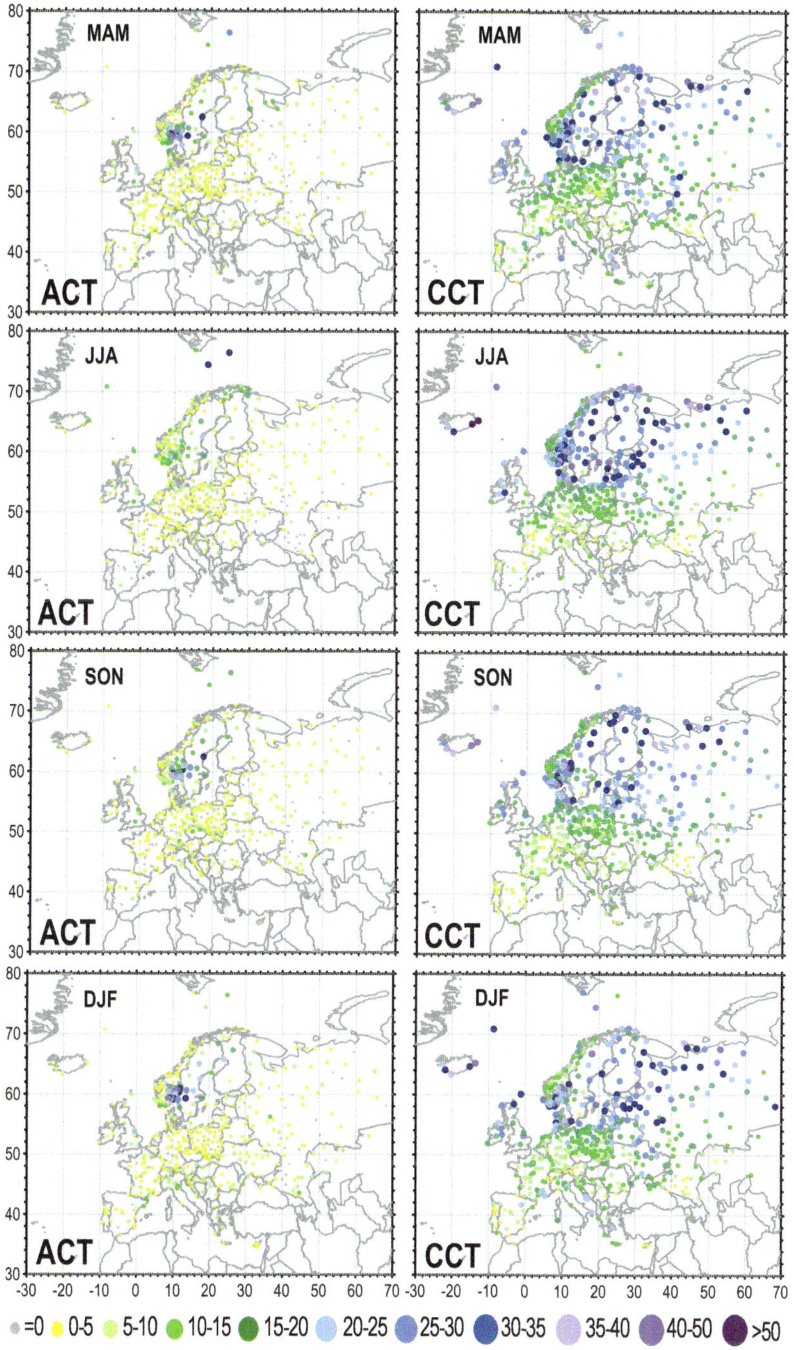

Fig. 6.37 Frequency of extreme precipitation associated with occlusion (type Fo) in anticyclones (*ACT*) and cyclones (*CCT*) for January 1951–February 2008. Right closed intervals

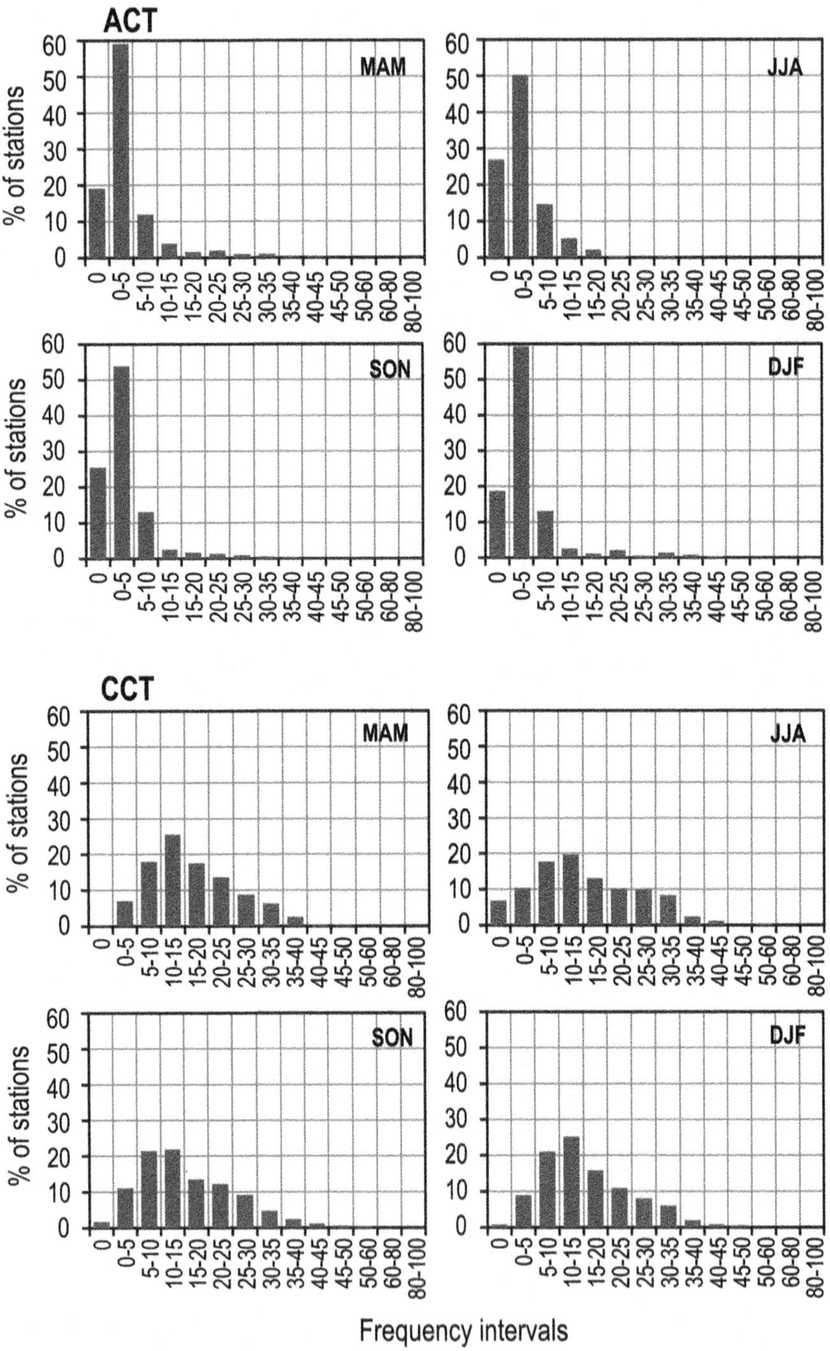

Fig. 6.38 Percentage of stations within the intervals of the frequency of extreme precipitation associated with occlusion (type Fo) in anticyclones (*ACT*) and cyclones (*CCT*) for January 1951–February 2008. Right closed frequency intervals

most stations does not exceed 10 % of ExP, and it is only in spring and winter, in the central part of the northern Mediterranean coasts, that it exceeds 20 % of ExP. The spatial distribution of the frequency of type Fo precipitation both in CCT and in ACT is associated with the course of the cyclogenesis during the passage of cyclones across Europe and the impact of orography on occluded front formation.

6.2.7.2 Frequency of Occluded Front Extreme Precipitation (Type Fo) in Anticyclonic Synoptic Situations

Extreme precipitation originating at occluded fronts in anticyclones is mainly recorded in the southern part of the Scandinavian Peninsula. The impact of atmospheric circulation on its occurrence in Europe is limited to that very area (Figs. 6.39 and 6.40).

Throughout the year, the highest frequency of type Fo precipitation is recorded in the southern part of the Scandinavian Peninsula on the lee side of the Scandinavian Mountains, and its occurrence is driven by advection of air from the southwest sector (situation S+SWa) (Fig. 6.40). In winter, type Fo precipitation in the situation represents more than 15 % of ExP, in spring and autumn 10 %, and in summer from 2.5 to 10 % of ExP.

Another advective situation that determines, although to a much lesser degree, the frequency of occluded front extreme precipitation in the Scandinavian Peninsula and several other stations located in different parts of the continent, is situation S+SEa. Type Fo occurs slightly more frequently than elsewhere in Europe, in the northernmost part of the Scandinavian Peninsula in summer in situation N+NEa (Fig. 6.39).

6.2.7.3 Frequency of Occluded Front Extreme Precipitation (Type Fo) in Cyclonic Synoptic Situations

The frequency of type Fo precipitation demonstrates the strongest links with atmospheric circulation in Scandinavia and Iceland. In the southern part of the Scandinavian Peninsula, the frequency of type Fo extreme precipitation is the highest (reaching approximately 30 % of ExP) during advection of air from the southwest sector (situation S+SWc). There is considerable type Fo precipitation in situation S+SWc, from 10 to 20 % of ExP, and also in the other areas of Scandinavia (spring and autumn), in Iceland (throughout the year), and on the British Isles (spring and winter). In winter, similar frequency of type Fo ExP in situation S+SWc (approximately 30 % of ExP) is observed in the areas of Jutland and Gulf of Riga (Fig. 6.42).

The frequency of type Fo precipitation in Northern Europe in spring, summer, and autumn also increases during inflow of air from the southeast sector (situation E+SEc). In each season, the situation also generates high (against the background of the continent) precipitation frequency on the British Isles and Iceland.

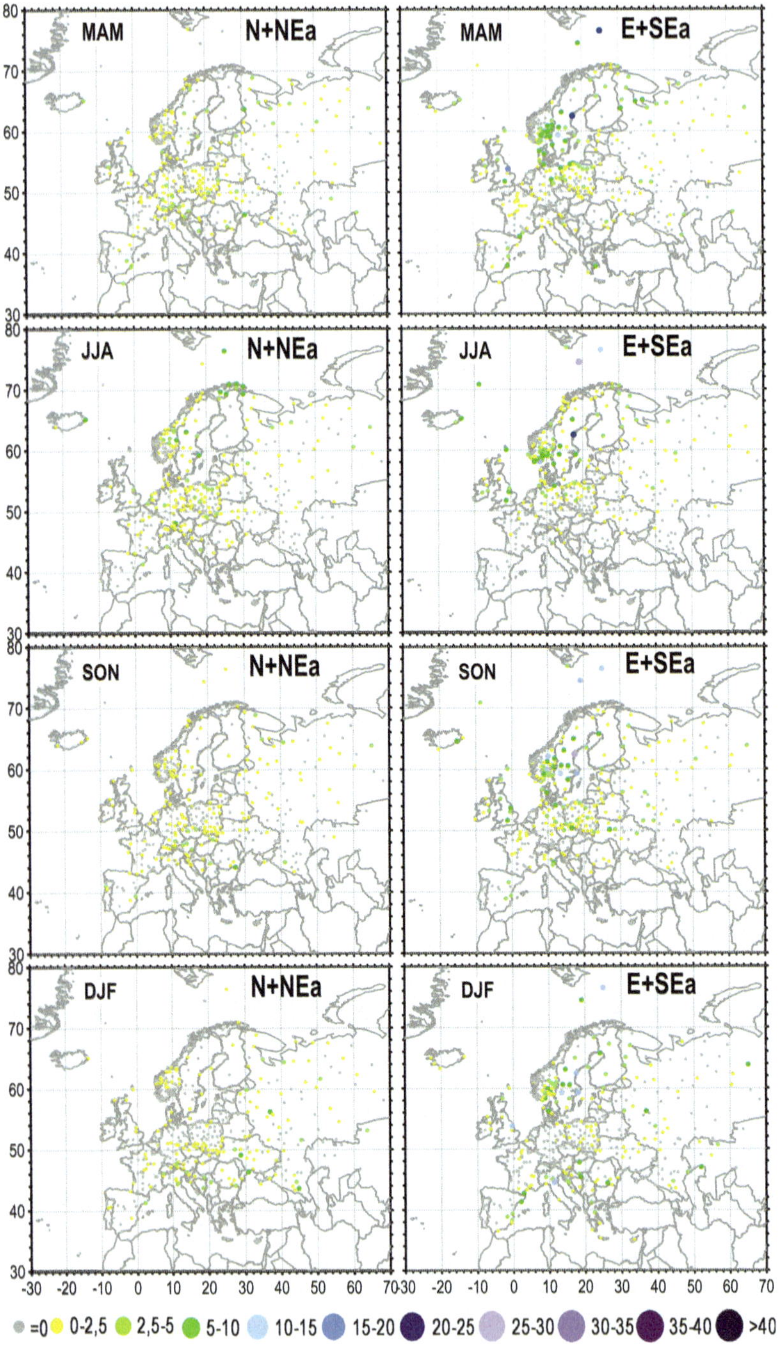

Fig. 6.39 Frequency of extreme precipitation associated occlusion (type Fo) in anticyclonic circulation types (N+NEa, E+SEa) for January 1951–February 2008. Right closed frequency intervals

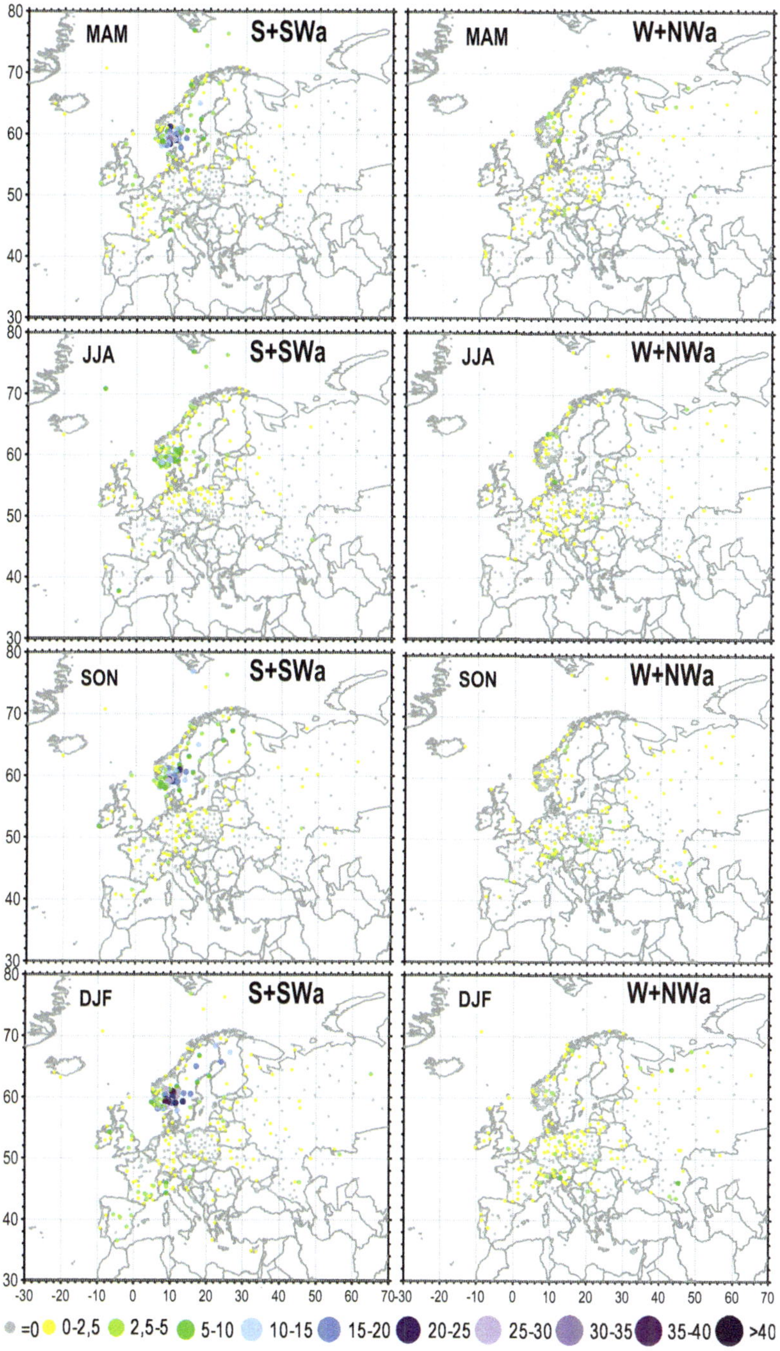

Fig. 6.40 Frequency of extreme precipitation associated with occlusion (type Fo) in anticyclonic circulation types (S+SWa, W+NWa) for January 1951–February 2008. Right closed frequency intervals

In spring and winter, the influence of situation E+SEc also covers the northern part of the Mediterranean Basin and the western and southwest part of Eastern Europe (Fig. 6.41). During inflow of air from the northeast sector (situation N+NEc), in summer, the frequency of type Fo precipitation exceeds 10 % of ExP mainly in the southern and northernmost part of the Scandinavian Peninsula, as well as in western and northern part of Eastern Europe.

In the latter area, type Fo precipitation occurs in situation N+NEc with a similar frequency also in spring, except that in this case, the number of stations where the frequency of type Fo precipitation exceeds 10 % of ExP, is markedly lower (Fig. 6.41).

Against the background of all advective types of circulation, type Fo precipitation is the least frequent during inflow of air from the west and northwest (situation W+NWc). In this situation, a slightly higher, as compared to the rest of the continent, frequency, within 2.5–5 % of ExP, is recorded by stations located mostly in the northern part of Western Europe and Central Europe (southern coasts of the North Sea and the Baltic), and, depending on the season, some stations in Eastern Europe (Fig. 6.42).

6.3 Conditional Probability of Air-Mass and Frontal Extreme Precipitation in Synoptic Situations

The analysis of the frequency of extreme precipitation types in synoptic situations presented in the previous chapters did not take into account the frequency of particular synoptic situations. Because their frequency varies during the year, the outcome of any frequency analysis of precipitation versus synoptic situation will, to a certain degree, depend on the occurrence of the situations.

Conditional probability is a method used to assess relationships between variables while taking into account the frequency of these variables. The statistic provides information about the degree to which each of the synoptic situations considered favours the occurrence of the extreme precipitation types and thus has a prognostic value. The probability of the occurrence of Event E_1 (a specific origin-based type of extreme precipitation) on condition of the occurrence of Event E_2 (a specific type of synoptic situation) is expressed as follows (Wilks 2006):

$$\Pr\{E_1 / E_2\} = \frac{\Pr\{E_1 \cap E_2\}}{\Pr\{E_2\}} \tag{6.1}$$

The conditional probability of the occurrence of ExPTs in the individual synoptic situations is low, with the highest values not exceeding 10 %, which is attributable to the relatively low frequency of extreme precipitation—a rare phenomenon by definition—and to the relatively high frequency of synoptic situations, which occur every day. Despite this, the results obtained provide valuable information, which can be used in modelling precipitation at various spatial and temporal scales.

The probabilistic analysis of the relationship between the occurrence of origin-based types of extreme precipitation and atmospheric circulation was carried out to complement the results of the analysis of ExPT frequency in synoptic situations.

6.3.1 Extreme Precipitation of All Origins (ExP)

6.3.1.1 Probability of Extreme Precipitation Occurrence (ExP) in Anticyclones (ACT) and Cyclones (CCT)

Throughout the year, the conditional probability of the occurrence of extreme precipitation in anticyclones in most areas of Europe does not exceed 1 %. Only in the area of high mountain ranges, that is, the Scandinavian Mountains (in particular in winter), the Alps, and the Caucasus, it is likely to occur more often (Fig. 6.43) as a result of favourable local conditions, the orographic barriers that force convection. The likelihood of ExP in anticyclones is clearly smaller than in cyclones also in the Alps, which was not demonstrated by the frequency analysis.

Throughout the year, the probability of the occurrence of ExP in cyclones exceeds 4 % on the British Isles and in the southern part of the Scandinavian Peninsula. Its probability is similar in the northern and western parts of Eastern Europe in autumn and winter, in Western Europe, and in the border area between Central Europe and Southern Europe (including the Alps) in summer, and more so in winter. The lowest, compared to the rest of the continent, likelihood of the occurrence of ExP in cyclonic situations, particularly in summer, characterises Southern Europe (Fig. 6.43).

6.3.1.2 Probability of Extreme Precipitation Occurrence (ExP) in Anticyclonic Synoptic Situations

In Europe, inflow of air from the northern sector as a whole (situations N+NEa and W+NWa) is conducive to the occurrence of extreme precipitation on the west coasts of the Scandinavian Peninsula, especially in autumn and winter, when the area records the highest seasonal precipitation totals. Advection of air from the northern sector, especially from the northwest (situation W+NWa), involves the highest, on a continental scale, likelihood of the occurrence of extreme precipitation also in the area of the Alps and the Caucasus. During inflow of air from the southeast sector (situation E+SEa), the possibility of extreme precipitation occurrence grows on the lee side of the Scandinavian Mountains and in the central area of the southernmost part of the Scandinavian Peninsula. In the southern part of the peninsula (Norway), extreme precipitation is fostered by inflow of air also from the other directions of the southern sector (situation S+SWa). In this situation, the probability of extreme

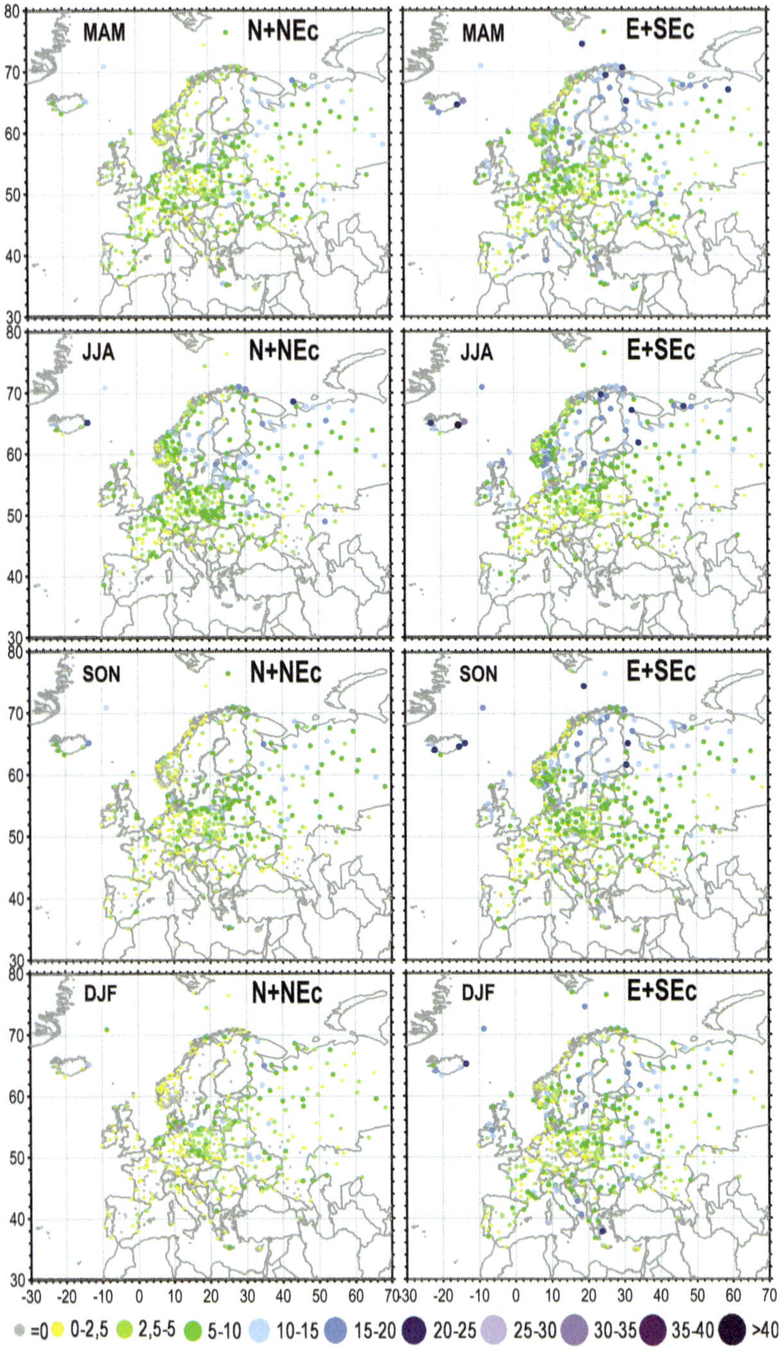

Fig. 6.41 Frequency of extreme precipitation associated with occlusion (type Fo) in cyclonic circulation types (N+NEc, E+SEc) for January 1951–February 2008. Right closed intervals

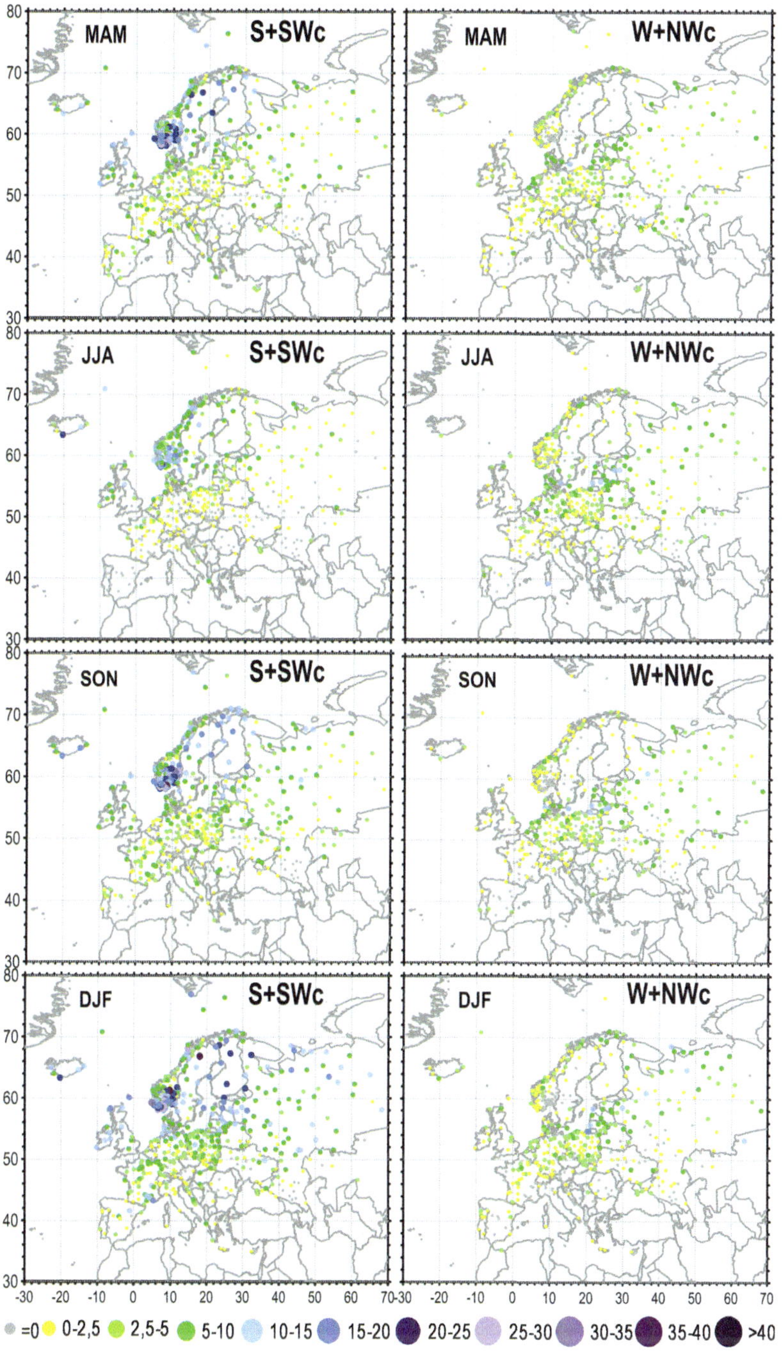

Fig. 6.42 Frequency of extreme precipitation associated with occlusion (type Fo) in cyclonic circulation types (S+SWc, W+NWc) for January 1951–February 2008. Right closed intervals

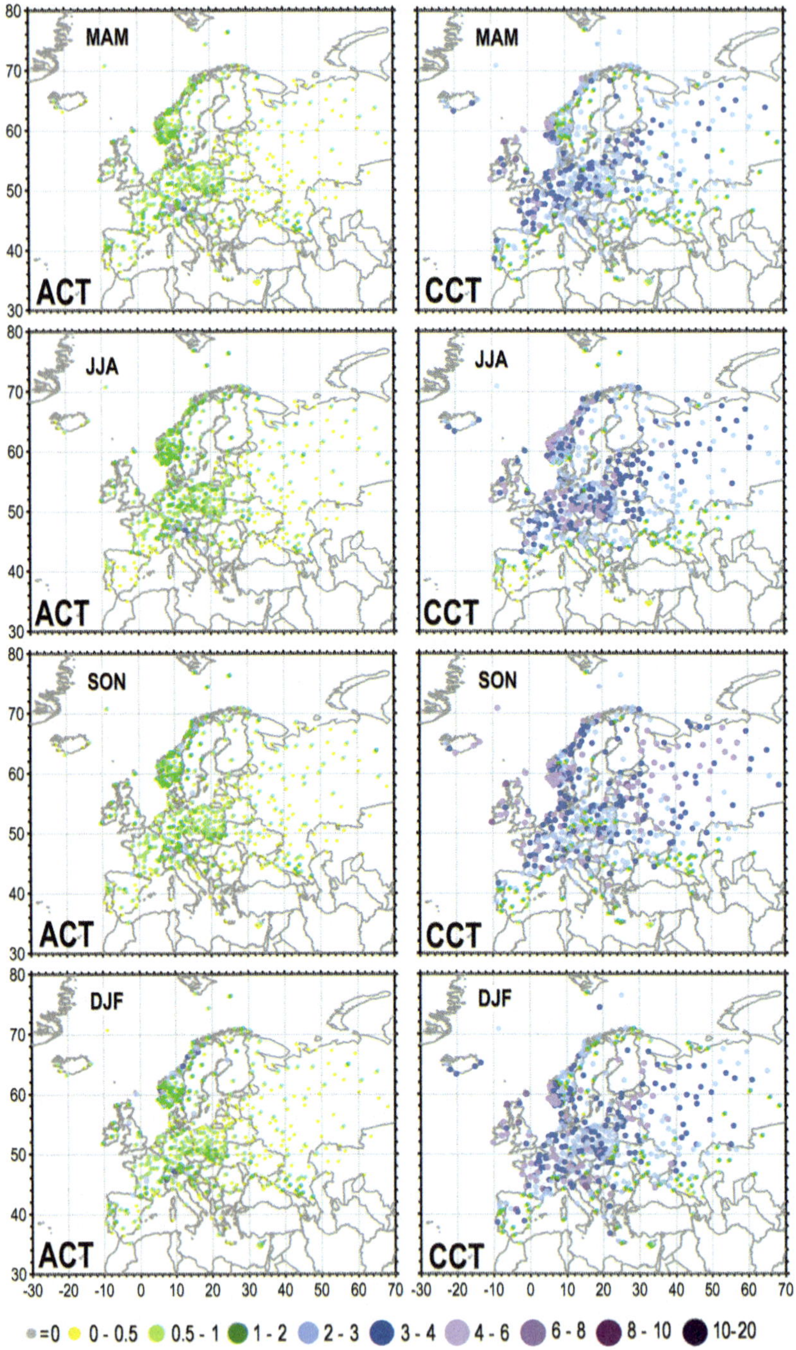

Fig. 6.43 Conditional probability of extreme precipitation occurrence in anticyclones (*ACT*) and cyclones (*CCT*) for January 1951–February 2008. Right closed intervals

precipitation occurrence is also slightly increased in southwest Europe (Iberian Peninsula, France) (Figs. 6.44 and 6.45).

6.3.1.3 Probability of Extreme Precipitation Occurrence (ExP) in Cyclonic Synoptic Situations

During advection of air from the northwest under an influence of a low-pressure system (situation W+NWc), extreme precipitation is most likely to occur in the western coasts of the Scandinavian Peninsula, in the Alps, in the Carpathians, in the southeast coasts of the North Sea, and in the Atlantic coasts of the Iberian Peninsula, as well as in France. In those areas (the Alps, the Carpathians), in addition to the atmospheric circulation, heavy precipitation is driven by the local conditions, that is, the high mountains. In Eastern and Northern Europe, on the lee side of the Scandinavian Mountains, extreme precipitation events are most often associated with inflow of air from the southeast sector (situation E+SEc). In the southern part of the Scandinavian Peninsula, the foregoing phenomena are most likely to be recorded in situation S+SWc; here, convection is also driven by the landform. The foregoing synoptic situation and situation E+SEc bring the highest amount of extreme precipitation to the British Isles (Figs. 6.46 and 6.47). It is also worth noting the nonadvective type Bc, for which the frequencies of extreme precipitation discussed in the previous chapter were very low. In most seasons, except spring, the conditional probability in the situation does not deviate significantly from the probability in situations characterised by a clear airflow direction. Thus, the low ExP frequency in type Bc is attributable to low frequency of this synoptic situation as such, which means that at many stations in Western Europe, Central Europe, and Eastern Europe the effect of situation Bc on ExP is comparable to the other circulation types. The spatial distribution of the conditional probability of extreme precipitation (irrespective of the type) does not change significantly during the year (Figs. 6.43, 6.46, and 6.47).

6.3.2 Air-Mass Precipitation (Type A)

6.3.2.1 Probability of Air-Mass Extreme Precipitation (Type A) in Anticyclones (ACT) and Cyclones (CCT)

The probability of the occurrence of air-mass extreme precipitation is very low both in anticyclones and in cyclones, which results, in the first place, from the low frequency of such precipitation. However, it must be pointed that in Europe, during anticyclones, air-mass precipitation is slightly more likely to occur in mountain areas (area of the Alps and the Caucasus), while during cyclones, it is more likely to be higher (on a continental scale) in the west coasts of the Scandinavian Peninsula,

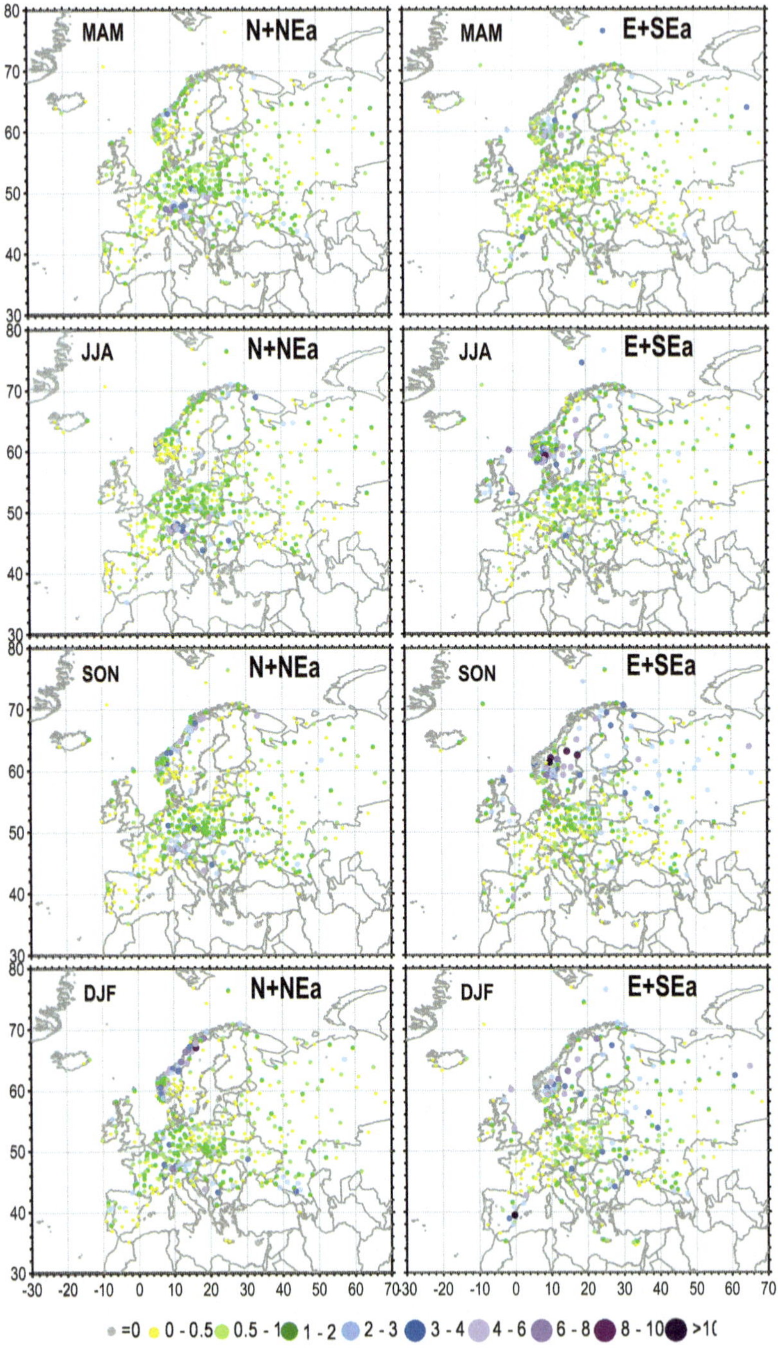

Fig. 6.44 Conditional probability of extreme precipitation occurrence in anticyclonic circulation types (N+NEa, E+SEa) for January 1951–February 2008. Right closed intervals

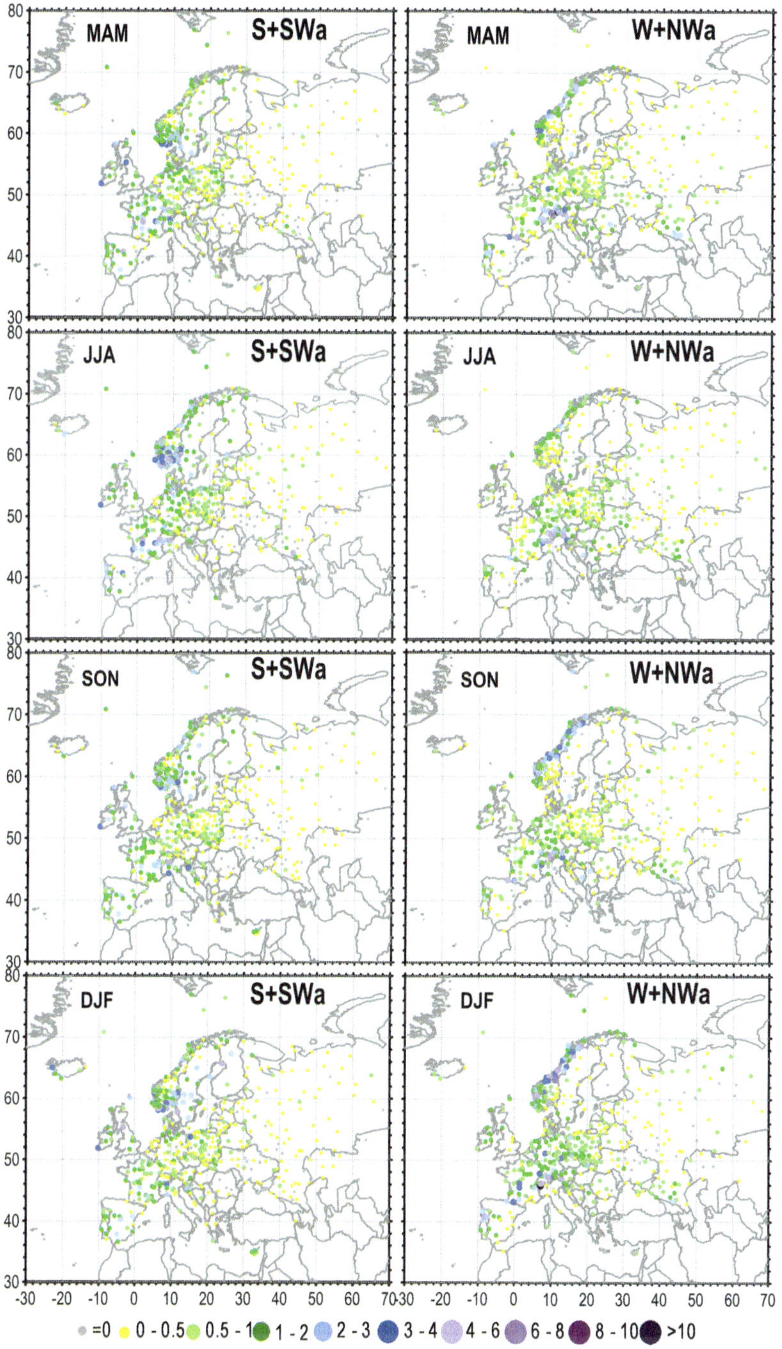

Fig. 6.45 Conditional probability of extreme precipitation occurrence in anticyclonic circulation types (S+SWa, W+NWa) for January 1951–February 2008. Right closed intervals

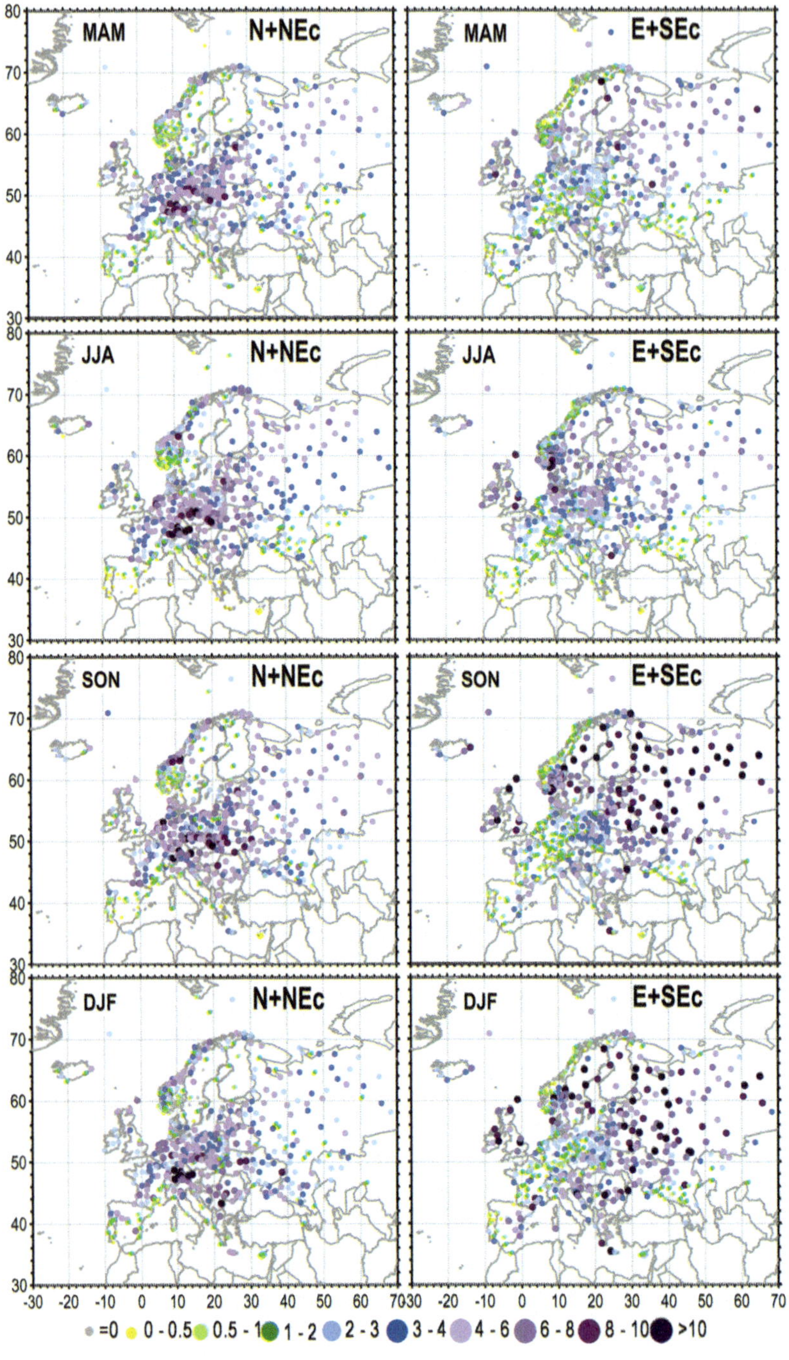

Fig. 6.46 Conditional probability of extreme precipitation occurrence in cyclonic circulation types (N+NEc, E+SEc) for January 1951–February 2008. Right closed intervals

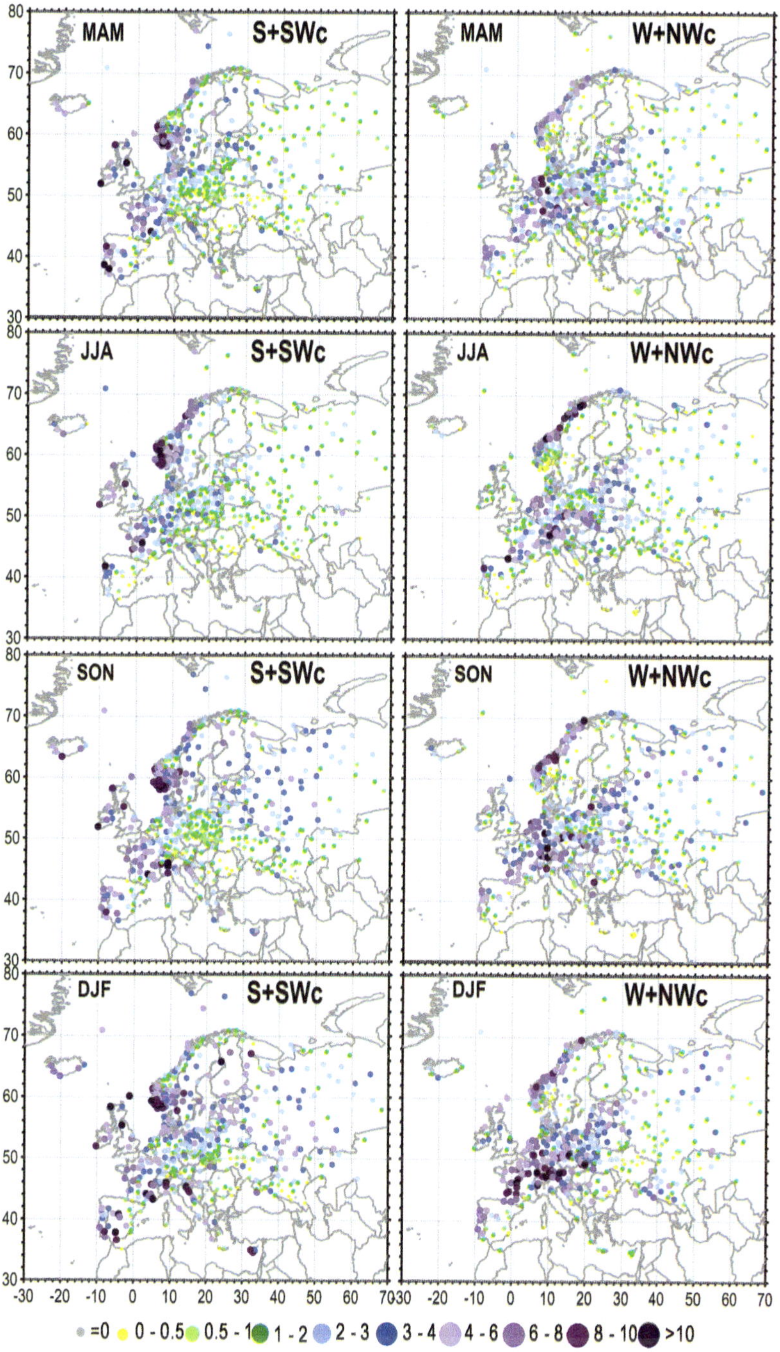

Fig. 6.47 Conditional probability of extreme precipitation occurrence in cyclonic circulation types (S+SWc, W+NWc) for January 1951–February 2008. Right closed intervals

at the southeast coasts of the North Sea, and in summer, also in elevated areas of Central Europe (the Alps, the Carpathians) (Fig. 6.48).

6.3.2.2 Probability of Air-Mass Extreme Precipitation (Type A) in Anticyclonic Synoptic Situations

In Europe, the occurrence of air-mass precipitation in anticyclones is mainly associated with advection of air from the eastern sector (situation N+NEa and E+SEa). In spring and summer during an inflow of the air from the northeast sector (situation N+NEa), the probability of extreme air-mass precipitation in high-mountain areas located inland (the Carpathians, the Alps, and the Caucasus) is slightly higher than elsewhere in Europe. Advection of air from the southeast sector (situation E+SEa) is a poor driver of extreme precipitation in Europe; in fact, its effect is limited to the northern part of the continent. In autumn and winter, in that situation, the likelihood of extreme type A precipitation is slightly higher in the northwest coasts of the Kola Peninsula and on the lee side of the Scandinavian Mountains (Figs. 6.49 and 6.50).

6.3.2.3 Probability of Air-Mass Extreme Precipitation (Type A) in Cyclonic Synoptic Situations

In Europe, the highest probability of air-mass extreme precipitation in cyclones is associated with the inflow of air from the northern sector (situations N+NEc and W+NWc). Similarly to anticyclonic situations, advection of air from the north-east sector (situation N+NEc) favours the occurrence of type A precipitation in the Alps (especially in summer and winter), and, depending on the season, in the Carpathian Mountains (spring and summer), in the southeast coasts of the North Sea (in summer and autumn), in the Polish Baltic coast (summer), and in some coastal areas of the Scandinavian Peninsula. The inflow of air from the northwest (situation W+NWc) mainly causes extreme precipitation risks in Western Europe, particularly on the coasts of the Scandinavian Peninsula, in the southeast coasts of the North Sea, in the area of the Bay of Biscay, and on the Iberian Peninsula. In the foregoing parts of Europe, the processes leading to the formation of extreme precipitation are intensified by orographic effects. In Eastern Europe, the probability of the occurrence of air-mass extreme precipitation at times when air inflows from the eastern sector (situations N+NEc and E+SEc) is slightly higher than in the other synoptic situations (Fig. 6.51). In the south of Europe, the links between the occurrence of type A precipitation and atmospheric circulation are the strongest in the Atlantic coasts of the Iberian Peninsula. In the area, the probability of the occurrence of type A precipitation in situation S+SWc in spring and summer exceeds 1 % (Fig. 6.52). The greatest likelihood of air-mass precipitation in situation Bc, which involves no advection, is recorded in summer. In the other seasons, air-mass precipitation in situation Bc occurs at few stations, but the likelihood of its occurrence is comparable, and even higher, than in advective situations.

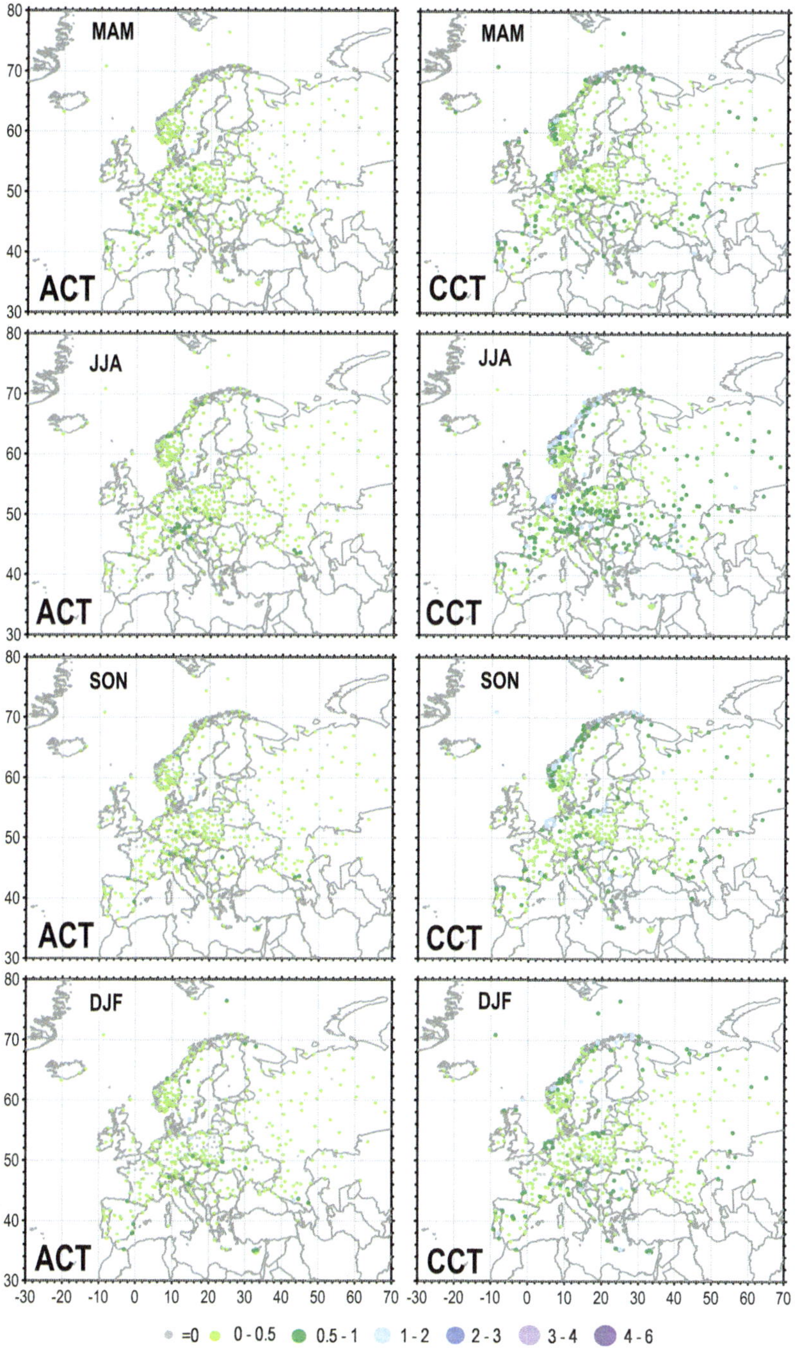

Fig. 6.48 Conditional probability of air-mass extreme precipitation (type A) occurrence in anticyclones (*ACT*) and cyclones (*CCT*) for January 1951–February 2008. Right closed probability intervals

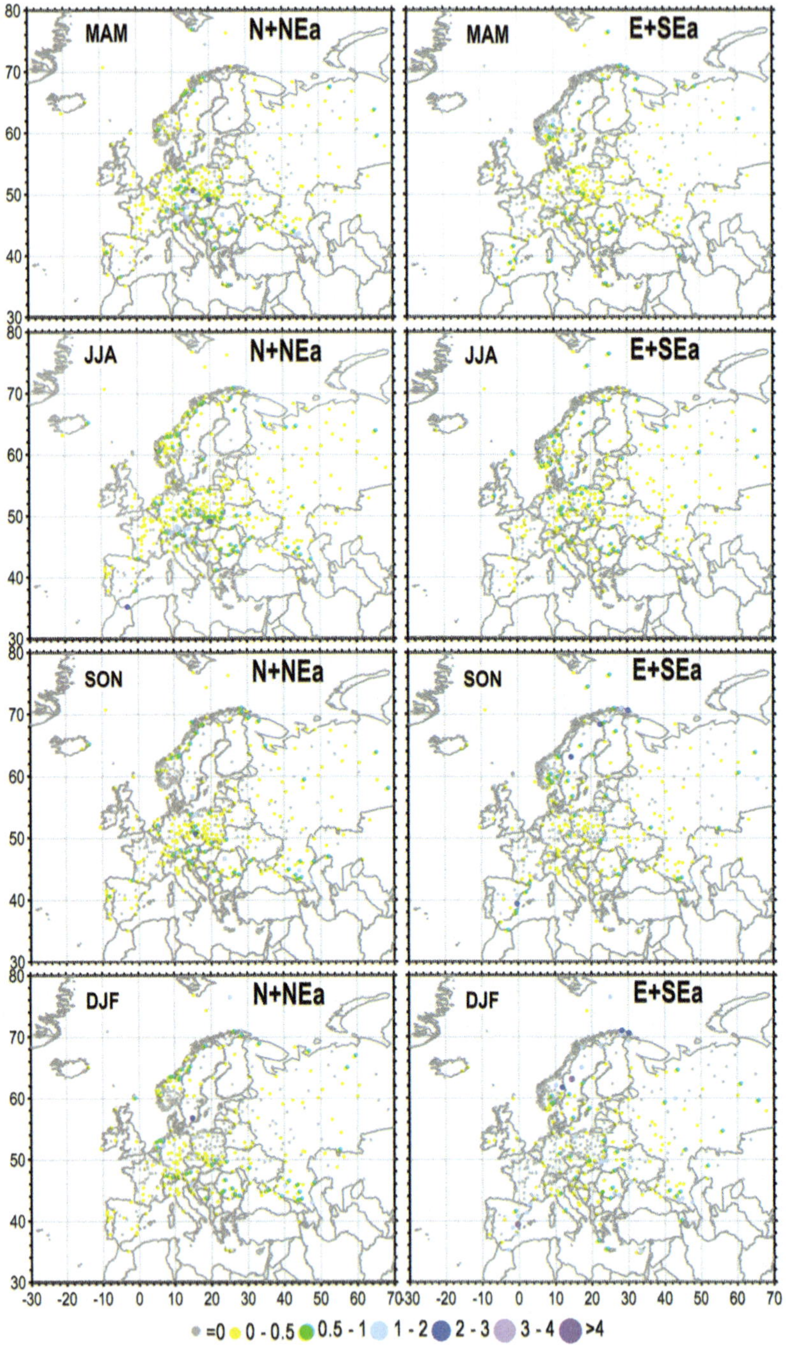

Fig. 6.49 Conditional probability of air-mass extreme precipitation (type A) occurrence in anticyclonic circulation types (N+NEa, E+SEa) for January 1951–February 2008. Right closed intervals

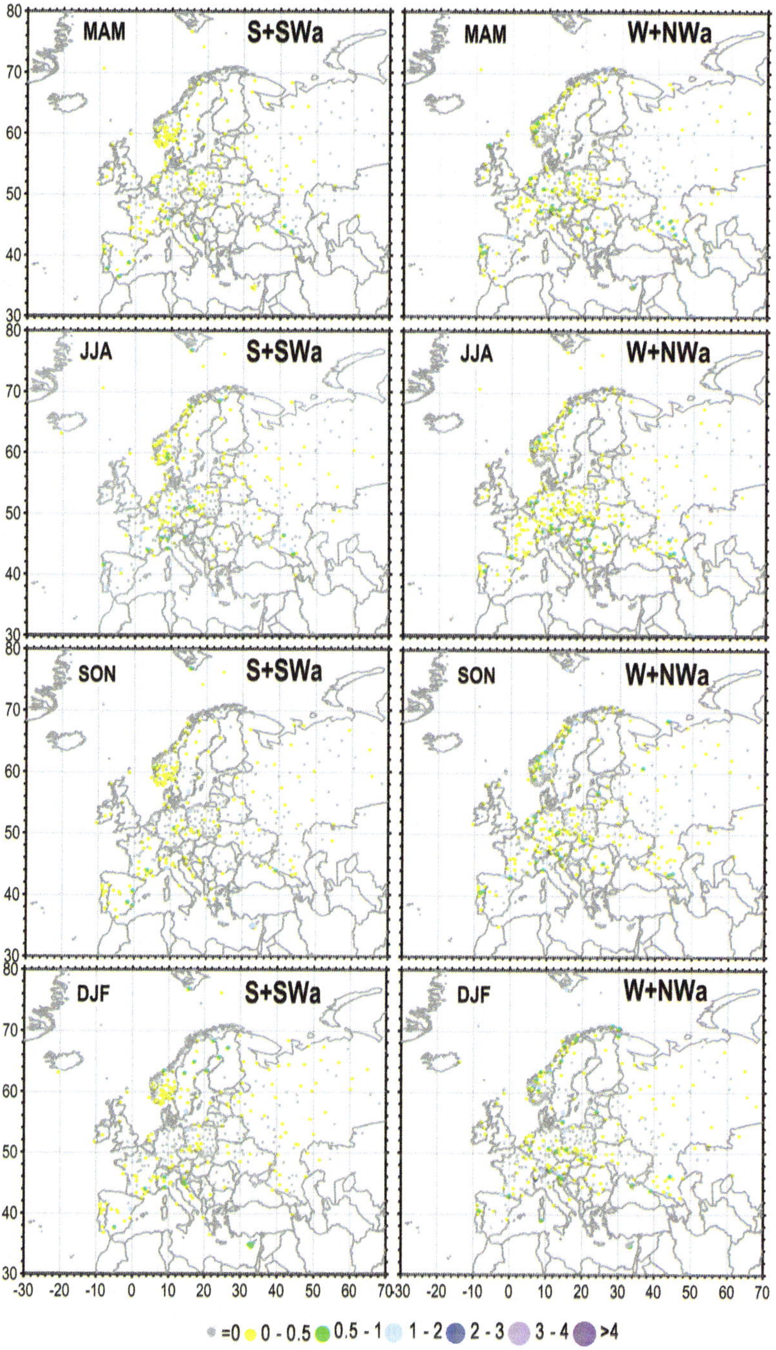

Fig. 6.50 Conditional probability of air-mass extreme precipitation (type A) occurrence in anticyclonic circulation types (S+SWa, W+NWa) for January 1951–February 2008. Right closed intervals

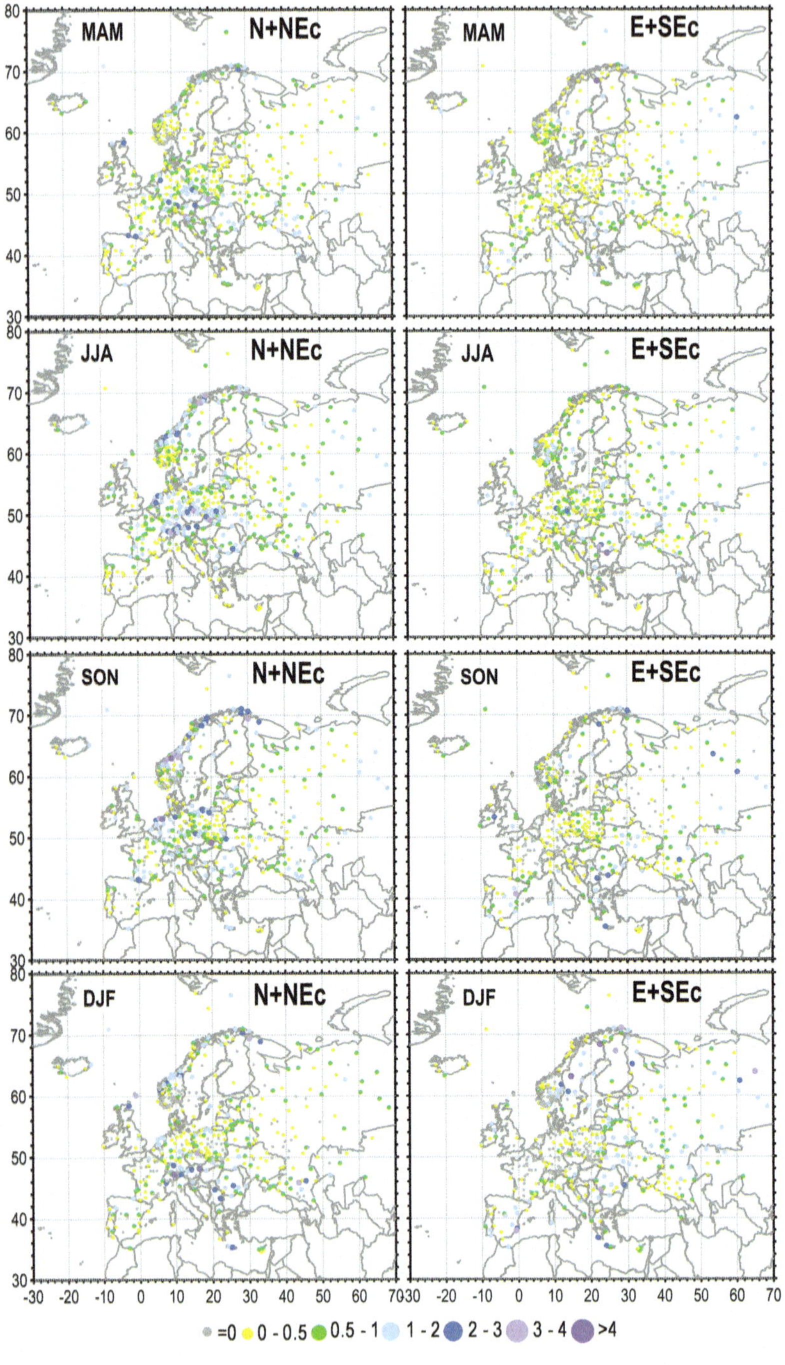

Fig. 6.51 Conditional probability of air-mass extreme precipitation (type A) occurrence in cyclonic circulation types (N+NEc, E+SEc) for January 1951–February 2008. Right closed intervals

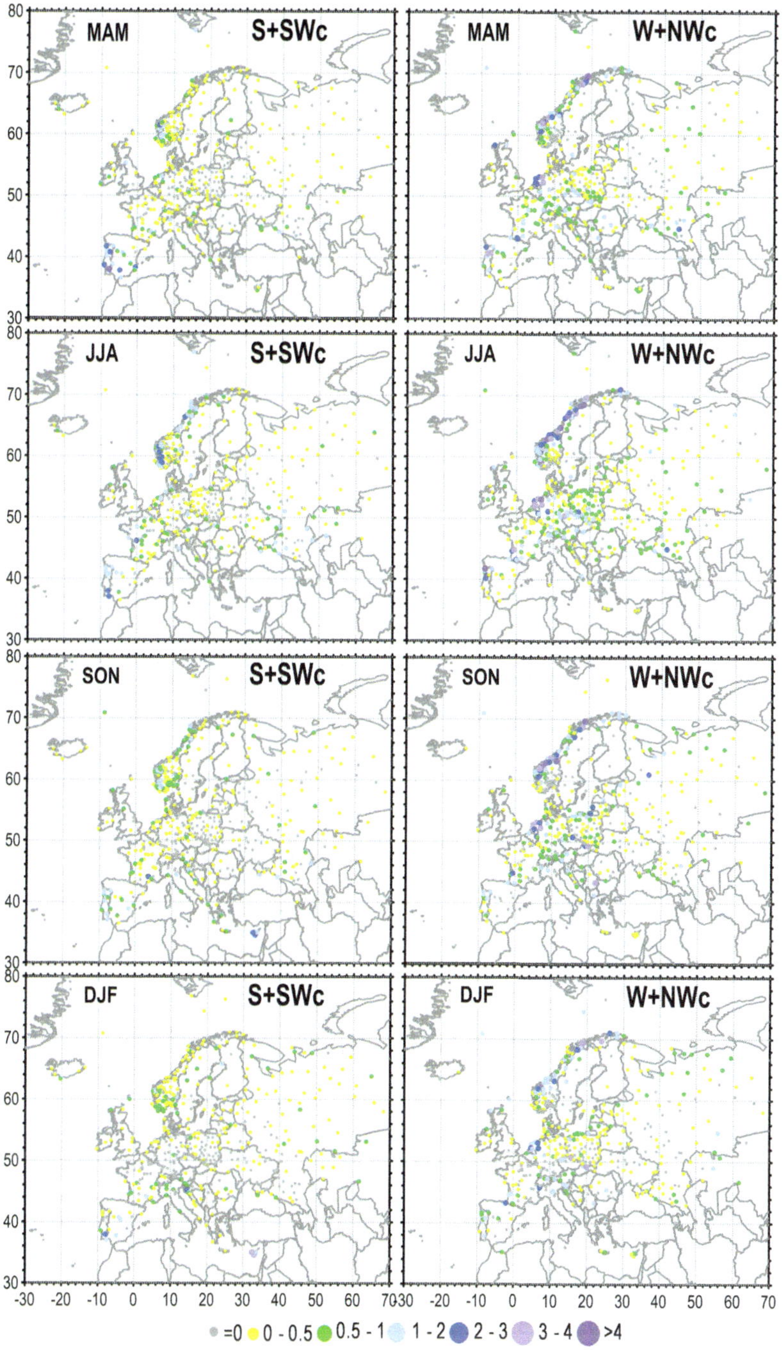

Fig. 6.52 Conditional probability of air-mass extreme precipitation (type A) occurrence in cyclonic circulation types (S+SWc, W+NWc) for January 1951–February 2008. Right closed probability intervals

6.3.3 *Frontal Precipitation (Type F)*

6.3.3.1 Probability of Frontal Extreme Precipitation (Type F) in Anticyclones (ATC) and Cyclones (CTC)

The highest frequency of frontal extreme precipitation in anticyclones characterises the western and central parts of the continent, notably the coasts of the Scandinavian Peninsula, the British Isles, and the Alps. In the Alps, throughout the year, the conditional probability of the occurrence of type F precipitation in anticyclonic situations is relatively high compared to the other parts of Europe (Fig. 6.53). In nearly all Europe, frontal precipitation is clearly more likely to occur in low-pressure systems than in high-pressure systems. The highest probability of type F precipitation in CCT, exceeding 4 %, occurs throughout the year on the British Isles, and, depending on the season, in Eastern Europe (mainly autumn), as well as in Western Europe and Alpine countries (winter) (Fig. 6.53).

6.3.3.2 Probability of Frontal Extreme Precipitation (Type F) in Anticyclonic Synoptic Situations

The occurrence of frontal precipitation (type F) in the Alps and western coasts of the Scandinavian Peninsula, as in the case of air-mass precipitation, is fostered by advection of air from the northern sector (situations N+NEa and W+NWa). In the Alps, the relationship is noticeable throughout the year, and in the west coasts of the Scandinavian Peninsula, it is the strongest in autumn and winter. This finding implies that the occurrence of extreme precipitation in anticyclones, irrespective of the precipitation types within areas where precipitation is fostered by orographic effect, is linked with northerly inflows. Higher—compared to the rest of the continent—frequencies of type F precipitation in situation N+NEa are also recorded in the south of Poland in winter. The occurrence of frontal precipitation is also strongly dependent on circulation in the Scandinavian Peninsula.

The occurrence of type F precipitation on the eastern side of the Scandinavian Mountains is enhanced by inflow of air from the southeast (situation E+SEa), and in the southernmost area of the Scandinavian Peninsula by inflow of air from the southwest sector (situation S+SWa). The probability of the occurrence of type F precipitation during situation S+SWa is relatively high also in the southwest part of Europe (France, Iberian Peninsula) (Figs. 6.54 and 6.55). At some stations in the central part of the continent, the occurrence of type F extreme precipitation is fostered by type Ba.

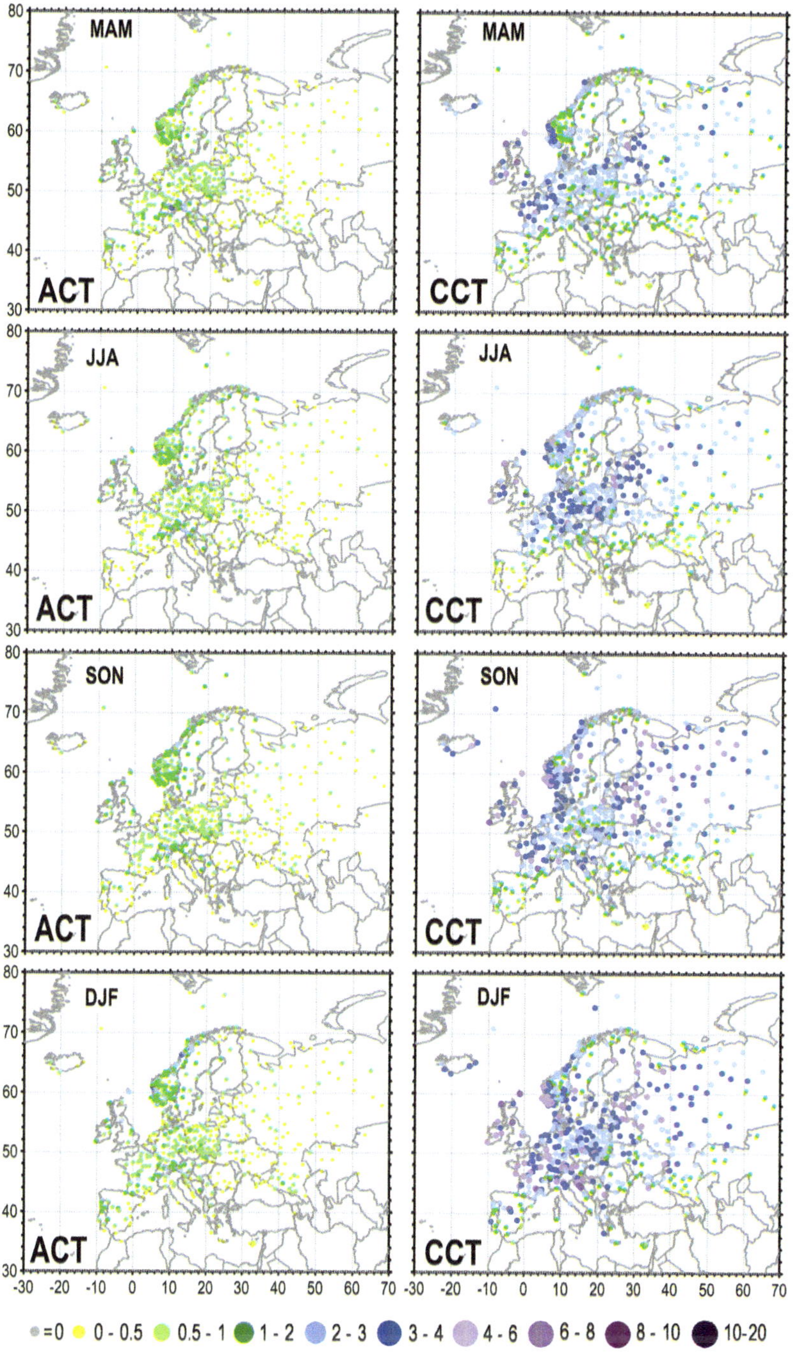

Fig. 6.53 Conditional probability of frontal extreme precipitation (type F) occurrence in anticyclones (*ACT*) and cyclones (*CCT*) for January 1951–February 2008. *Right* closed intervals

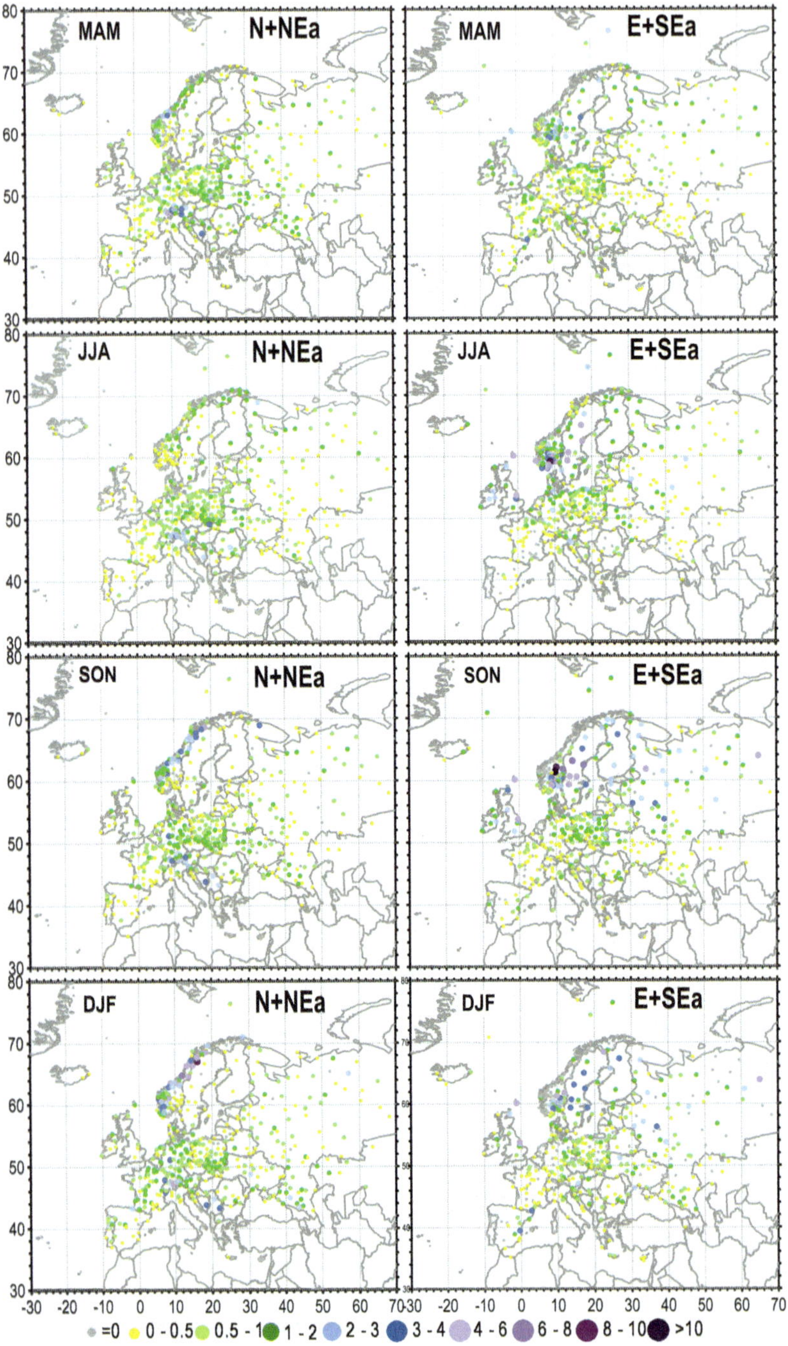

Fig. 6.54 Conditional probability of frontal extreme precipitation (type F) occurrence in anticyclonic circulation types (N+NEa, E+SEa) for January 1951–February 2008. Right closed probability intervals

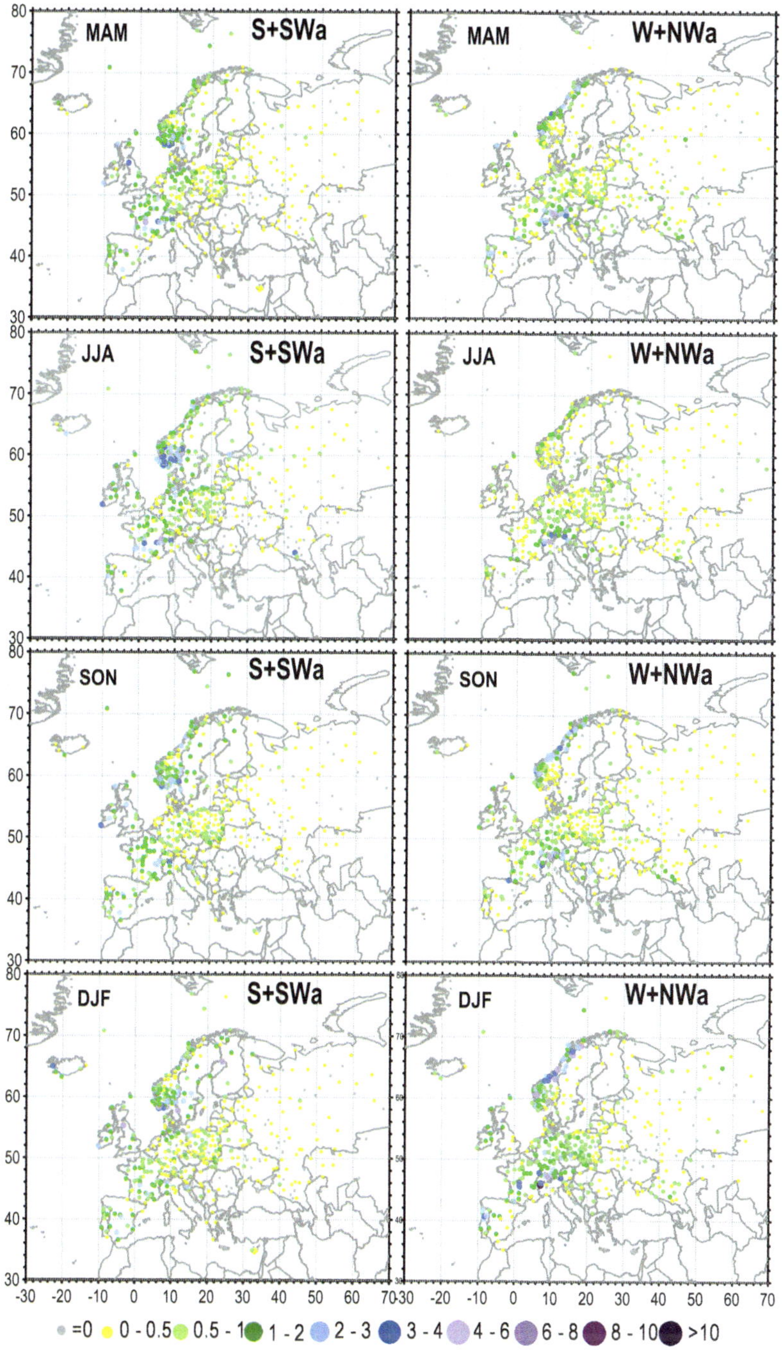

Fig. 6.55 Conditional probability of frontal extreme precipitation (type F) occurrence in anticyclonic circulation types (S+SWa, W+NWa) for January 1951–February 2008. Right closed probability intervals

6.3.3.3 Probability of Frontal Extreme Precipitation (Type F) in Cyclonic Synoptic Situations

The spatial distribution of the conditional probability of frontal extreme precipitation occurrence demonstrates a pronounced diversity depending on the synoptic situation. Throughout the year, inflow of air from the northern sector (situation N+NEc) drives type F precipitation in Central Europe and the northern part of Eastern Europe.

In autumn and winter, the zone where the probability of the precipitation in question exceeds 4 % extends to stations in Western Europe and various parts of Eastern Europe. Inflow of air from the northwest sector (situation W+NWc) fosters frontal extreme precipitation to the greatest extent in Western Europe, from the Iberian Peninsula to the southern part of Germany, and along the Atlantic coastline of the Scandinavian Peninsula. The relationships are particularly pronounced in winter (Fig. 6.57). In the other seasons, in the Iberian Peninsula and in France, the possibility of the occurrence of type F extreme precipitation is the greatest during advection of air from the southwest sector. Situation S+SWc is most conducive to the occurrence of type F precipitation, also in the southernmost part of the Scandinavian Peninsula and at some stations on the British Isles.

The probability of the occurrence of type F precipitation in situation E+SEc exceeds 6 %, and at some stations even 8 % on the British Isles, on the western side of the Scandinavian Mountains, in the eastern part of Iceland, and in winter, also in southern Europe (Figs. 6.56 and 6.57).

6.3.4 Cold Front Precipitation (Type Fc)

6.3.4.1 Probability of Cold Front Extreme Precipitation (Type Fc) in Anticyclones (ACT) and Cyclones (CCT)

In Europe, cold front extreme precipitation in anticyclones is rare. Its occurrence probability is slightly higher only in the Alps, especially in spring and summer, as well as on the western coasts of the Iberian Peninsula in winter (Fig. 6.58). Throughout the year, the probability of type Fc precipitation occurrence in cyclones exceeds 0.5 % in Western Europe, and in summer, also in Central Europe. In summer and autumn, similar probability is also observable at some stations in Eastern Europe (Fig. 6.58).

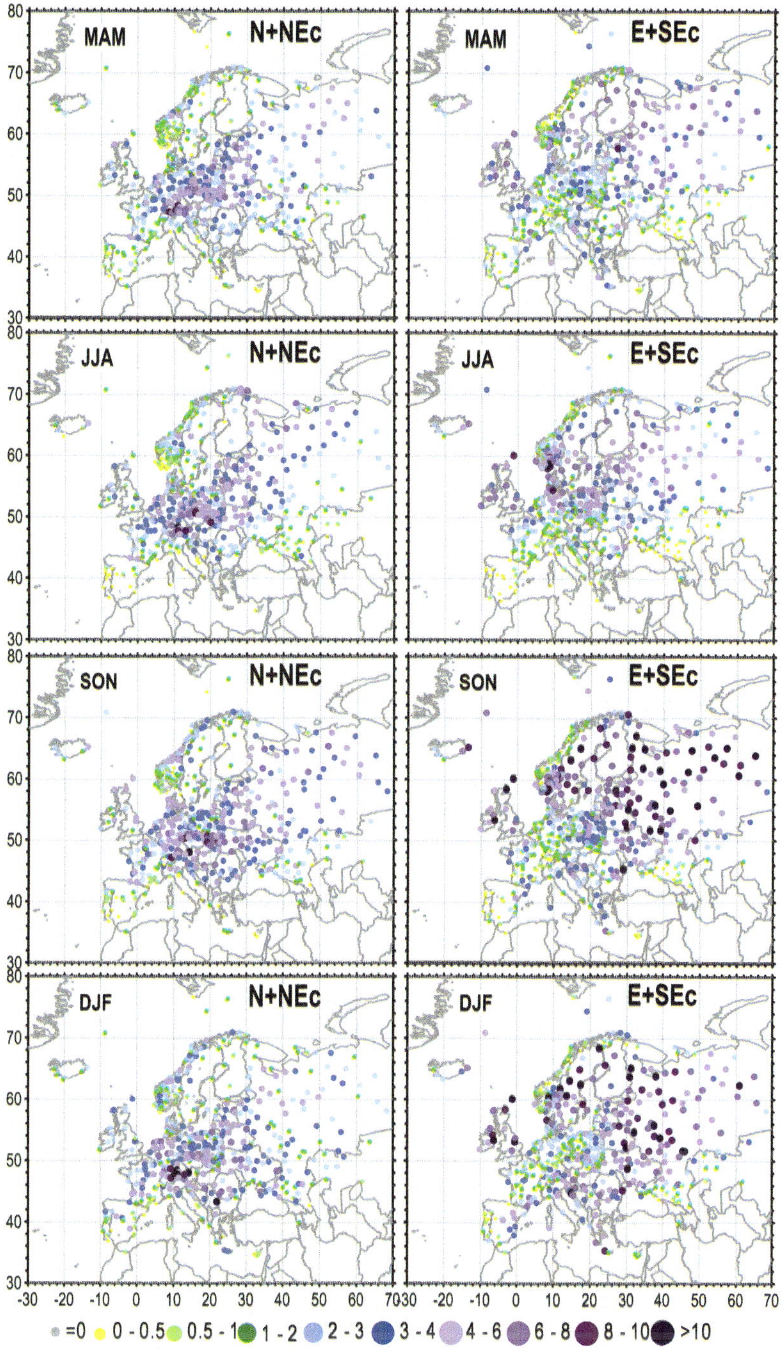

Fig. 6.56 Conditional probability of frontal extreme precipitation (type F) occurrence in cyclonic circulation types (N+NEc, E+SEc) for January 1951–February 2008. Right closed intervals

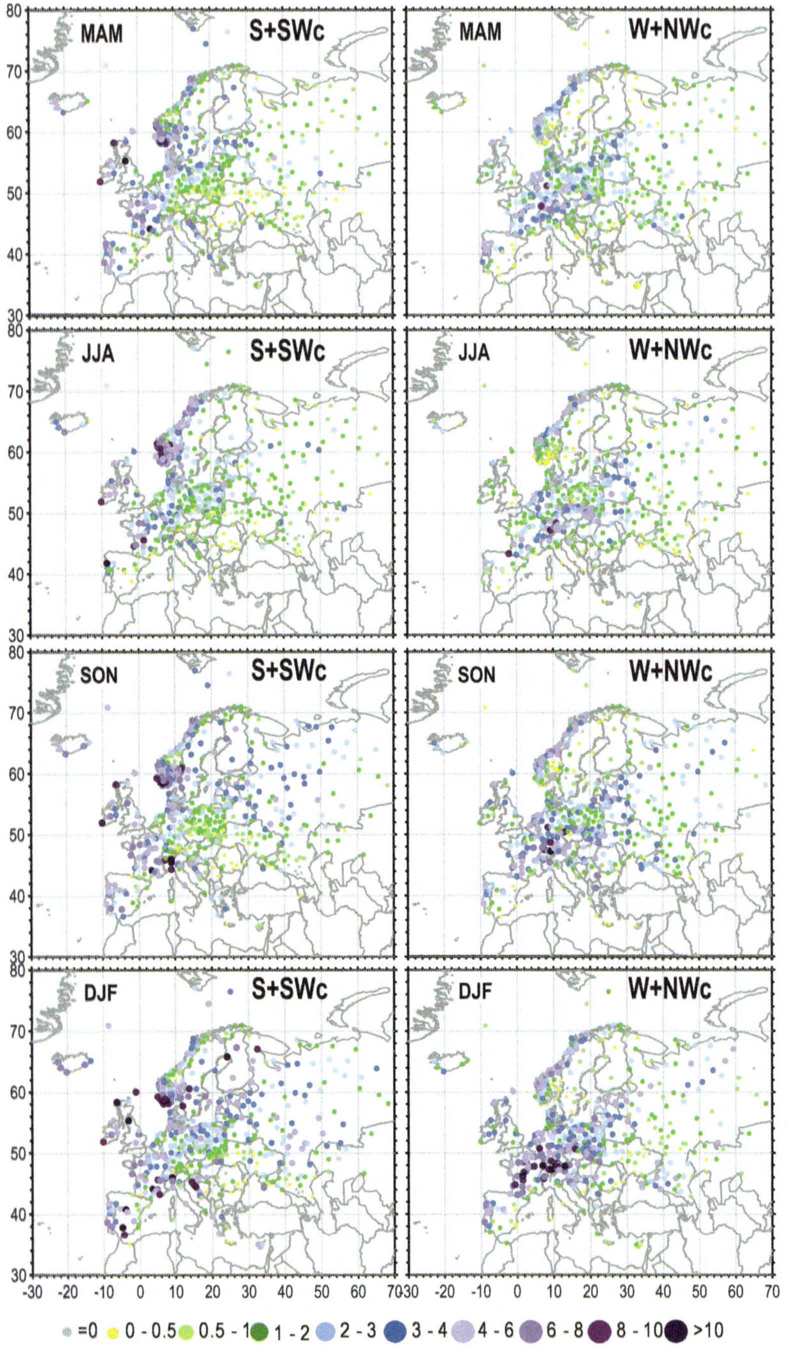

Fig. 6.57 Conditional probability of frontal extreme precipitation (type F) occurrence in cyclonic circulation types (S+SWc, W+NWc) for January 1951–February 2008. Right closed intervals

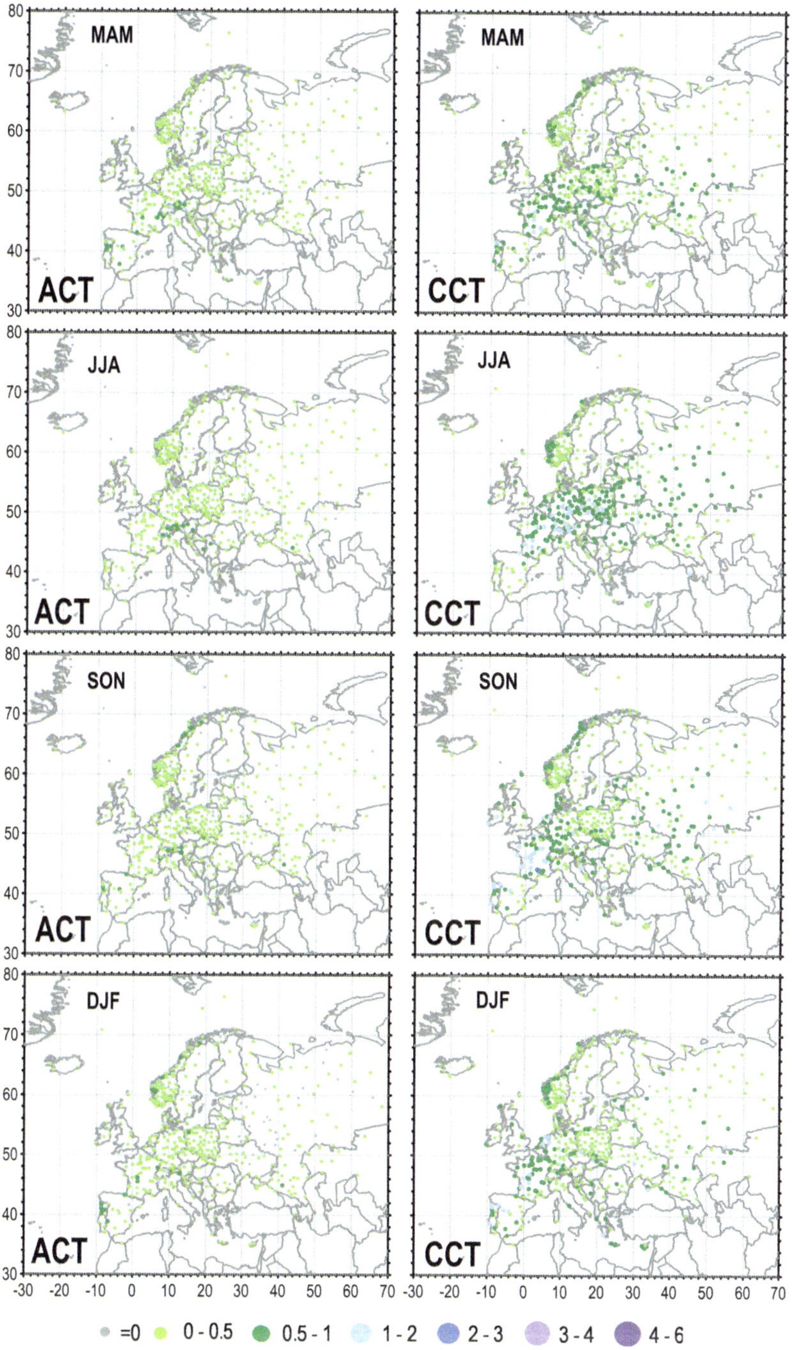

Fig. 6.58 Conditional probability of cold front extreme precipitation (type Fc) occurrence in anticyclones (*ACT*) and cyclones (*CCT*) for January 1951–February 2008. Right closed probability intervals

6.3.4.2 Probability of Cold Front Extreme Precipitation (Type Fc) in Anticyclonic Synoptic Situations

The occurrence of extreme precipitation associated with the passage of a cold front in high-pressure systems is fostered, in most seasons, except for summer, by advection of air from the northeast sector, and in autumn and winter, also from the northwest sector (situations N+NEa and W+NWa) (Figs. 6.59 and 6.60). In each season, inflow of air from the entire northern sector translates into an increased likelihood of type Fc precipitation in the Alps and in the foreland of the Caucasus. In France, the relationship between the occurrence of type Fc and the direction of advection is strong only in spring and summer; then, its occurrence is fostered by situation S+SWa. In this situation, type Fc precipitation is more frequent than in other synoptic types also at different, depending on the season, stations in the Scandinavian Peninsula (Fig. 6.60). In highland and mountain areas of Southern Poland, type Fc precipitation in summer and spring is most probable in situation Ba, which involves no advection, and in summer, also during advection of air from the northeast sector (situation N+NEa).

6.3.4.3 Probability of Cold Front Extreme Precipitation (Type Fc) in Cyclonic Synoptic Situations

Similarly to anticyclonic types, also in cyclonic situations, cold front precipitation is most likely, throughout the year, during inflow of air from the north (situations N+NEc and W+NWc) in the Alps, in the Caucasus area, and in Southern Poland (Figs. 6.61 and 6.62). In southwestern Europe (the Iberian Peninsula, France), the occurrence of such precipitation is linked to advection of air from the western sector (situations S+SWc and W+NWc). Such advection also fosters type Fc precipitation on the coasts of the Scandinavian Peninsula. In spring, autumn, and winter, cold front precipitation is more likely in situation W+NWc, while in summer, especially in the southernmost part of the peninsula, in situation S+SWc (Fig. 6.62). It is worth noting that in Central Europe, in summer, when the frequency of extreme precipitation associated with the passage of a cold front is the highest, its occurrence in the northern part of the area is fostered by situation E+SEc, while in the southern part by situation S+SWc (Figs. 6.61 and 6.62). The likelihood of the occurrence of type Fc precipitation in a nonadvective situation, irrespective of the season, is higher than 1 %, while in autumn, at many stations in Central Europe, it falls between 2 and 4 %.

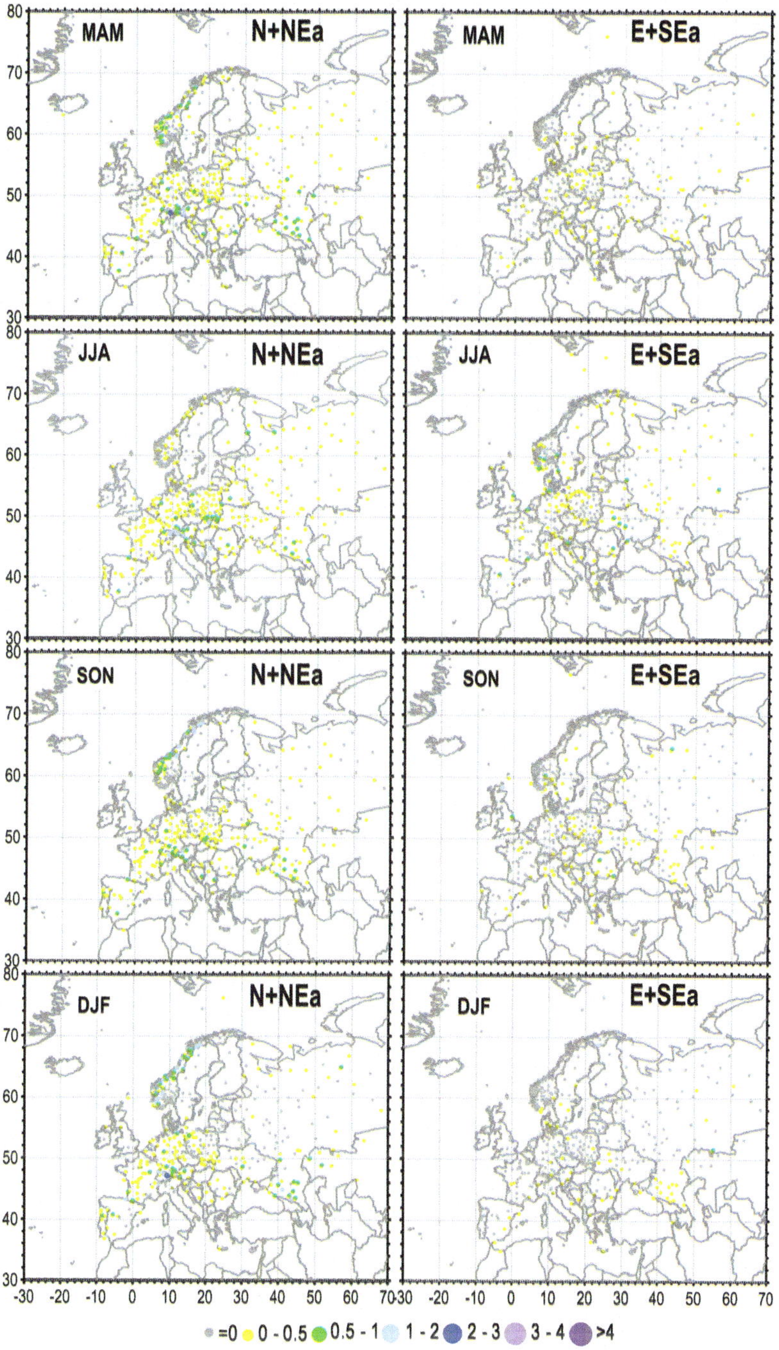

Fig. 6.59 Conditional probability of cold front extreme precipitation (type Fc) occurrence in anticyclonic circulation types (N+NEa, E+SEa) for January 1951–February 2008. Right closed intervals

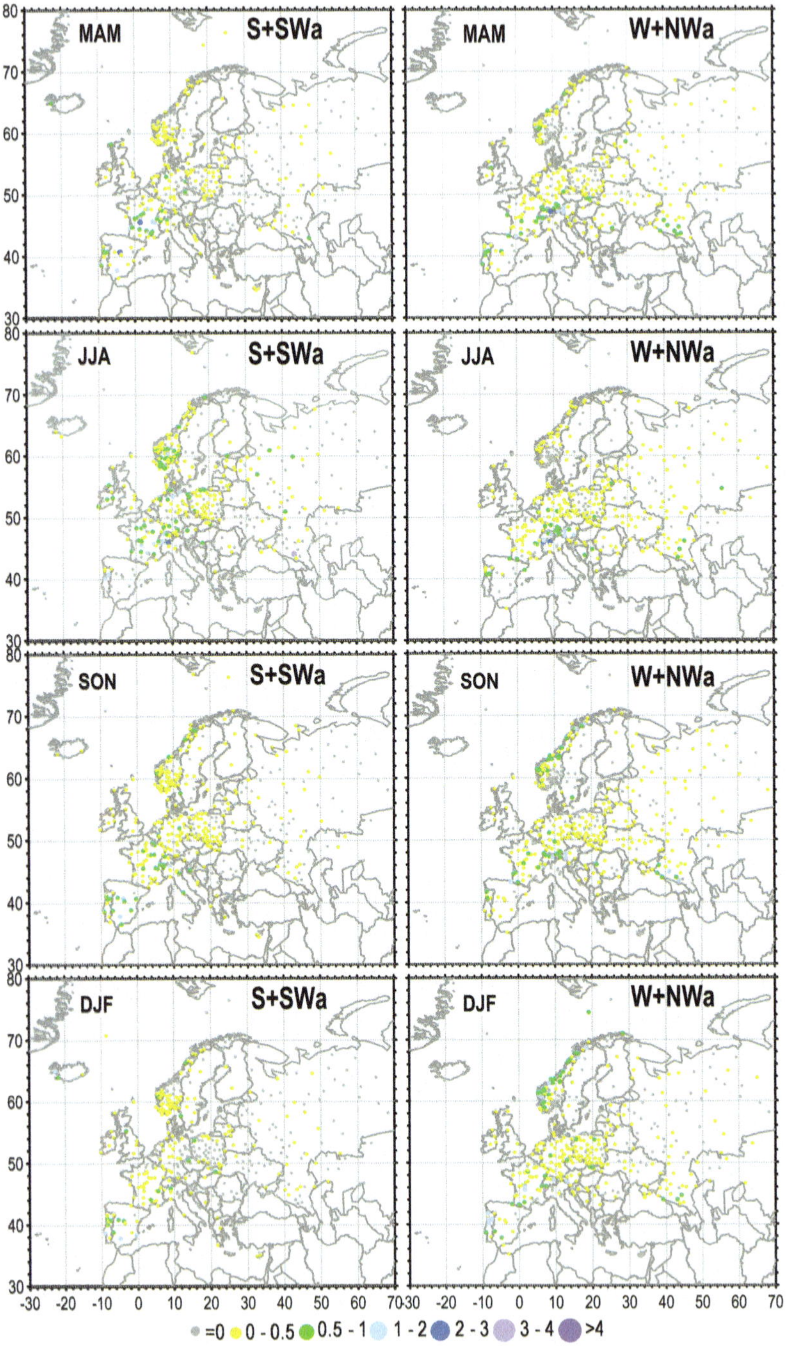

Fig. 6.60 Conditional probability of cold front extreme precipitation (type Fc) occurrence in anticyclonic circulation types (S+SWa, W+NWa) for January 1951–February 2008. Right closed probability intervals

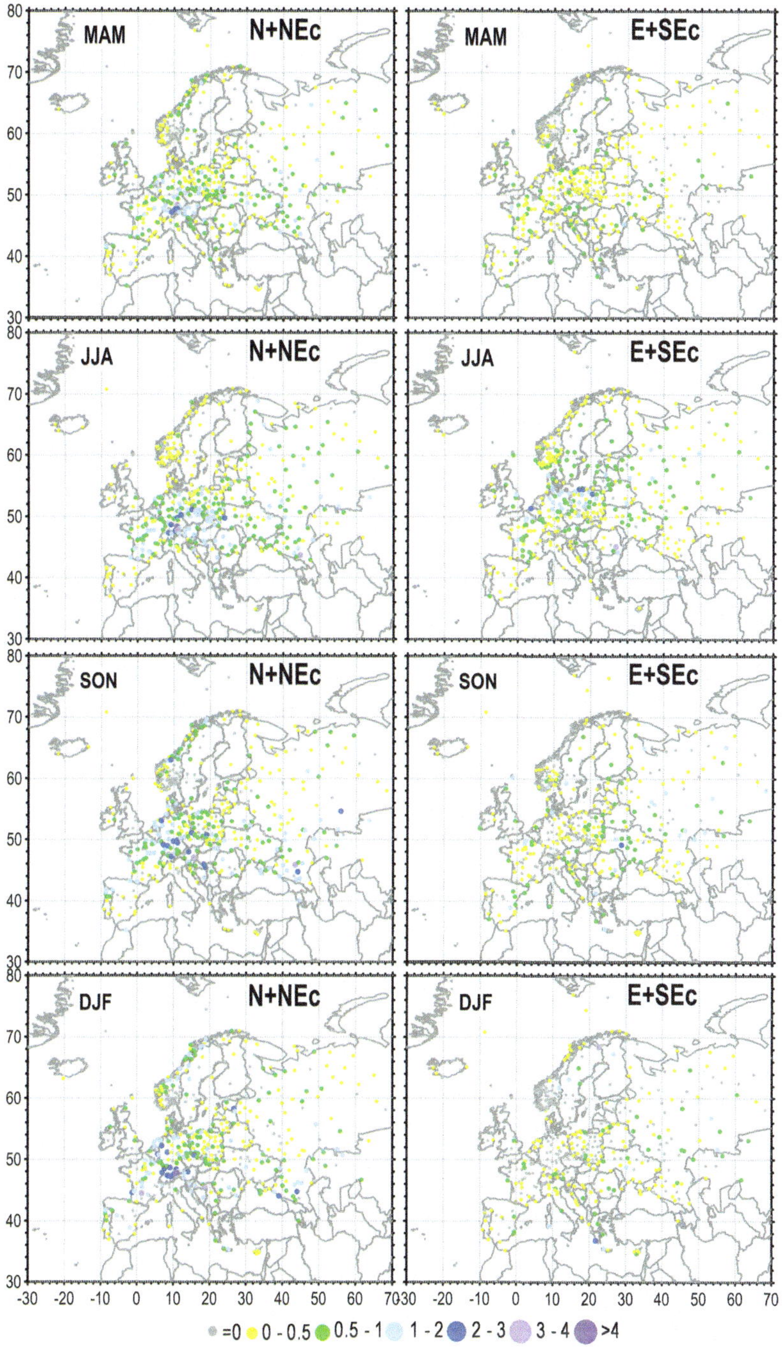

Fig. 6.61 Conditional probability of cold front extreme precipitation (type Fc) occurrence in cyclonic circulation types (N+NEc, E+SEc) for January 1951–February 2008. Right closed intervals

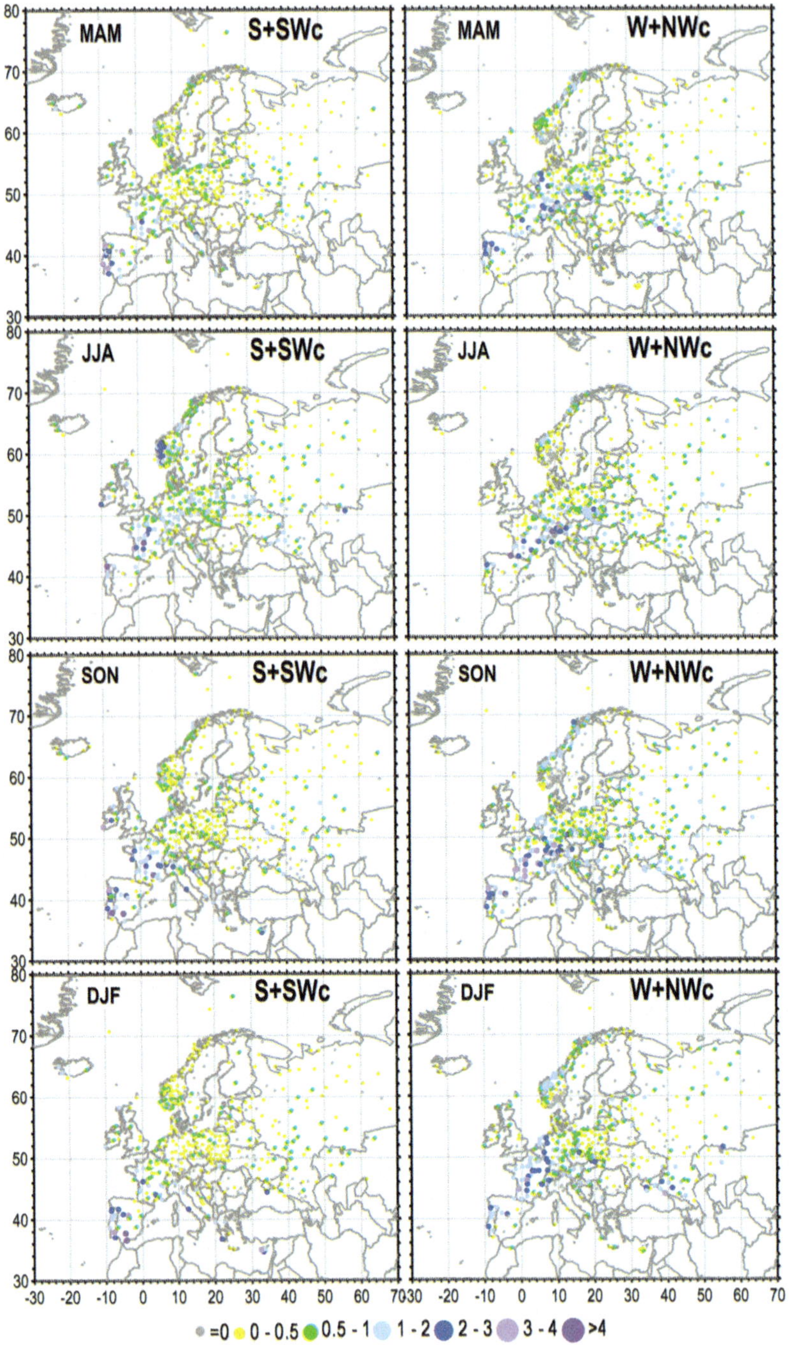

Fig. 6.62 Conditional probability of cold front extreme precipitation (type Fc) occurrence in cyclonic circulation types (S+SWc, W+NWc) for January 1951–February 2008. Right closed intervals

6.3.5 *Warm Front Precipitation (Type Fw)*

6.3.5.1 Probability of Warm Front Precipitation (Type Fw) in Anticyclones (ATC) and Cyclones (CTC)

The likelihood of warm front precipitation in anticyclones is very low in all seasons. Only in winter does it fall between 0.5 and 1 % at some stations located in the west coasts of the Scandinavian Peninsula. The probability of the occurrence of the precipitation in question in cyclones is equally low, exceeding 0.5 % only at isolated stations, mainly in Eastern Europe. Only in winter, when the frequency of warm front precipitation is the highest, does the likelihood of such precipitation in CCT grow slightly compared to the other seasons in Eastern Europe, in the Alpine area, and at some sites on the British Isles and Iceland (0.5–1 %) (Fig. 6.63). The low probability of warm front extreme precipitation in both anticyclones and cyclones is generally attributable to its low overall frequency (Fig. 6.63).

6.3.5.2 Probability of Warm Front Precipitation (Type Fw) in Anticyclonic Synoptic Situations

The occurrence of extreme precipitation associated with the passage of a warm front in high-pressure systems demonstrates a perceivable relationship with the type of synoptic situation in the western coasts of the Scandinavian Peninsula (in winter) and at some stations in Central Europe (mainly in winter) and in Northern Europe (autumn and winter). The occurrence of type Fw precipitation in the Atlantic coasts of the Scandinavian Peninsula is perceptibly connected with advection from the north (situations N+NEa and W+NWa). In Central Europe such precipitation is most likely in situation Ba, and in Northern Europe in situation E+SEa (Figs. 6.64 and 6.65).

6.3.5.3 Probability of Warm Front Precipitation (Type Fw) in Cyclonic Synoptic Situations

The strongest relationships between the occurrence of precipitation associated with a warm front (type Fw) and the direction of air advection were found in Eastern Europe in winter and autumn. In this area, the occurrence of type Fw precipitation both in winter and autumn is most strongly associated with advection of air from the southeast sector (E+SEc). In winter, in the western part of Eastern Europe, the values of the index in question reach 2–4 %. In autumn, the probability of type Fw in situation E+SEc is higher than 2 % at stations in the eastern part of Eastern Europe. In Central Europe, and in the area stretching from the Alps to the Black Sea, the occurrence of type Fw is fostered the most by situation N+NEc across the seasons (Fig. 6.66). A slightly higher probability of the occurrence of type Fw precipitation

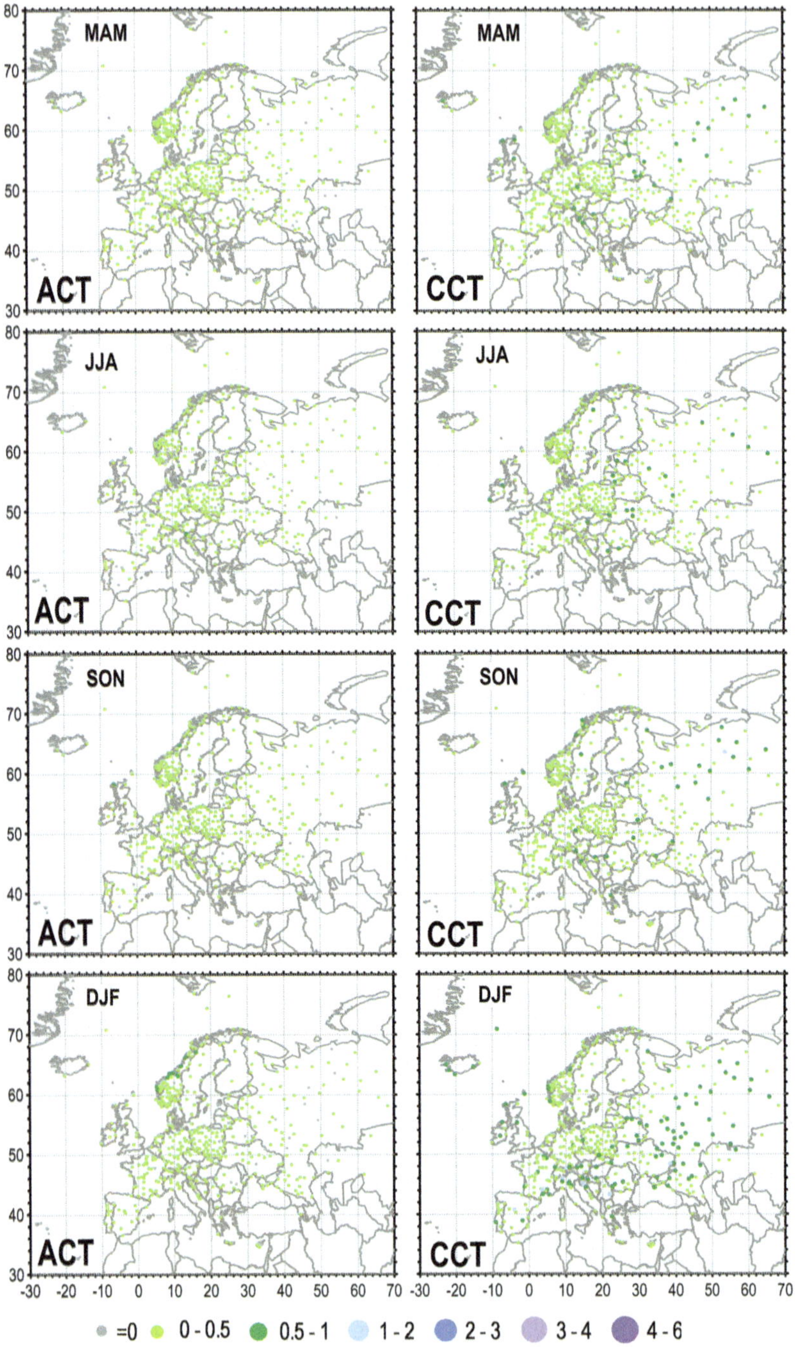

Fig. 6.63 Conditional probability of warm front extreme precipitation (type Fw) occurrence in anticyclones (*ACT*) and cyclones (*CCT*) for January 1951–February 2008. Right closed intervals

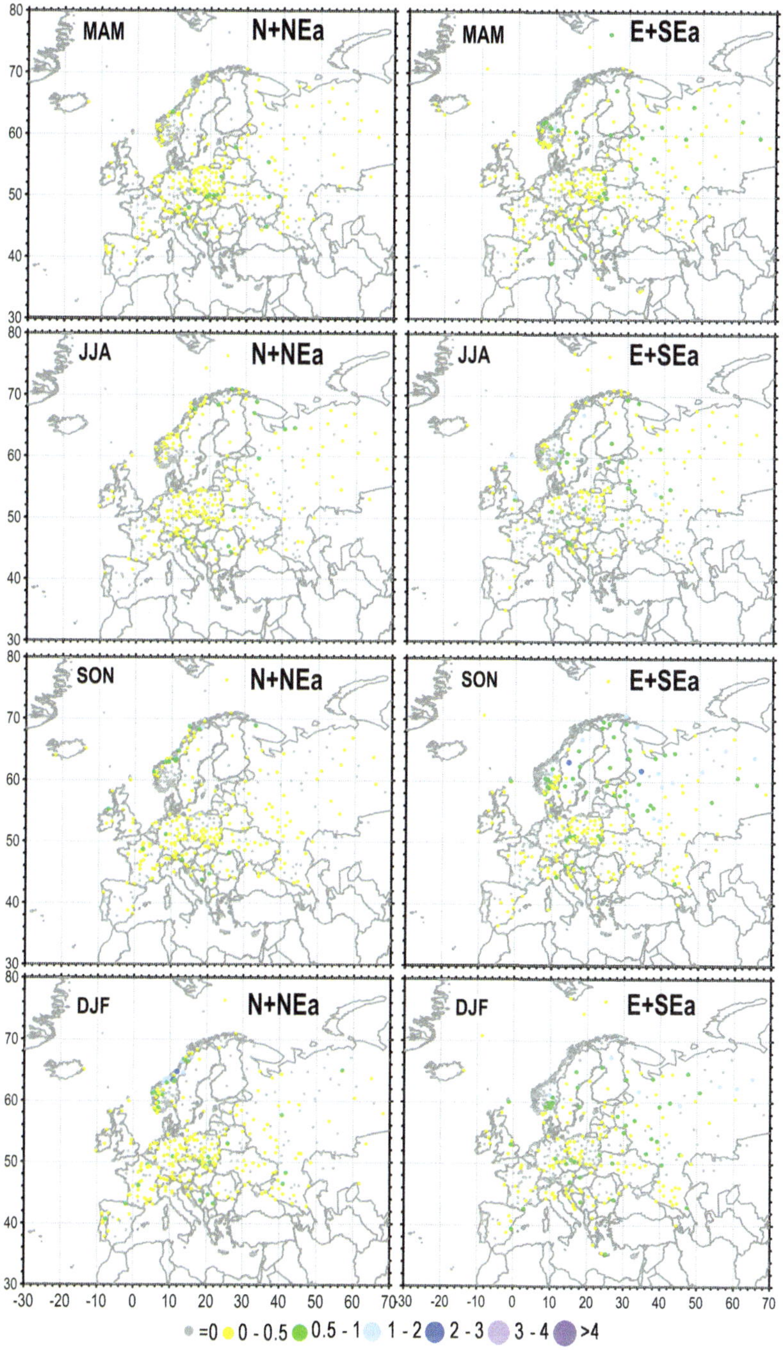

Fig. 6.64 Conditional probability of warm front extreme precipitation (type Fw) occurrence in anticyclonic circulation types (N+NEa, E+SEa) for January 1951–February 2008. Right closed intervals

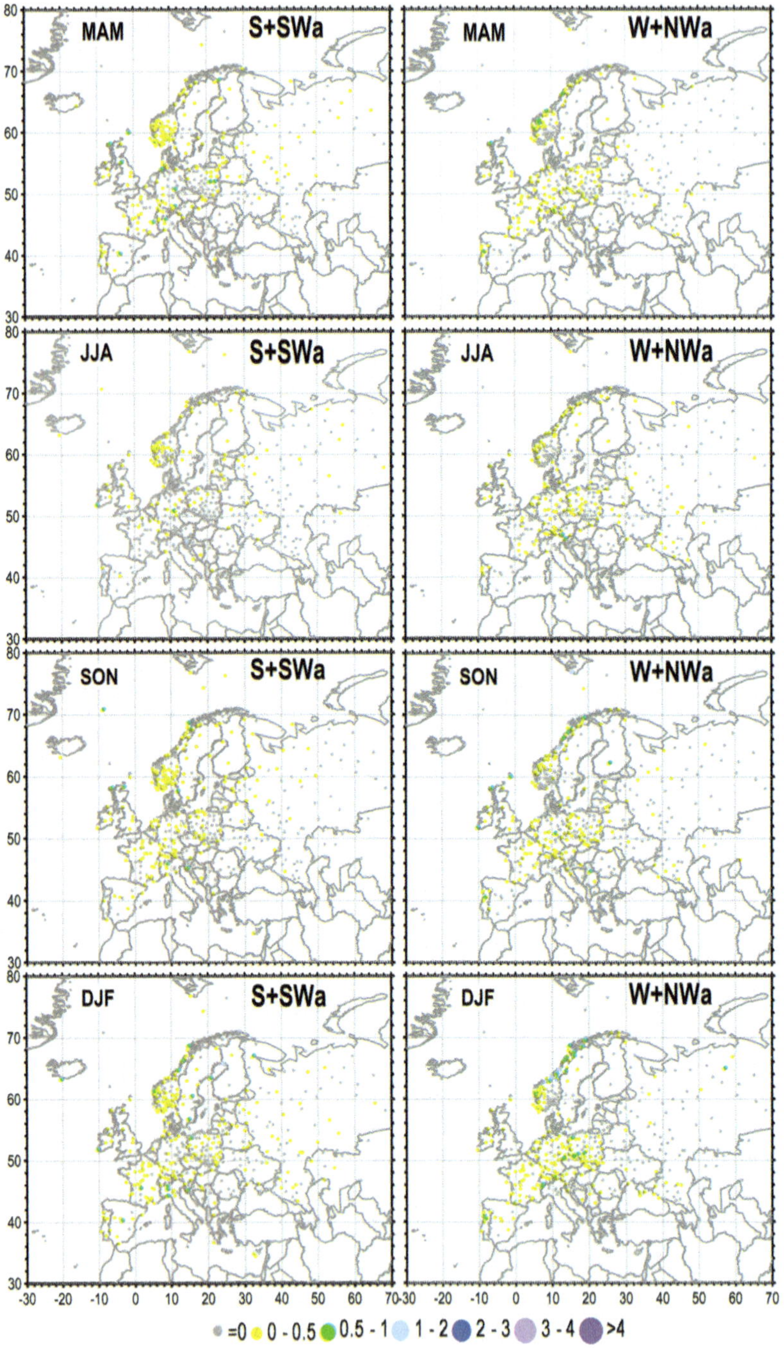

Fig. 6.65 Conditional probability of warm front extreme precipitation (type Fw) occurrence in anticyclonic circulation types (S+SWa, W+NWa) for January 1951–February 2008. Right closed probability intervals

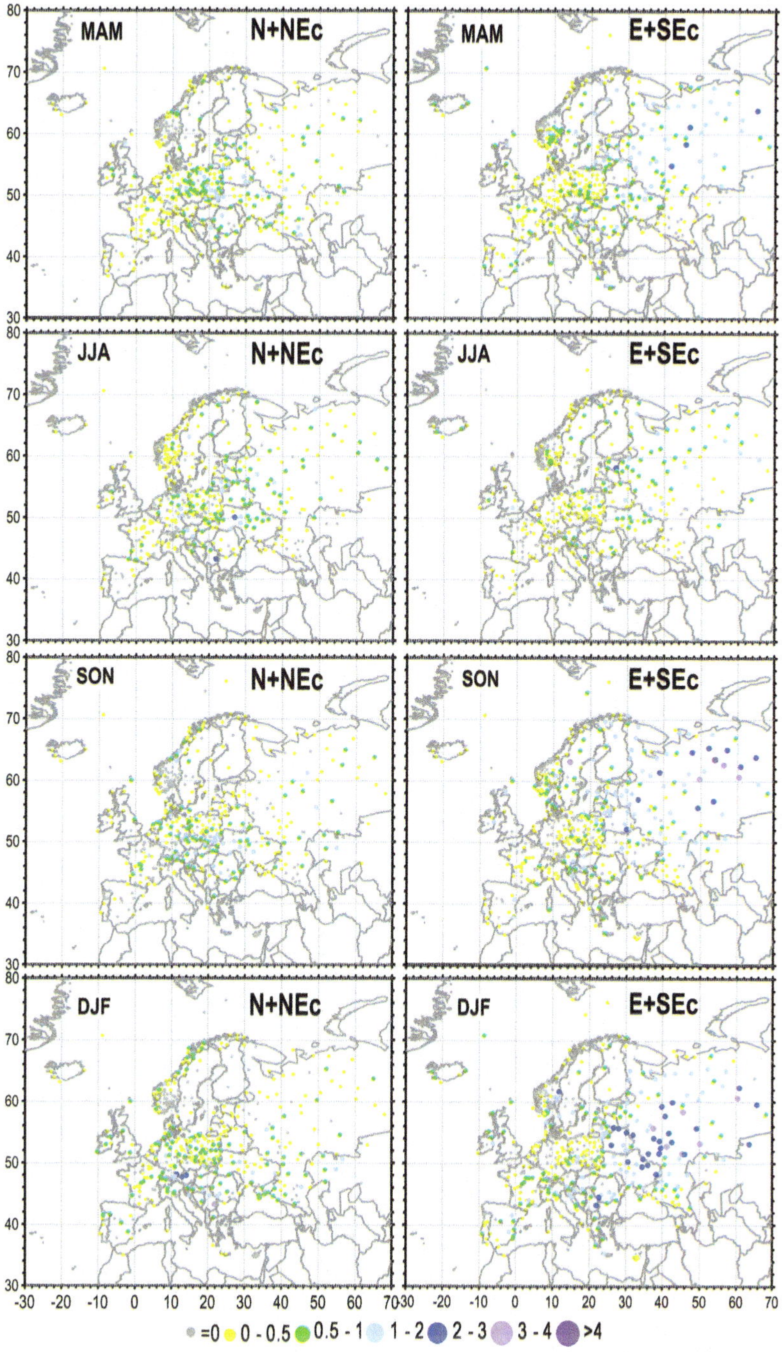

Fig. 6.66 Conditional probability of warm front extreme precipitation (type Fw) occurrence in cyclonic circulation types (N+NEc, E+SEc) for January 1951–February 2008. Right closed intervals

in winter characterises the western and central parts of Southern Europe, at times of advection from the southwest (situation S+SWc) and the Alpine area, as well as some stations in southern Poland in situation W+NWc (Fig. 6.67).

6.3.6 Precipitation Associated with the Passage of Different Fronts (Type Ff)

6.3.6.1 Probability of Extreme Precipitation Associated with the Passage of Different Fronts (Type Ff) in Anticyclones (ACT) and Cyclones (CCT)

In Europe, the probability of the occurrence of precipitation associated with the passage of several fronts in high-pressure systems is low, up to 0.5 % at most stations. Type Ff precipitation appears slightly more frequently in CCT (1 % probability) at the Atlantic coasts of the Scandinavian Peninsula and at Alpine stations (Fig. 6.68).

Throughout the year, the possibility of the occurrence of type Ff in low-pressure systems is the highest in Western Europe and Central Europe, mainly in the area stretching from the Bay of Biscay to the Gulf of Finland, as well as in the Atlantic coasts of the Scandinavian Peninsula. In autumn, and more so in winter, type Ff precipitation is recorded at many stations on 3–4 % of the days with cyclonic situations. The southern part of Europe demonstrates the lowest likelihood of the occurrence of type Ff precipitation (Fig. 6.68).

6.3.6.2 Probability of Extreme Precipitation Linked to the Passage of Diffrent Fronts (Type Ff) in Anticyclonic Synoptic Situations

In Europe, the highest probability of the occurrence of extreme precipitation associated with the passage of several fronts in anticyclonic situations characterises: the Scandinavian Peninsula, the Alps, and, in some seasons, also France. As in the case of the earlier discussed types, type Ff precipitation on the west coasts of the Scandinavian Peninsula is fostered by inflow of air from the north (situations N+NEa and W+NWa) (Figs. 6.69 and 6.70). These synoptic situations, in particular W+NWa, create the greatest probability of such precipitation in the Alps. In the southernmost part of the Scandinavian Peninsula, type Ff precipitation is most likely to occur during advection of air from the south (situations E+SEa and S+SWa). Situation S+SWa demonstrates a clear relationship with the frequency of type Ff precipitation in the southern coast of the peninsula, while the occurrence of such precipitation on the lee, that is, eastern, side of the Scandinavian Mountains is most enhanced by easterly flows (situation E+SEa). In situation S+SWa type Ff precipitation occurs frequently also at many French stations (Fig. 6.69). The increased probability of the occurrence of many types of extreme precipitation in

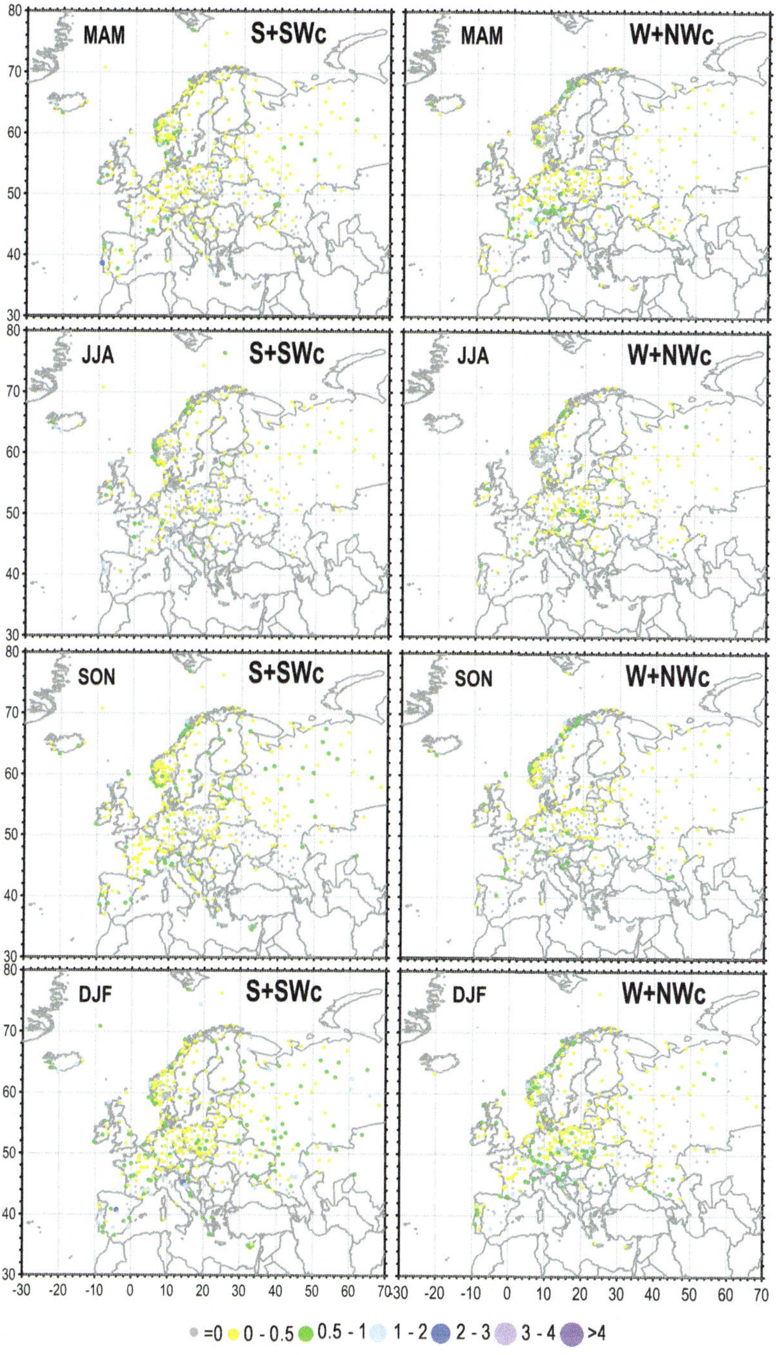

Fig. 6.67 Conditional probability of warm front extreme precipitation (type Fw) occurrence in cyclonic circulation types (S+SWc, W+NWc) for January 1951–February 2008. Right closed intervals

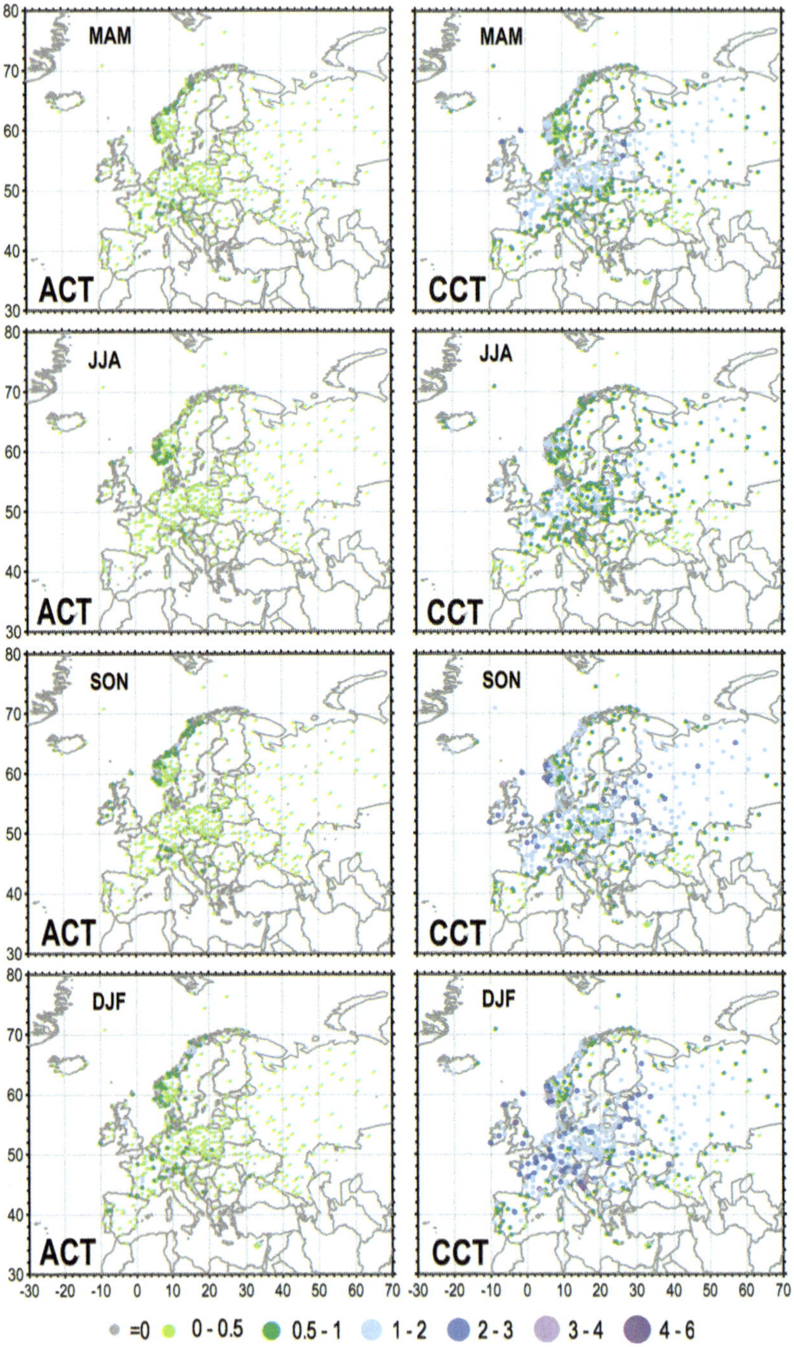

Fig. 6.68 Conditional probability of the occurrence of extreme precipitation associated with the passage of different fronts (type Ff) in anticyclones (*ACT*) and cyclones (*CCT*) for January 1951–February 2008. Right closed probability intervals

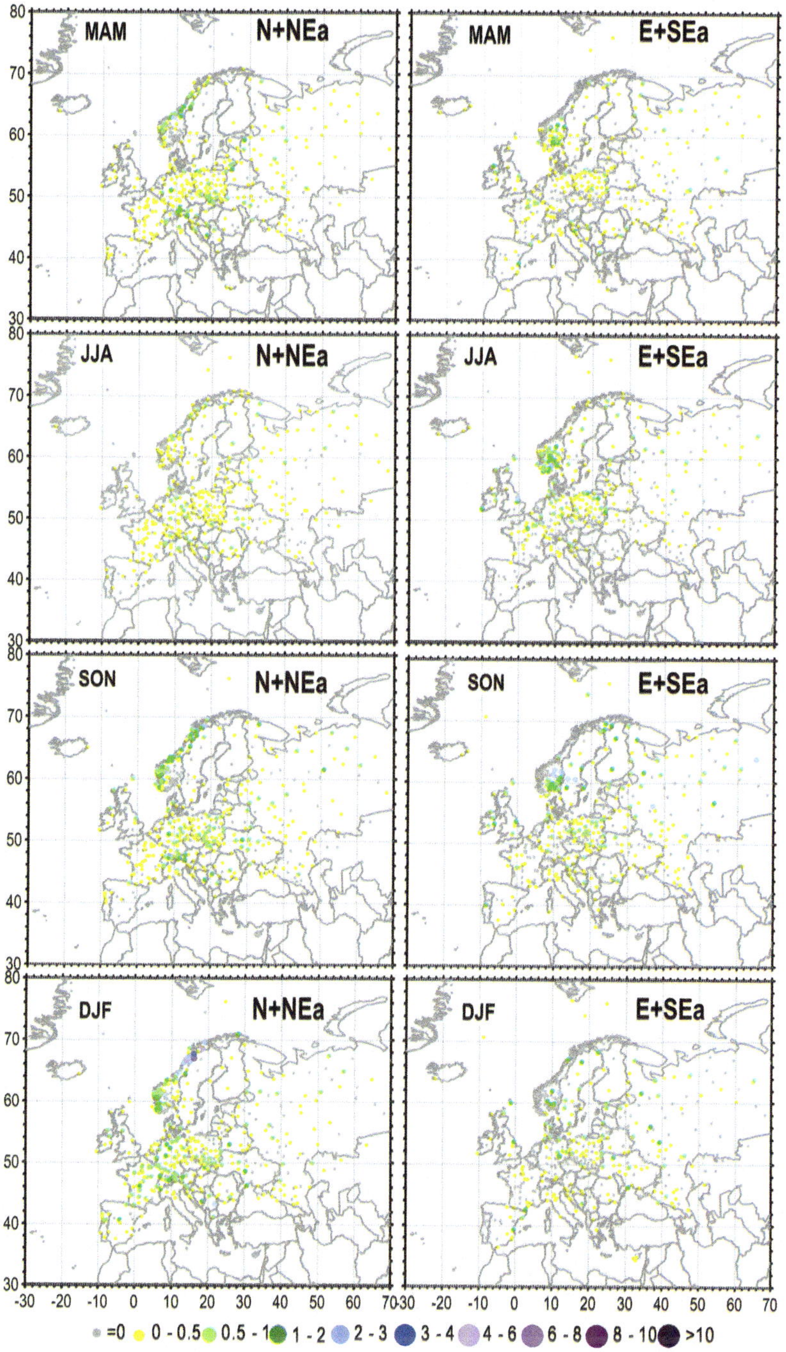

Fig. 6.69 Conditional probability of the occurrence of extreme precipitation associated with the passage of different fronts (type Ff) in anticyclonic circulation types (N+NEa, E+SEa) for January 1951–February 2008. Right closed probability

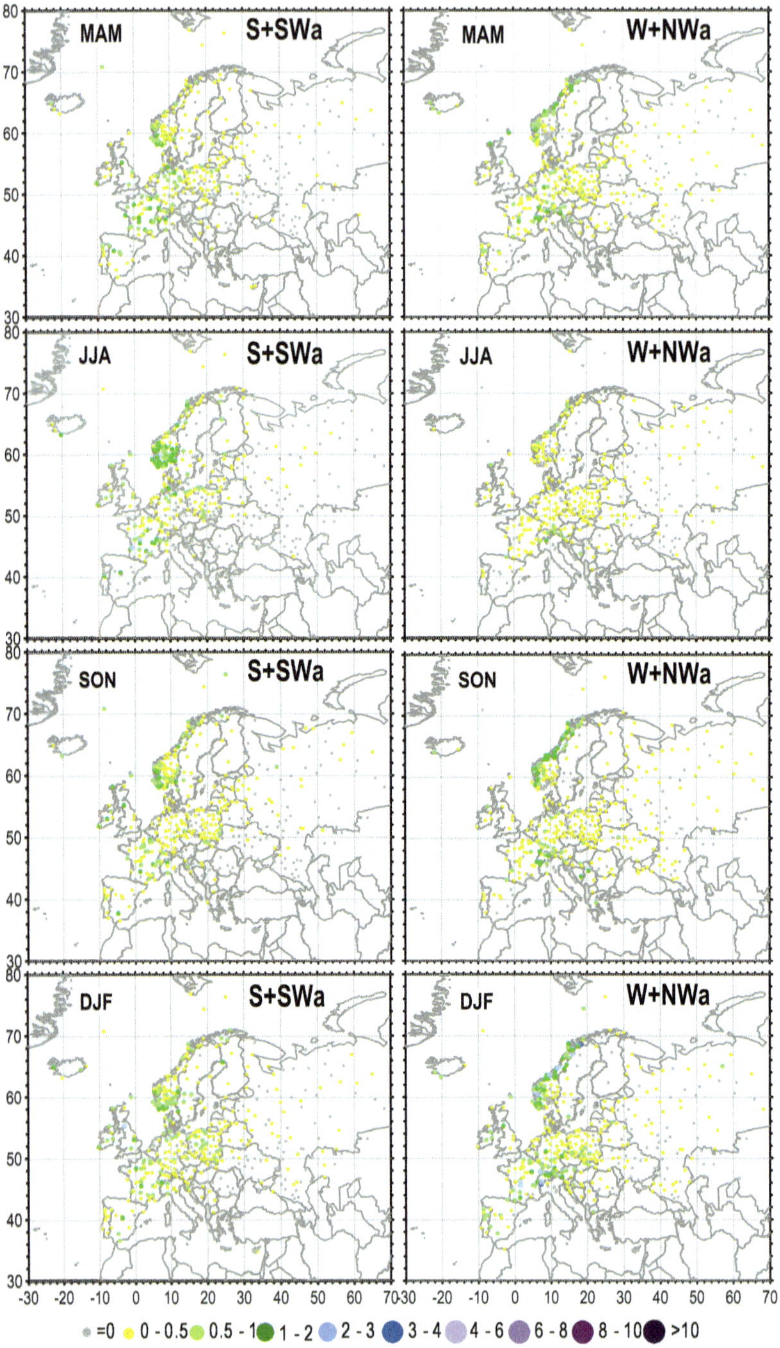

Fig. 6.70 Conditional probability of the occurrence of extreme precipitation associated with the passage of different fronts (type Ff) in anticyclonic circulation types (S+SWa, W+NWa) for January 1951–February 2008. Right closed intervals

the mountains and their foreland is attributable, partly, to the generally higher number of days with precipitation in those parts of Europe.

6.3.6.3 Probability of Extreme Precipitation Associated with the Passage of Different Fronts (Type Ff) in Cyclonic Synoptic Situations

Precipitation associated with the passage of several fronts is the most frequent in autumn and winter, when the dynamics of the atmosphere is greater than in the warmer seasons, and the rate at which low-pressure systems along with weather fronts move over the continent grows (Figs. 6.71 and 6.72).

Across the seasons, the occurrence of type Fr precipitation in cyclones at stations in Western Europe and Central Europe, in particular those located between the Bay of Biscay and the Gulf of Finland, is favoured by inflow of air from the north. In winter, during advection of air from the northwest sector (situation W+NWc), the probability of the occurrence of such events at some stations in France and the Alps exceeds 4 %.

Inflow of air from the north increases the likelihood of Ff precipitation also in highland and mountain areas of southern Poland and Germany, and on the coasts of the Scandinavian Peninsula. The occurrence of such precipitation in the southern part of Central Europe is driven particularly strongly by situation N+NEc, whereas in the Alps (mainly in winter) and in the Atlantic coasts of the Scandinavian Peninsula (especially in autumn and winter) type Ff extreme precipitation is associated, for the most part, with situation W+NWc.

With inflow of air from the southwest sector (situation S+SWc), the likelihood of the precipitation in question increases in Iceland, in the southernmost part of the Scandinavian Peninsula, on the British Isles, and in most seasons, also in France. On the British Isles, the probability of the occurrence of type Ff precipitation in situation S+SWc is only slightly lower than in situation E+SEc, which means that in that part of Europe type Ff precipitation is strongly related to air advection from the entire southern sector (Figs. 6.71 and 6.72).

In Eastern Europe, similar to the other parts of the continent, extreme precipitation originating during the passage of several fronts is the most probable in winter and autumn. It is most likely to occur in those seasons during advection of air from the southeast (situation E+SEc). At many stations in Europe, extreme precipitation associated with the passage of several fronts does not occur at all in situation Bc, that is, one that involves no advection; however, if it does appear, the probability of its occurrence is relatively high compared to advective situations, especially in autumn and winter.

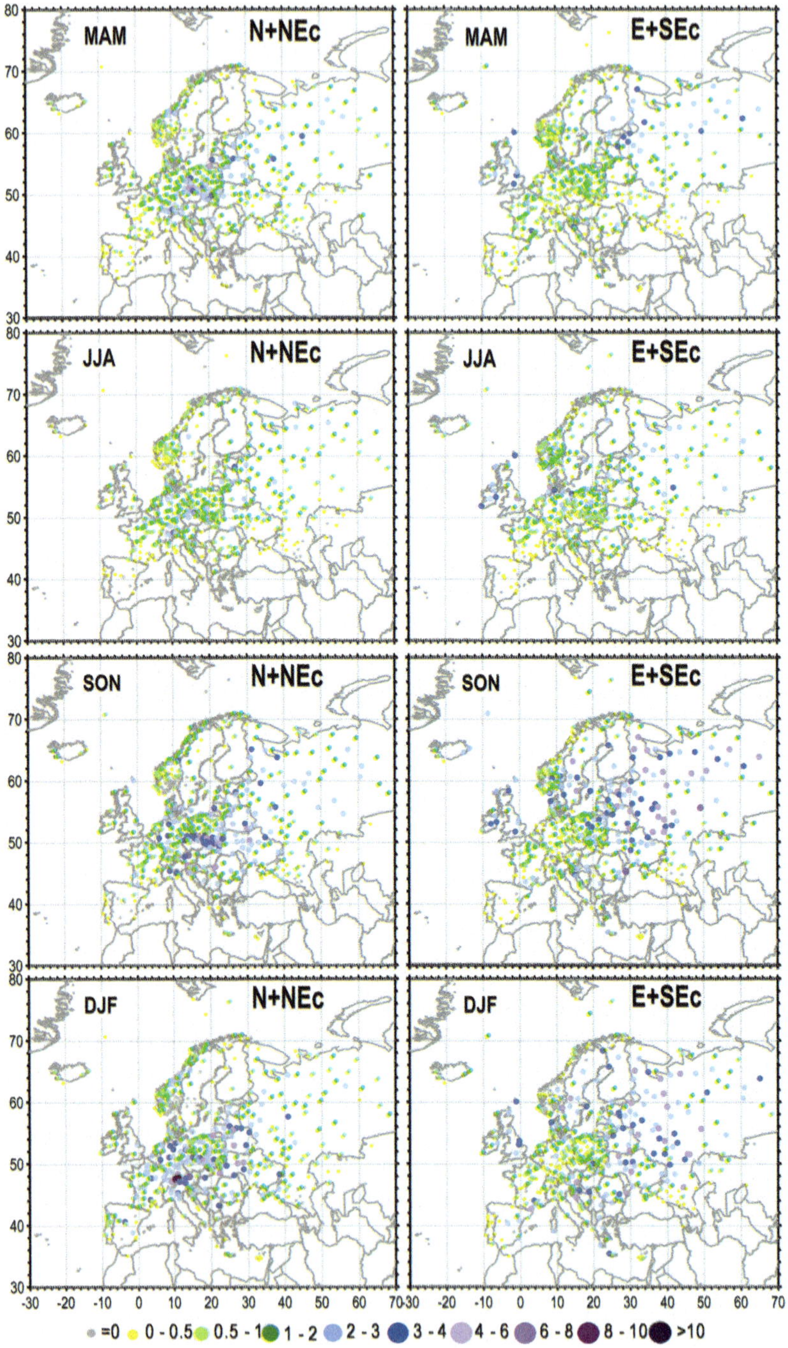

Fig. 6.71 Conditional probability of the occurrence of extreme precipitation associated with the passage of different fronts (type Ff) in cyclonic circulation types (N+NEc, E+SEc) for January 1951–February 2008. Right closed intervals

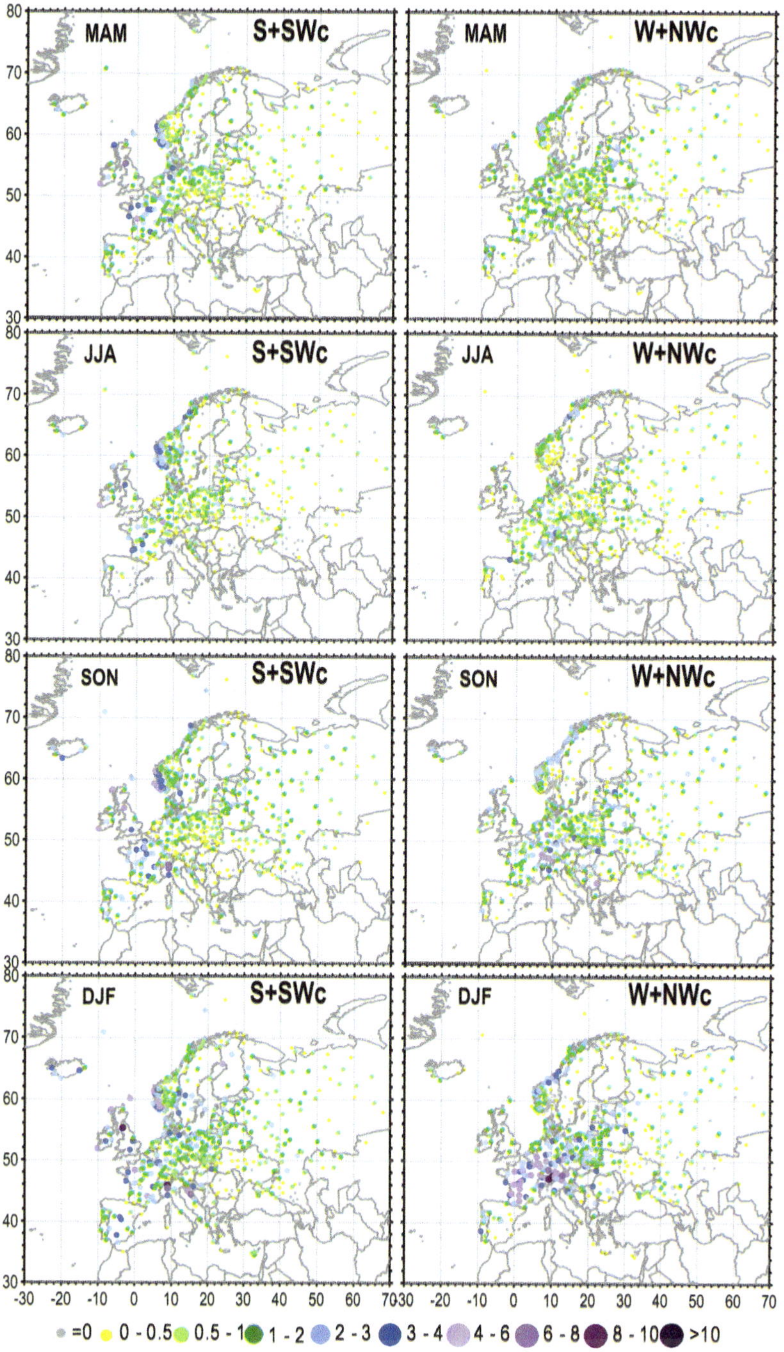

Fig. 6.72 Conditional probability of the occurrence of extreme precipitation associated with the passage of different fronts (type Ff) in cyclonic circulation types (S+SWc, W+NWc) for January 1951–February 2008. Right closed intervals

6.3.7 Occluded Front Precipitation (Type Fo)

6.3.7.1 Probability of Occluded Front Precipitation (Type Fo) in Anticyclones (ACT) and Cyclones (CCT)

The probability of the occurrence of occluded front extreme precipitation in anticyclones falls between 0.5 and 1 % only in the central part of the southernmost area of the Scandinavian Peninsula, and is smaller elsewhere in Europe (Fig. 6.73). The corresponding distribution of the probability in cyclonic situations does not change considerably from season to season. As far as CCT is concerned, similarly to ACT, the highest probability, exceeding 2 % at some stations, characterises the central part of the southern end of the Scandinavian Peninsula, especially in autumn and winter. A slightly lower, but still high (against the background of the continent as a whole) probability of type Fo precipitation characterises Northern Europe. The values of the index in question decline in the north–south direction (Fig. 6.73).

6.3.7.2 Probability of Occluded Front Precipitation (Type Fo) in Anticyclonic Synoptic Situations

The occurrence of Fo precipitation in anticyclones on the eastern side of the Scandinavian Mountains, and especially in the central part of the southern end of the Scandinavian Peninsula, in autumn and summer, is favoured by air advection from the southeast (situation E+SEa). In the southern part of the Scandinavian Peninsula, occluded front extreme precipitation forms, more often than elsewhere in Europe, also in situation S+SWa. In summer, at mountain stations in the Alpine area, in the southern part of Poland and in the Caucasus area, the likelihood of type Fo extreme precipitation increases at times of advection from the northeast (situation N+NEa) (Figs. 6.74 and 6.75).

6.3.7.3 Probability of Occluded Front Precipitation (Type Fo) in Cyclonic Synoptic Situations

Occluded front precipitation occurs most frequently in the north of the continent. The highest likelihood of its occurrence, in particular on the lee side of the Scandinavian Mountains, in nearly all seasons, except summer, is related to situation E+SEc. The situation is also highly conductive to the occurrence of type Fo precipitation on the British Isles (Fig. 6.76). In the southernmost part of the Scandinavian Peninsula, the probability of type Fo precipitation during advection of air from the southwest sector (situation S+SWc) is comparable, and even higher than in situation E+SEc. The direction of advection in question (situation S+SWc) increases the likelihood of type Fo occurrence also at the other stations in Northern Europe, but the probability of type Fo precipitation in this situation is lower than in

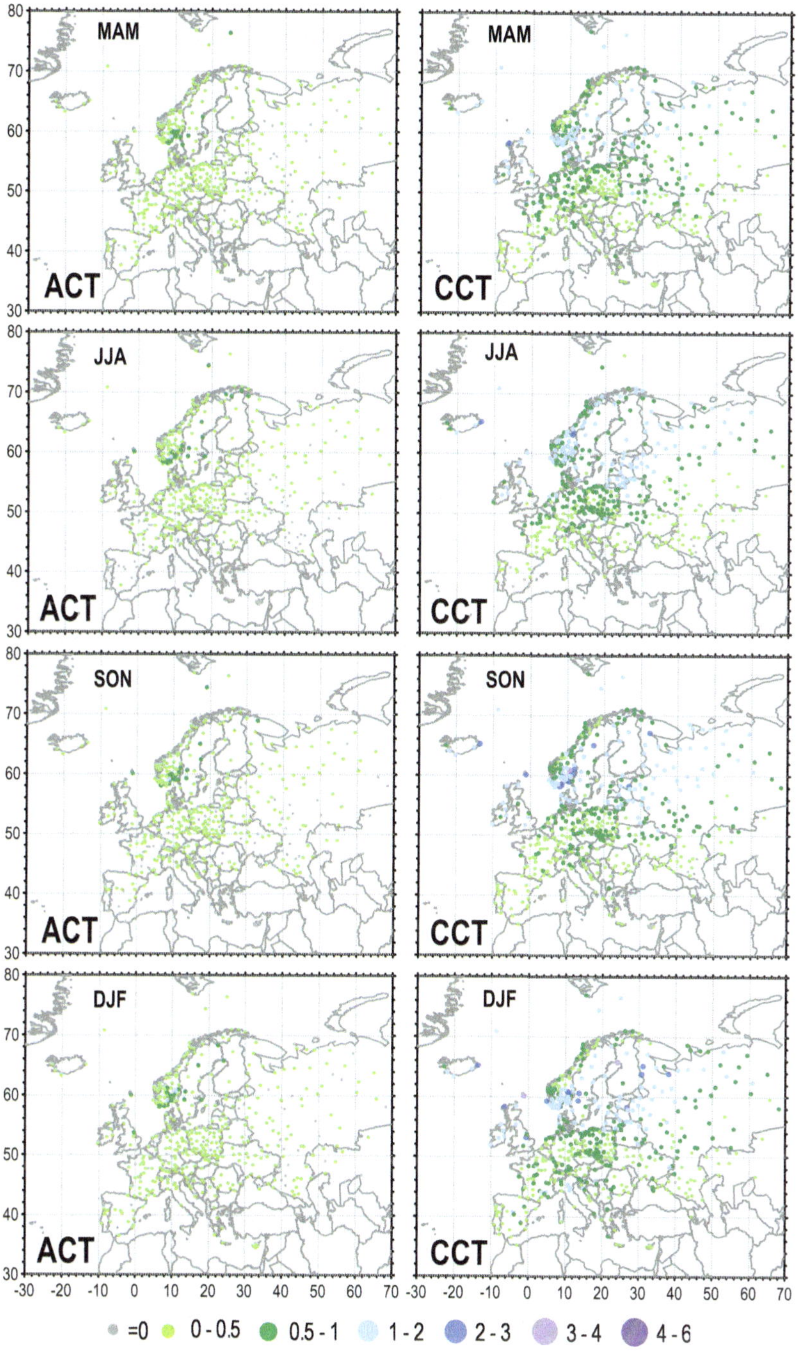

Fig. 6.73 Conditional probability of the occurrence of extreme precipitation associated with occlusion (type Fo) in anticyclones (*ACT*) and cyclones (*CCT*) for January 1951–February 2008. Right closed intervals

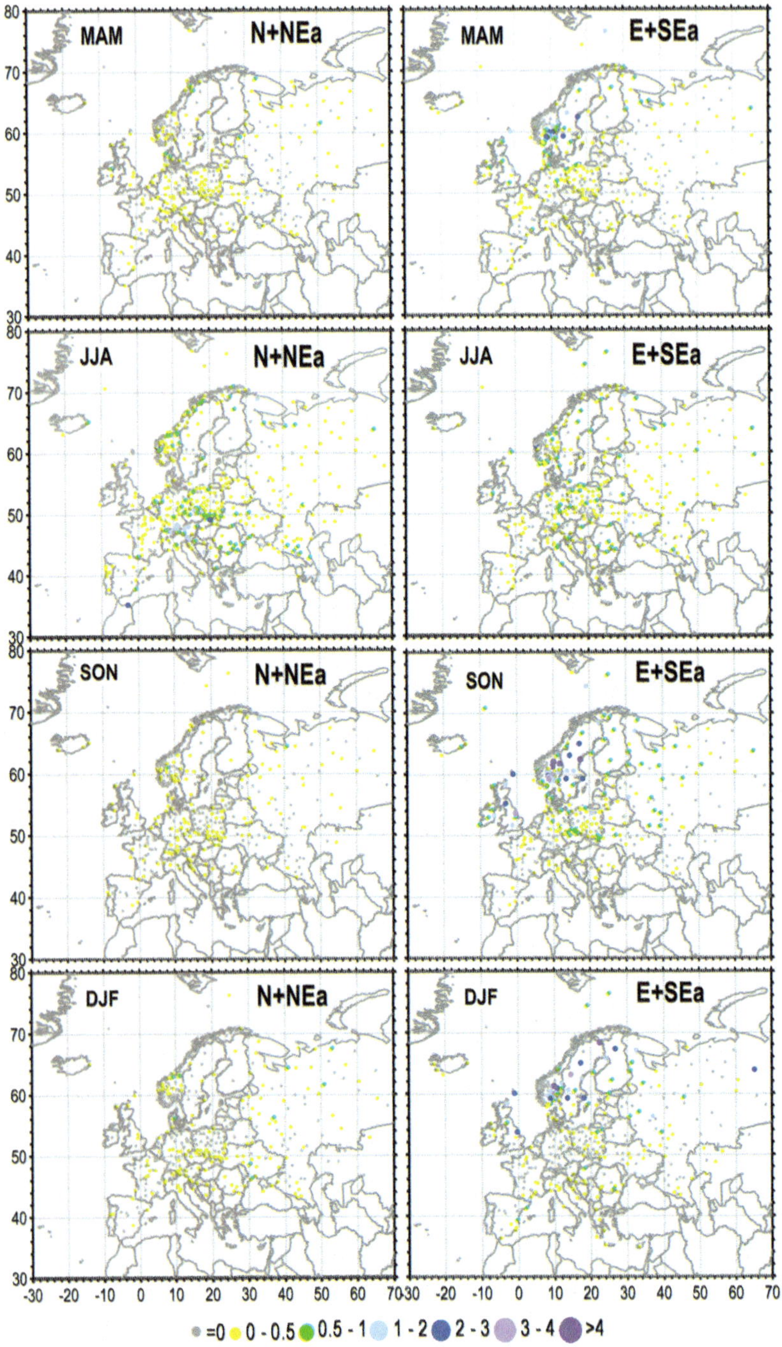

Fig. 6.74 Conditional probability of the occurrence of extreme precipitation associated with occlusion (type Fo) in anticyclonic circulation types (N+NEa, E+SEa) for January 1951–February 2008. Right closed probability intervals

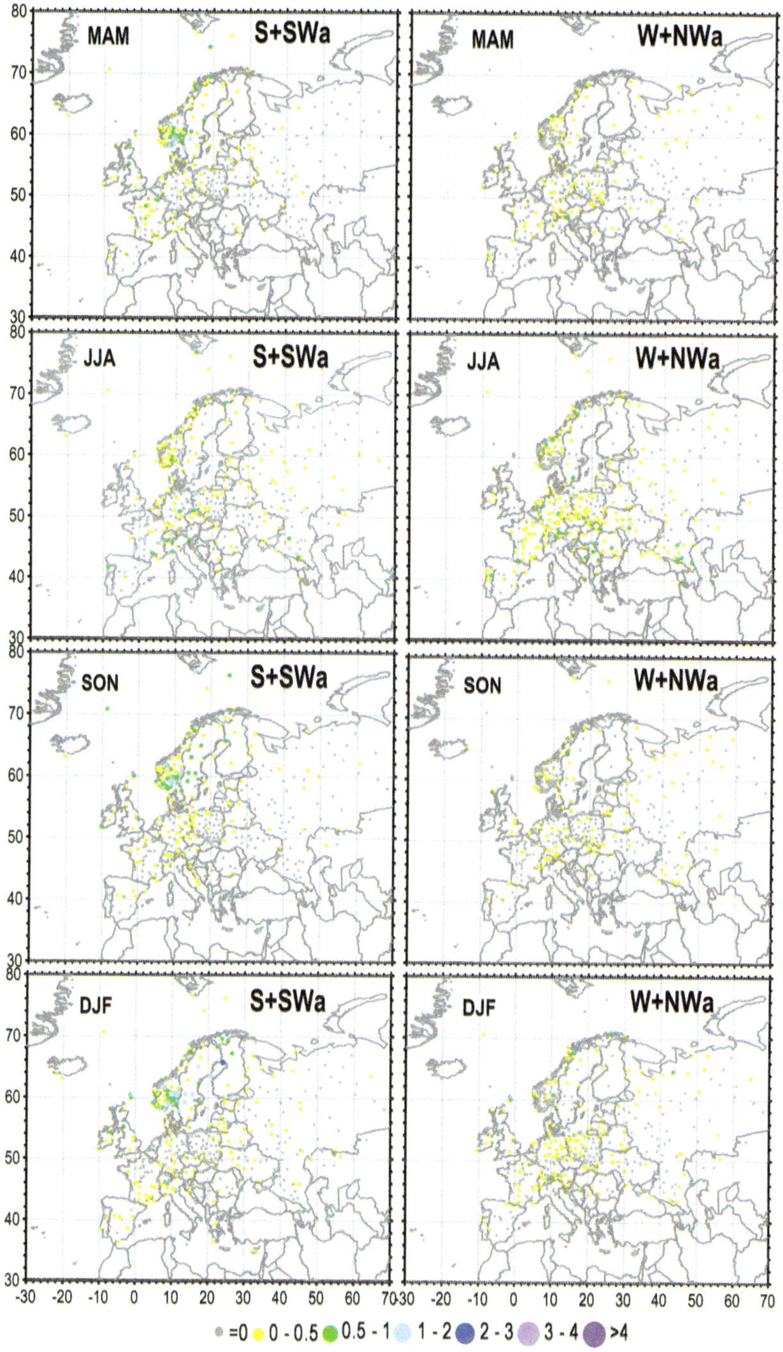

Fig. 6.75 Conditional probability of the occurrence of extreme precipitation associated with occlusion (type Fo) in anticyclonic circulation types (S+SWa, W+NWa) for January 1951–February 2008. Right closed probability intervals

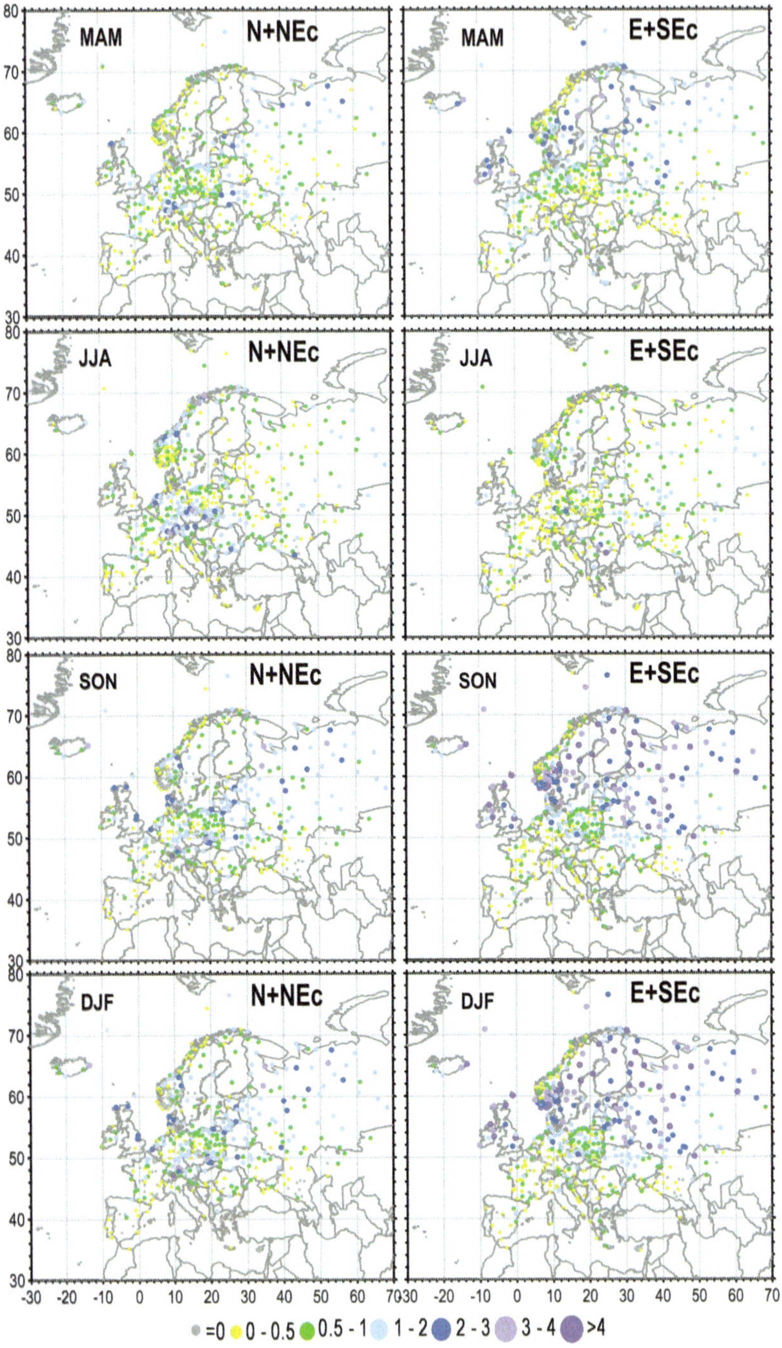

Fig. 6.76 Conditional probability of the occurrence of extreme precipitation associated with occlusion (type Fo) in cyclonic circulation types (N+NEc, E+SEc) for January 1951–February 2008. Right closed intervals

E+SEc type. In the central part of Europe, type Fo is the most probable in mountain areas (the Alps, Southern Poland), where its occurrence is associated, in the first place, with northeasterly inflows, especially in summer (situation N+NEc). In this season, the inflow of air from the north sector (situations N+NEc and W+NWc) contributes to the occurrence of type Fo ExP in the western coasts of the Scandinavian Peninsula (Figs. 6.76 and 6.77).

A high—compared to the advective synoptic situations—probability of the occurrence of type Fo precipitation is also that in nonadvective situation, that is, Bc, especially in autumn and winter.

6.4 Conclusions

The influence of mesoscale circulation on the occurrence of extreme precipitation varies among precipitation types. The frequency of various origin-based ExP types in Europe is linked to the direction of air advection and to the type of the pressure system, but varies from region to region. All origin-based types of extreme precipitation (both air-mass and frontal) occur in both cyclonic and anticyclone situations, with the former accounting for a higher frequency.

Relationships between air-mass precipitation and atmospheric circulation are characterised by less regional variability than those between circulation and the occurrence of frontal precipitation. Extreme air-mass precipitation displays the strongest link with atmospheric circulation in areas including Southern Europe, mountainous areas, the southeastern North Sea coast, and the western slopes of the Scandinavian Mountains. Type A precipitation in southern Europe occurs during air advection from the southeastern sector (situation E+SEc) during the whole year. Elsewhere in Europe this type is normally observed during air advection from the northern sector. Along the western Scandinavian coast its occurrence is particularly favoured by situations with advection from the northwestern sector (situation W+NWc), whereas in mountainous areas deeper in the continent it occurs most frequently during air advection from the northwestern sector (situation N+NEc), especially in summer.

The occurrence of extreme frontal precipitation (type F) is associated with both the direction of air advection and the type of the pressure system. This relationship remains stable between seasons in anticyclone systems, although in cyclones there is visible seasonality. The strongest dependencies between the frequency of type F precipitation and anticyclone situations was found in the Alps, in the central part of the southern tip of the Scandinavian Peninsula, and on the lee side of the Scandinavian Mountains. In the Alps the occurrence of type F precipitation, just as type A, is favoured by the northern advection, especially situation W+NWa. In the central part of the southern tip of the Scandinavian Peninsula and on the lee side of the Scandinavian Mountains, type F precipitation displays the strongest connection with southern advection, especially from the southeast (situation S+SEa). A conditional probability analysis of the occurrence of type F precipitation also suggests

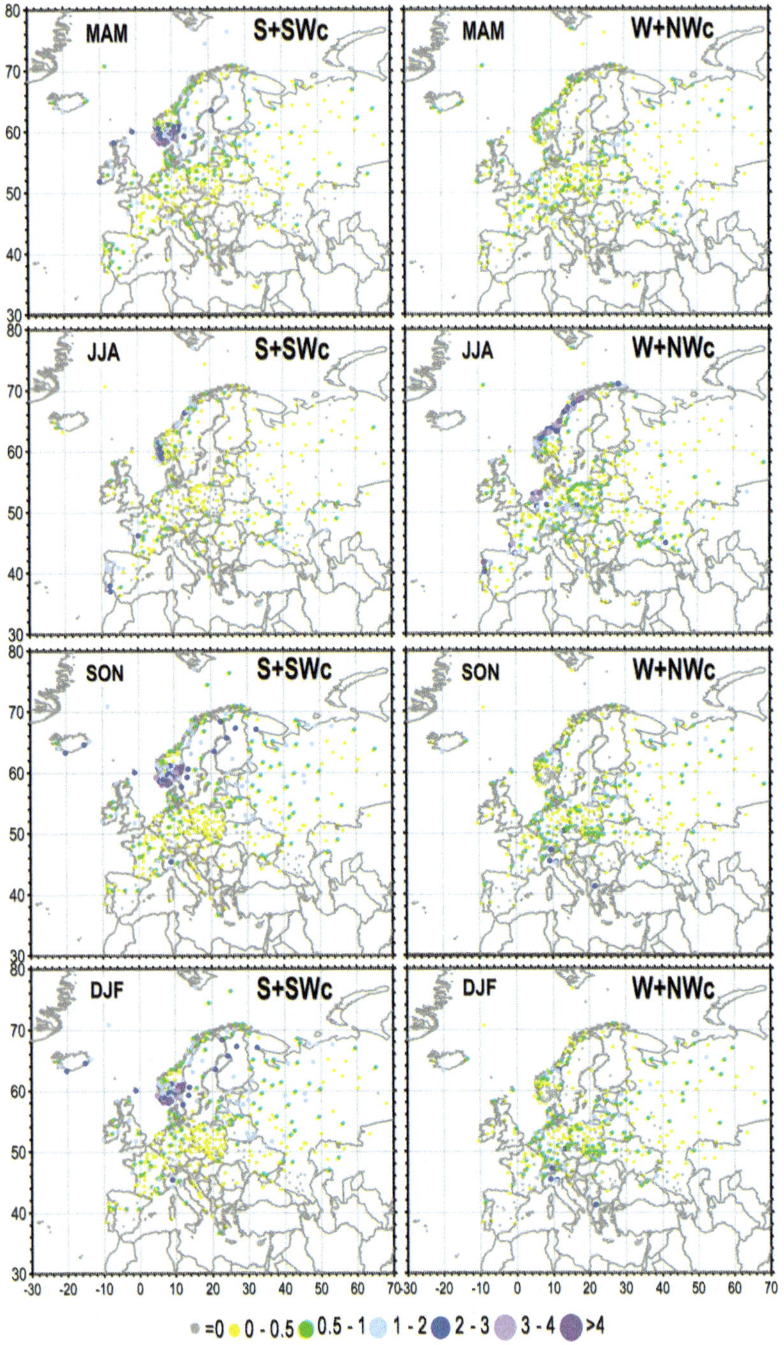

Fig. 6.77 Conditional probability of the occurrence of extreme precipitation associated with occlusion (type Fo) in cyclonic circulation types (S+SWc, W+NWc) for January 1951–February 2008. Right closed intervals

that in autumn and in winter along the western Scandinavian coast this type is the most likely in situations N+NEa and W+NWa, whereas in the southern part of Norway during the whole year it is in situation S+SWa.

The relationship between frontal precipitation and cyclonic situations is much more complicated. In Northern Europe (except the western coast of the Scandinavian Mountains) and in Eastern Europe, extreme frontal precipitation has the strongest link with air advection from the southeast (situation E+SEc). In Central Europe the highest frequency and the highest probability of the occurrence of type F precipitation coincide with air advection from the southeastern sector (situation N+NEc). In Western Europe its occurrence is favoured by air advection from the northwestern sector (situation W+NWc). All these relationships gain in strength in autumn and in winter. Throughout the year, type F precipitation in the British Isles and in the southern part of the Scandinavian Peninsula coincides mostly with air advection from the southeast (situation S+SWc). In the British Isles, the probability of type F precipitation in situation E+SEc is comparable to that in situation S+SWc.

The breaking down of extreme frontal precipitation into the types of fronts by which it was produced helped to better define the relationship between its occurrence and atmospheric circulation, especially in cyclonic synoptic situations. This relationship between precipitation associated with a cold front (type Fc) and anticyclonic situations is weak, but discernible in specific areas, such as the Alps and, in certain seasons, also in the Iberian Peninsula and along the western Scandinavian coast. The Alpine precipitation associated with a cold front is favoured by northern air advection, especially situation W+NWa, regardless of the season. Along the western Scandinavian coast the frequency of type Fc precipitation remains low in all anticyclone situations, but the analysis of conditional probability demonstrated that its occurrence in situation N+NEa is more likely than that in other anticyclone situations, especially in winter and in autumn. Along the western Iberian coast the situation W+NWa (in summer and in winter) coincides with increased frequencies of type Fc precipitation, although this relationship is not confirmed by the results of conditional probability analysis.

Extreme precipitation of type Fc displays a stronger relationship with cyclonic situations. In southeastern Europe the occurrence of this type is normally favoured by advection from the western sector. Around France this type coincides with advection from the northwestern sector (situation W+NWc, especially in winter and in autumn) whereas along the western Iberian coast its probability is similar or, at some stations, slightly higher in situation S+SWc. Advection from the northern sector favours the occurrence of type Fc extreme precipitation in the southern part of Eastern Europe. In the Alps, type Fc is linked to air advection from the northern sector in both cyclonic and anticyclonic situations, although its probability in lows is higher.

In Europe precipitation generated by warm fronts (type Fw) is relatively rare, and it displays a clear relationship with atmospheric circulation in cyclonic situations. This occurrence is the strongest in Eastern Europe and is slightly weaker in a broadly defined Central Europe that reaches up to the Alps and the western part of Eastern Europe. In Eastern Europe it coincides with southeastern advection (situation E+SEc) throughout the year and especially in winter. In the Central part of Europe this type is normally linked with air flowing from the northeast (situation N+NEc), but with seasonal variations: in spring, this is true of the territory of Poland, in summer in western part of Eastern Europe, and in autumn in southern Poland and in areas to the south of it between the Alps and the Black Sea.

Precipitation associated with the passage of various fronts (type Ff) is the most frequently occurring origin-based precipitation type, and for this reason the relationship between its frequency and atmospheric circulation follows a similar pattern to that of frontal precipitation (type F), especially in cyclonic situations. However, because type Ff precipitation accounts for the largest proportion of extreme precipitation in Western Europe, its occurrence in winter is favoured by advection from the northwestern sector (situation W+NWc). The frequency of type Ff precipitation in anticyclone situations is lower than that of type F precipitation, but its spatial variability is similar. The only difference concerning Ff type is a lack of the relationship between northern air advection and this type in the Alps in autumn and in spring.

Precipitation generated on occluded fronts (type Fo) accounts for the largest proportion of extreme precipitation in northern Europe. The area the most prone to occurrence of this precipitation type is the central part of the southern tip of the Scandinavian Peninsula. There, type Fo precipitation is favoured by air advection from the southeastern sector in both cyclonic and anticyclone situations (situations S+SWc and S+SWa). Elsewhere in Europe type Fo is only related to cyclonic situations. In Northern Europe, especially on the lee side of the Scandinavian Mountains, in the British Isles, and in Eastern Europe, the occurrence of ExO type Fo, especially in autumn and in winter, is linked to southwestern air advection. In summer, along the western Scandinavian coast and the southeastern North Sea coast, most of the extreme precipitation is related to occluded fronts during air advection from the northern sector (situations N+NEc and W+NWc). Extreme precipitation of type Fo displays the strongest relationship with the type of the pressure system, which most frequently is the cyclonic system.

This study provides the first pan-European comprehensive synoptic and climatic analysis of extreme precipitation to take into account both circulation types and atmospheric front types. Six regional groups featuring seasonally different patterns of the occurrence of origin-based extreme precipitation types were identified. Regional and seasonal variability of the relationship between the occurrence of origin-based types of extreme precipitation and atmospheric circulation was also found.

References

Barry RG, Perry AH (1973) Synoptic climatology. Methods and applications. Methuen, London
Djurić D (1994) Weather analysis. Prentice Hall, Englewood Cliffs
Goldreich Y (2003) The climate of Israel: observation, research and application. Springer, Israel

Goldreich Y, Moses H, Rosenfeld D (2004) Radar analysis of cloud systems and their rainfall yield in Israel. Isr J Earth Sci 53:63–76
Hare FK (1964) Dynamiczne aspekty klimatologii (Dynamic sides of Cimatology, in Polish). Przegląd Zagranicznej Literatury Geograficznej, PAN IG, Warszawa 2:826–840
Harmann JR, Winkler JA (1991) Synoptic climatology: themes, applications and prospects. Phys Geogr 12(3):220–230
Kalnay E, Kanamitsu M, Kistler R, Collins W, Deaven D, Gandin L, Iredell M, Saha S, White G, Woollen J, Zhu Y, Leetmaa A, Reynolds R (1996) The NCEP/NCAR 40-year reanalysis project. Bull Am Meteorol Soc 77:437–471
Kundzewicz Z, Dobrowolski A, Lorenc H, Niedźwiedź T, Pińskwar I, Kowalczak P (2012) Floods in Poland. In: Kundzewicz Z (ed) Changes in flood risk in Europe. IAHS Press, Wallingford
Lorenzo MN, Taboadab JJ, Gimenoa L (2008) Links between circulation weather types and teleconnection patterns and their influence on precipitation patterns in Galicia (NW Spain). Int J Climatol 28:1493–1505
Millán MM, Estrela MJ, Miró J (2005) Rainfall components: variability and spatial distribution in a Mediterranean area (Valencia region). J Climate 18:2682–2705
Niedźwiedź T (1981) Sytuacje synoptyczne i ich wpływ na zróżnicowanie przestrzenne wybranych elementów klimatu w Dorzeczu Górnej Wisły (Synoptic situations and their influence on spatial variability selected climate element in the upper Vistula basine). Uniwersytet Jagielloński, Rozprawy habilitacyjne 58, Kraków
Niedźwiedź T (2003a) Extreme precipitation in Central Europe and its synoptic background. Global Change 10:15–29
Niedźwiedź T (2003b) Extreme precipitation events on the northern side of the Tatra mountains. Geogr Pol 76(2):15–23
Pastor F, Estrela MJ, Peñarrocha D, Millán MM (2001) Torrential rains on the Spanish Mediterranean coast: modeling the effects of the sea surface temperature. J Appl Meteorol 40:1180–1195
Piotrowski P (2009) Obiektywna metoda klasyfikacji cyrkulacji atmosferycznej dla Polski (The objective classification of atmospheric circulation for Poland), Folia Geographica Physica 10. Wydawnictwo Uniwersytetu Łódzkiego, Łódź
Saaroni H, Halfon N, Ziv B, Alpertd P, Kutielb H (2010) Links between the rainfall regime in Israel and location and intensity of Cyprus lows. Int J Climatol 30:1014–1025
Schneider SH (1996) Encyclopedia of climate and weather, vols 1 & 2. Oxford University Press, New York, p 875
Strangeways I (2007) Precipitation. Theory, measurements and distribution. Cambridge University Press, New York, p 290
Sumner G (1996) Daily precipitation patterns over Wales: towards a detailed precipitation climatology. Trans Inst Br Geogr NS 21:157–176
Tolika K, Anagnostopoulou CHR, Maheras P, Kutiel H (2007) Extreme precipitation related to circulation types for four case studies over the Eastern Mediterranean. Adv Geosci 12:87–93
Ustrnul Z (1997) Zmienność cyrkulacji atmosfery na półkuli północnej w XX wieku (Variability of atmospheric circulation over northern hemisphere in the 20th century), Materiały Badawcze, Seria Meteorologia 27. IMGW, Warszawa
Wilby RL (1998) Modelling low-frequency rainfall events using airflow indices, weather patterns and frontal frequencies. J Hydrol 212-213:380–392
Wilks DS (2006) Statistical methods in the atmospheric sciences, International Geophysics Series. Elsevier, San Diego
Yarnal B (2000) Synoptic climatology in environmental analysis. A primer. Belhaven Press, London
Zangvil A, Karas S, Sasson A (2003) Connection between eastern Mediterranean seasonal mean 500 hPa height and sea-level pressure patterns and the spatial rainfall distribution over Israel. Int J Climatol 23:1567–1576

Index

E. Łupikasza, *The Climatology of Air-Mass and Frontal Extreme Precipitation*,
Springer Atmospheric Sciences, DOI 10.1007/978-3-319-31478-5

The manufacturer's authorised representative in the EU is Springer Nature Customer Service Centre GmbH, Europaplatz 3, 69115 Heidelberg, Germany. If you have any concerns regarding our products, please contact ProductSafety@springernature.com

Printed and bound by CPI Group (UK) Ltd, Croydon, CR0 4YY
01/10/2026
02225901-0003